Lecture Notes in Computer Science 6263

Torben Bach Pedersen Mukesh K. Mohania
A Min Tjoa (Eds.)

Data Warehousing and Knowledge Discovery

12th International Conference, DaWaK 2010
Bilbao, Spain, August/September 2010
Proceedings

Volume Editors

Torben Bach Pedersen
Aalborg University Selma
Department of Computer Science
Lagerløfs Vej 300
9220 Aalborg, Denmark
E-mail: tbp@cs.aau.dk

Mukesh K. Mohania
IBM India Research Lab
4, Block C, Institutional Area, Vasant Kunj
New Delhi 110 070, India
E-mail: mkmukesh@in.ibm.com

A Min Tjoa
Vienna University of Technology
Institute of Software Technology andInteractive Systems
Favoritenstr. 9/188
1040 Wien, Austria
E-mail: amin@ifs.tuwien.ac.at

Library of Congress Control Number: 2010931871

CR Subject Classification (1998): H.2, H.2.8, H.3, H.4, J.1, H.5

LNCS Sublibrary: SL 3 – Information Systems and Application, incl. Internet/Web and HCI

ISSN 0302-9743
ISBN-10 3-642-15104-3 Springer Berlin Heidelberg New York
ISBN-13 978-3-642-15104-0 Springer Berlin Heidelberg New York

springer.com

Printed in Germany

Typesetting: Camera-ready by author, data conversion by Scientific Publishing Services, Chennai, India
Printed on acid-free paper 06/3180

Preface

Data warehousing and knowledge discovery has been widely accepted as a key technology for enterprises and organizations to improve their abilities in data analysis, decision support, and the automatic extraction of knowledge from data. With the exponentially growing amount of information to be included in the decision-making process, the data to be considered become more and more complex in both structure and semantics. New developments such as cloud computing add to the challenges with massive scaling, a new computing infrastructure, and new types of data.

Consequently, the process of retrieval and knowledge discovery from this huge amount of heterogeneous complex data forms the litmus test for research in the area.

In the last decade, the International Conference on Data Warehousing and Knowledge Discovery (DaWaK) has become one of the most important international scientific events bringing together researchers, developers, and practitioners to discuss the latest research issues and experiences in developing and deploying data warehousing and knowledge discovery systems, applications, and solutions.

This year's conference, the 12th International Conference on Data Warehousing and Knowledge Discovery (DaWaK 2010), continued the tradition by discussing and disseminating innovative principles, methods, algorithms, and solutions to challenging problems faced in the development of data warehousing, knowledge discovery, the emerging area of "cloud intelligence," and applications within these areas. In order to better reflect novel trends and the diversity of topics, the conference was organized in four tracks: Cloud Intelligence, Data Warehousing, Knowledge Discovery, and Industry and Applications.

The papers presented at DaWaK 2010 covered a wide range of topics within cloud intelligence, data warehousing, knowledge discovery, and applications. The topics included data warehouse modeling, spatial data warehouses, mining social networks and graphs, physical data warehouse design, dependency mining, business intelligence and analytics, outlier and image mining, pattern mining, and data cleaning and variable selection.

It was encouraging to see that many papers covered emerging important issues such as social network data, spatio-temporal data, streaming data, non-standard pattern types, complex analytical functionality, multimedia data, as well as real-world applications. The wide range of topics bears witness to the fact that the data warehousing and knowledge discovery field is dynamically responding to the new challenges posed by novel types of data and applications.

From 112 submitted abstracts, we received 89 papers from 16 countries in Europe, North and South America, Asia, Africa, and Oceania. The Program Committee finally selected 26 papers, yielding an acceptance rate of 29%.

We would like to express our most sincere gratitude to the members of the Program Committee and the external reviewers, who made a huge effort to review the papers in a timely and thorough manner. Due to the tight timing constraints and the high number of submissions, the reviewing and discussion process was a very challenging task, but the commitment of the reviewers ensured that a very satisfactory result was achieved. We would like to thank Alfredo Cuzzocrea for his tireless contributions as Track Chair and Publicity Chair. We would also like to thank all authors who submitted papers to DaWaK 2010, for their contribution to making the technical program so excellent.

Finally, we send our warmest thanks to Gabriela Wagner for delivering an outstanding level of support within all aspects of the practical organization of DaWaK 2010. We also thank Amin Anjomshoaa for his support with the conference management software.

August 2010

Torben Bach Pedersen
Mukesh Mohania
A Min Tjoa

Organization

Program Chairs

Torben Bach Pedersen	Aalborg University, Denmark
Mukesh Mohania	IBM India Research Lab, India
A Min Tjoa	Vienna University of Technology, Austria

Publicity Chair

Alfredo Cuzzocrea	ICAR CNR & University of Calabria, Italy

Program Committee

Alberto Abello	Universitat Politecnica de Catalunya, Spain
Ira Assent	Aalborg University, Denmark
Elena Baralis	Politecnico di Torino, Italy
Ladjel Bellatreche	ENSMA, France
Petr Berka	University of Economics, Prague, Czech Republic
Jorge Bernardino	ISEC - Instituto Superior de Engenharia de Coimbra, Portugal
Mokrane Bouzeghoub	CNRS - Université de Versailles SQY, France
Stephane Bressan	National University of Singapore, Singapore
Peter Brezany	University of Vienna, Austria
Robert Bruckner	Microsoft, USA
Jesús Cerquides	Universitat de Barcelona, Spain
Zhiyuan Chen	University of Maryland Baltimore County, USA
Sunil Choenni	The Netherlands Ministry of Justice, The Netherlands
Frans Coenen	University of Liverpool, UK
Bruno Cremilleux	Université de Caen, France
Alfredo Cuzzocrea	ICAR-CNR & University of Calabria, Italy
Agnieszka Dardzinska	Bialystok University of Technology, Poland
Karen Davis	University of Cincinnati, USA
Kevin Desouza	University of Washington, USA
Curtis Dyreson	Utah State University, USA
Todd Eavis	Concordia University, Canada
Johann Eder	University of Klagenfurt, Austria
Tapio Elomaa	Tampere University of Technology, Finland
Roberto Esposito	Università di Torino, Italy
Vladimir Estivill-Castro	Griffith University, Australia
Christie Ezeife	School of Computer Science, University of Windsor, Ontario, Canada
Jianping Fan	UNC-Charlotte, USA
Ling Feng	Tsinghua University

Eduardo Fernandez Medina	University of Castilla-La Mancha, Spain
Dragan Gamberger	Ruder Boskovic Institute, Croatia
Gyözö Gidófalvi	Royal Institute of Technology (KTH), Sweden
Matteo Golfarelli	University of Bologna, Italy
Eui-Hong (Sam) Han	Sears Holdings Corporation, USA
Wook-Shin Han	Kyungpook National University, Korea
Jaakko Hollmén	Aalto University School of Science and Technology, Finland
Jimmy Huang	York University, Canada
Farookh Hussain	Curtin University of Technology, Australia
Ryutaro Ichise	National Institute of Informatics, Japan
Mizuho Iwaihara	Waseda University, Japan
Murat Kantarcioglu	University of Texas at Dallas, USA
Jinho Kim	Kangwon National University, Korea
Sang-Wook Kim	Hanyang University, Korea
Jörg Kindermann	Fraunhofer Institute IAIS, Germany
Jens Lechtenboerger	Westfälische Wilhelms-Universität Münster, Germany
Wolfgang Lehner	Dresden University of Technology, Germany
Sanjay Kumar Madria	University of Missouri-Rolla, USA
Anirban Mondal	University of Tokyo, Japan
Jose-Norberto Mazón	University of Alicante, Spain
Ullas Nambiar	IBM Research, India
Jian Pei	Simon Fraser University, Canada
Evaggelia Pitoura	University of Ioannina, Greece
Stefano Rizzi	University of Bologna, Italy
Alkis Simitsis	HP Labs
Koichi Takeda	Tokyo Research Laboratory, IBM Research, Japan
Dimitri Theodoratos	New Jersey Institute of Technology, USA
Christian Thomsen	Aalborg University, Denmark
Juan-Carlos Trujillo Mondéjar	University of Alicante, Spain
Vincent Shin-Mu Tseng	National Cheng Kung University, Taiwan
Panos Vassiliadis	University of Ioannina, Greece
Wolfram Woess	University of Linz, Austria
Robert Wrembel	Poznan University of Technology, Poland
Man Lung Yiu	Hong Kong Polytechnic University, Hong Kong
Qiankun Zhao	Telefonica, Spain
Xiaofang Zhou	University of Queensland, Australia
Esteban Zimányi	Université Libre de Bruxelles, Belgium

External Reviewers

Timo Aho
Annalisa Appice
Ryan Bissell-Siders
Rajkumar Bondugula
Panos Bouros
Giulia Bruno

Peggy Cellier
Tania Cerquitelli
Eugenio Cesario
Fabio Fassetti
Christina Feilmayr
Alessandro Fiori
Paolo Garza
Teemu Heinimäki
Christian Koncilia
Jussi Kujala
Stefano Lodi
Jose-Norberto Mazón
Francois Rioult
Paolo Serafino
Jose Jacobo Zubcoff

Table of Contents

Data Warehouse Modeling and Spatial Data Warehouses

Mining Social Networks and Graphs

Physical Data Warehouse Design

Dependency Mining

Business Intelligence and Analytics

Outlier and Image Mining

Pattern Mining

Data Cleaning and Variable Selection

Logic Programming for Data Warehouse Conceptual Schema Validation

Carlo dell'Aquila, Francesco Di Tria, Ezio Lefons, and Filippo Tangorra

Dipartimento di Informatica
Università degli Studi di Bari "Aldo Moro"
Via Orabona 4, 70125, Bari, Italy
{dellaquila,francescoditria,lefons,tangorra}@di.uniba.it

Abstract. The current lack of a standard methodology for data warehouse design has led to have many possible lifecycles. In some of them, the validation of the data warehouse conceptual schema is a specific process that precedes the translation of such a schema into a logical one. This activity must ensure that the data warehouse to be implemented effectively allows all the analytical queries to be executed correctly. To accomplish this, the validation process takes the preliminary workload into account, that is, a set of queries defined from user requirements to obtain the typical information the users are interested in. The methodologies that perform such a validation process define some guidelines that must be manually executed by an expert. In this paper, we introduce a logic program to automate this activity, by checking a set of predefined issues with an inferential engine.

Keywords: logic programming; workload; conceptual schema.

1 Introduction

Companies are devoting more and more attention to the benefits emerging from the exploitation of data warehouses (DWs) in the scope of Business Intelligence (BI) systems. Indeed, DWs are used as data sources for On-Line Analytical Processing (OLAP) and Machine Learning [1], in order to produce information and knowledge useful in decision making processes. Therefore, the design of a DW requires methodologies quite different from those adopted for On-Line Transactional Processing (OLTP) systems; such methodologies must satisfy precise quality factors, such as believability of data in terms of their completeness and consistency [2].

The basic lifecycle of a DW comprises: (a) *analysis and requirements definition*, where end-users needs are investigated in order to understand what kind of information they are interested in; (b) *conceptual design*, based on the user requirements, the schemata, and the documentation of the source databases; (c) *logical design*, where the conceptual schema of the DW is traduced into the logical schema; (d) *implementation*, where the DW is implemented according to the logical model (ROLAP or MOLAP) supported by the DBMS; (e) *ETL design*, producing a plan to feed and to periodically update the DW; (f) *refreshing*, that consists of the execution of the ETL, repeated at regular intervals of time, depending on the refresh necessity; and

T.B. Pedersen, M.K. Mohania, and A M. Tjoa (Eds.): DaWaK 2010, LNCS 6263, pp. 1–12, 2010.

(g) *BI applications development*, consisting of traditional reports, analytical processing, and data mining applications [3].

A popular methodological framework to design DWs [4], establishes that, in the requirements definition step, the designer must first define a *preliminary workload* that consists of a set of queries, expressed according to a high level language. These queries represent the typical analytical queries that the business users will perform on the DW and they help the designer to identify facts, dimensions, and measures during the next conceptual design step. The conceptual model adopted is the Dimensional Fact Model (DFM) [5], which produces facts schemata according to the Multidimensional Model [6]. Before proceeding with the logical design, the designer must be sure that the designed conceptual schema supports the preliminary workload. This step is the so-called *DW conceptual schema validation* and its aim is to verify whether the multidimensional schema properly accords to the preliminary workload. In particular, the designer must be sure that all the measures, useful to produce business information, have been identified, and that all the hierarchies are well-structured to perform data aggregation. Only if all the queries in the workload can be effectively expressed on such a conceptual schema, then the designer can safely translate it into a logical one. On the other hand, if the designer realizes that one or more queries are not executable against that conceptual schema, then s/he has the possibility to go back to the conceptual design step, in order to produce a multidimensional model that satisfies the user requirements and that allows business users to obtain all the needed information. In detail, the schema validation is executed by re-writing the preliminary workload via a simple language that allows defining a query in accordance with the DFM. In this context, a query is represented by an expression, describing a measure to be retrieved, an aggregation pattern, and a selection clause. Currently, the validation is a made-by-human work and consists of a manual mapping of each expression to the graphical representation of the conceptual schema. Of course, this test represents, in some cases, a waste of time and can easily produce misunderstandings, oversights, and human errors, due to the difficulty to check a very large set of queries on complex conceptual schemata.

As the conceptual schema design represents the most crucial step to capture user requirements such to be error-free [7], nowadays, it is emerged the necessity to support the designers, by providing them with new methodologies to obtain objective evaluations about the quality of the conceptual schemas [8], to create strong formal models of user requirements [9, 10], to automate their activities [11, 12], and to extend the existing ones with more powerful features [13]. In this context, the validation of the conceptual schema is the only mean to produce a final DW that is as close to the user needs as possible [14]. In our opinion, the validation phase can be effectively replaced by an automatic process, based on an inferential engine, whose knowledge base is composed of metadata representing a multidimensional schema. The aim of this paper is to describe the architecture and the functionality of such an inferential engine, able to validate the conceptual schema automatically, effectively replacing the activity that a human expert makes.

The paper is organized as follows. In Section 2, we report an overview of the related work about conceptual schema validation. Section 3 introduces the metadata to be used to represent a conceptual schema. Section 4 describes the inferential process for the validation and the goal used to start the inferential process. Section 5 illustrates the compiler that translates a workload into goals for the logical program. In Section 6, we report the testing scenario of the methodology. Finally, Section 7 contains our conclusions.

2 Related Work

The conceptual schema is the result of the conceptual design and represents the most important step of the design of both relational databases [15] and DWs [16]. In the logical design step, the conceptual schema must be translated into a logical one. However, in order to produce an effective logical schema, the conceptual schema must be first validated. While in relational databases the validation is devoted to verify whether the conceptual schema satisfies a set of constraints [17], such as cardinality constraints [18] for example, in the scope of data warehousing, it is well-known that the validation consists of verifying whether the preliminary workload, defined on the basis of the user requirements, can be supported by the designed conceptual schema [4]. For the sake of simplicity, validation means controlling whether each query of the analytical workload can be effectively executed on the designed schema.

A similar approach is used in [11], where a conceptual schema is chosen, among a set of conceptual schemas designed by an algorithm, provided it is able to accomplish an answer to each query included in the workload.

A more general methodology allows designers to verify the correctness of a conceptual schema, by checking some desirable properties (such as satisfiability, non-redundancy of integrity constraints, and executability of operations) according to a plan, expressed using the first-order logic. For each property, opportune initial state and goal are defined, and the designer assumes the property is satisfied if there exists a sequence of derivations to accomplish the given goal [19].

In our opinion, the first-order logic is a very powerful language to perform logical deductions on the basis of a semantic level, such as the understanding of a conceptual schema.

3 Metadata Modelling

There are several kinds of metadata associated to a DW [20]. As an example, there are metadata describing the refresh status of data. However, the most important class of metadata is the one describing the multidimensional model of the DW. These metadata are usually used in ROLAP systems to generate SQL queries [21]. Currently, the standard language for the representation of DW metadata is described by the Common Warehouse Metamodel (CWM) [22]. In this paper, we adopt this standard representation. In fact, according to the CWM, we model the main concepts and relationships via the Predicate Calculus (PC) [23], in order to define a set of metadata to be used as a knowledge base for a logical program able to perform the validation of a conceptual schema. This metadata modelling defines a set of predicates, able to represent the conceptual schema of a DW. The predicates are listed in Table 1.

Table 1. Predicates of the Metamodel

Predicate	Semantics
cube(C)	*C* is a cube.
measure(M, C)	*M* is a measure of *C*. *C* must be a cube.
dimension(D)	*D* is a dimension.
hierarchy(H, D)	*H* is a hierarchy of *D*. *D* must be a dimension.
level(L, N, H)	*L* is the level number *N* of *H*. *H* must be a hierarchy. *N* must be a natural number.
cube_dim(C, L)	*L* is one of the first levels of aggregation of *C*. *C* must be a cube.
attribute(L, A, T)	*A* is an attribute of *L*. *L* must be a level of a dimension. *T* value is *id* (identifier) or *desc* (descriptive).

4 Conceptual Schema Validation

According to the traced guidelines in [19], we define the following issues related to the validation of a conceptual schema in reference to the queries included into the preliminary workload:

- a query involves a cube that has not been defined as such;
- a query requires a measure that is not an attribute of the given cube;
- a query presents an aggregation pattern on levels that are unreachable from the given cube;
- a query requires an aggregation on a field that has not been defined as a dimensional attribute.

In reference to these issues, a set of tests to be performed has been designed, as explained in the next Sub-section.

4.1 Inferential Engine

The Inferential Engine (IE) is a logic program, composed of a set of rules, expressed according to the PC. The conceptual schema validation is executed by the IE, via an inferential process that allows verifying the issues pointed out in the previous Sub-section. At the end of the inferential process, the IE states whether the conceptual schema is valid or not, on the basis of a given preliminary workload and a set of multidimensional metadata. The logic program has been developed in Prolog [24] and it performs a set of tests. Notice that, for simplicity, the first rule has been entirely reported, while only the head of the other rules is shown.

- Cube test:

  verify_cube(C):- cube(C), write(C), writeln(' is a cube.').
  verify_cube(C):- not(cube(C)), write(C), writeln(' is not a cube.'), fail.
  If *C* is a cube, then IE shows a validation message. On the contrary, if *C* is not a cube, then IE shows an error message.
- Measure test: *verify_measure(M, C)*. This rule verifies whether *M* is a measure of the cube *C*. If *M* is not a measure of the cube *C*, then IE shows an error message.

- Attribute test: *verify_attribute(A, L)*. This rule verifies whether there exists a level *L*, where *A* is an attribute. If *A* is not an attribute of any level, then IE checks if *A* is a measure of the *C* cube.
- Path test: *verify_path(C, D)*. This rule checks whether *D* is part of the primary aggregation pattern of the *C* cube or belongs to the same hierarchy of the level representing the primary aggregation pattern of the *C* cube (*viz.*, it checks whether there is an aggregation path from *C* to *D*).
- Aggregation test: *verify_level(*[*Head* | *Tail*]*)*. The rule scans a list recursively on the tail of the list (the ending condition is represented by an empty list). It checks whether all the elements of the list are dimensional attributes (*i.e.,* level identifiers). If an element of the list is not a dimensional attribute, then IE shows an error message.

4.2 Goal

In order to start the inferential process, the IE needs a goal. All the goals are generated by a compiler (*see*, Section 5) using the queries included into the workload. So, each goal corresponds to a query to be tested. The goal is represented by the predicate:

$$goal(C, V, A),$$

where *C* is the cube on which the query is based, *V* is a list of dimensional attributes on which to perform data aggregation, and *A* is an attribute which can be a measure or a descriptive attribute.

The main goal is divided into the following three sub-goals:

- *fact(C, A)*. This goal performs both the cube and the attribute tests.
- *aggregation(V)*. This goal performs the aggregation test. *V* is a list of dimensional attributes.
- *path(C, V)*. This goal performs the path test on each element *D* of the list *V*.

5 Compiler

The general workflow of the validation process is the following. The Compiler translates the workload into goals for the IE. The workload is written according to a high-level language, as explained in [5]. In particular, it generates a goal for each query in the workload. Then, the IE uses both the goals and the metadata to check whether the schema is valid or not. The Compiler is based on a Syntactical Analyzer, that, on turn, uses a Lexical Analyzer for string pattern recognition.

5.1 Syntactical Analyzer

The Syntactical Analyzer (SA) is a parser that verifies the syntactical structure of a statement. The SA has been developed using Bison [25], which is a tool that (a) reads a grammar-file, and (b) generates a C-code program. This C-code program represents the SA. In particular, the grammar-file contains the declaration of a set of terminal symbols and a set of grammar rules, expressed according to the Backus Naur Form (BNF).

First of all, the terminal symbols (tokens) of the grammar must be defined. The tokens include literals (*i.e.*, string constants), identifiers (*i.e.*, string variables), and numeric values. The tokens defined for the SA are the following:

```
%token VAR                              %token AND "AND"
%token DIGIT                            ...
%token COMMA ","                        %token OPEN "("
%token SEMICOL ";"                      %token CLOSE ")"
%token DOT "."                          %token OPSQ "["
%token EQ "="                           %token CLOSESQ "]"
%token GT ">"                           %token SUPS "'"
```

The first and the second tokens represent string variables and numeric values, respectively. The other tokens represent string constants, as language keywords. Once all tokens have been defined, the rules of the grammar follow.

A query against a conceptual schema is a statement expressed according to the following BNF grammar:

```
<query>          ::= <expression>.<measure> |
                     <expression>.<attribute name>
<expression>     ::= <fact name> <aggr. clause>
<aggr. clause>::= [<pattern>] | [<pattern>; <sel. clause>]
<pattern>        ::= <attribute name> | <pattern>, <attribute name>
<sel. clause>    ::= <predicate> |
                     <sel. clause> <logical operator> <predicate>
<predicate>      ::= <attribute name> <comparison operator> <value>
```

This grammar is composed of a set of rules and defines all the well-formed phrases of the language to express queries against the conceptual schema.

As an example, the string

"*sales*[*day, product; city='Rome' and product='milk'*].*quantity*"

is a correct phrase, while the string

"*sales*[*day, product; city='Rome' and product='milk'*],*quantity*"

generates a syntax error, due to the comma instead of the dot, before the *quantity* attribute.

In Bison, each rule has the form:

<result>: *<components>* { *<statement>* };

where *<result>* is a non-terminal symbol, *<components>* is a set of terminal and/or non-terminal symbols, and *<statement>* is the C-code statement to be executed when the rule is applied. In order to implement the grammar, the rules defined for the SA are the following:

```
query:      express DOT attribute { printf(", %s) \n", $3); };
express:    fact aggreg;
fact:       VAR { printf("goal(%s",$1); };
aggreg:     OPSQ pattern CLOSESQ { printf("],"); } |
            OPSQ pattern SEMICOL seq_sel CLOSESQ { printf("]"); };
pattern:    attribute { printf(",[%s",$1); } |
            pattern COMMA attribute { printf(",%s", $3); };
seq_sel:    selection | seq_sel logic selection;
selection: attribute operator value;
value:      SUPS VAR SUPS | DIGIT;
operator:  EQ | GT | LT | GT EQ | LT EQ;
logic:      AND | OR;
attribute: VAR;
```

The first rule, the tagged *query* one, is applied when the SA recognizes a string like "*<a>.<b>*", where *<a>* is a valid *<express>* non-terminal symbol, and *<b>* is a *VAR* token, *i.e.*, when the string is syntactically correct. In this case, the SA prints: (a) a comma, (b) the third parameter of the *<components>* (*i.e.*, the attribute), (c) a closed round bracket, (d) a carriage return, and (e) a line feed. Then, the SA ends with no error message generation.

Example 1. The query string "*sales*[*day, product; city='Rome' and product='milk'*]. *quantity*" is translated into the following goal to be submitted to the IE:

goal(sales, [*day, product*], *quantity)*

expressed according to the PC. Note that the *selection clause* is ignored. (This issue will be addressed in future works.) □

5.2 Lexical Analyzer

The Lexical Analyzer (LA) is the component used by the SA, in order to obtain an ordered sequence of tokens. The tokens are recognized by the LA inside a string (pattern matching on text) and, then, passed to the SA. The LA has been developed using Flex [26], which is a tool that (a) uses the tokens defined for the SA program, (b) reads a rule-file, and (c) generates a C-code program. This C-code program represents the LA. In particular, the rule-file is composed of two sections: (a) definition, and (b) rules. The definition section includes the tokens defined with Bison, plus further identifiers. The identifiers define how to perform the matching between a sequence of alphanumeric characters and a token. The identifiers contained in the definition section of the rule-file are the following:

```
UVAR [a-z][a-z0-9]*
UDIGIT [0-9]*
```

Correct instance of UVAR is any string that starts with an alphabetic character, followed by an arbitrary number of alphabetic characters or digits 0 to 9 (for example, a1, qr55, m5n9, abbbddd, ...). Correct instance of UDIGIT is any numeric value, composed of an arbitrary number of digits 0 to 9. The rule section includes a set of rules, defining the action to perform when a matching happens. The rules defined for the LA are the following:

```
[=] {return EQ;}                {UVAR} {return VAR;}
[<] {return LT;}                {UDIGIT} {return DIGIT;}
[>] {return GT;}                ...
                                <<EOF>> { yyterminate();}
```

The first rule states that, whenever the constant "=" is recognized inside the input string, the token EQ must be returned. In fact, when the LA recognizes an identifier, it returns the corresponding token to the SA. When the LA encounters the end-of-file symbol, it stops the string scanning (last rule).

Example 2. In analyzing the string "*sales*[*day, product; city='Rome' and product= 'milk'*].*quantity*", the LA returns the following sequence of tokens: VAR OPSQ VAR COMMA VAR SEMICOL VAR EQ SUPS VAR SUPS AND VAR EQ SUPS VAR SUPS CLOSESQ DOT VAR. At last, the tokens are passed from the LA to the SA for the syntactical control. □

6 Testing Scenario

Figure 1 shows the conceptual schema of two cubes: *sales* and *shipments*.

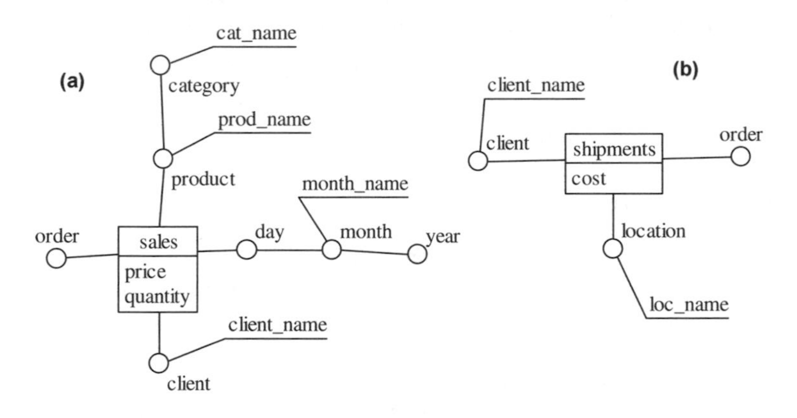

Fig. 1. Conceptual schemas. (a) *sales* cube. (b) *shipments* cube.

Here, *sales* is a four-dimensional cube. The four dimensions are *clients*, *time*, *orders*, and *products*. *Time* is a one-hierarchy dimension. This hierarchy is formed by three levels: *days*, *months*, and *years*. Each level has at least one attribute (the dimensional attribute, denoted by a circle), representing the identifier of the level. Some levels can also have descriptive attributes (represented by emphasized names). In the example, *days* and *years* levels have only the *day* and *year* dimensional attributes, while *months* has the *month* dimensional attribute and also the *month_name* descriptive attribute. *Orders* is a one-hierarchy dimension. This hierarchy is formed by the one level *order*. This level has no descriptive attributes. *Clients* is a one-hierarchy dimension. This hierarchy is formed by the one level *client*. This level has the *client_name* descriptive attribute. *Products* is a one-hierarchy dimension. This hierarchy is formed by the two levels *product* and *category*. Each of these levels has its own descriptive attributes. *Day*, *product*, *order*, and *client* levels represent the primary aggregation pattern of the *sales cube*.

The *shipments* cube is a three-dimensional cube. It has the client and order dimensions in common with *sales* and has location as geographical dimension.

Let us suppose the following preliminary workload (composed of five queries) has been defined from user requirements using the high-level language introduced in [5]:

```
sales[day, product].price,        orders[day, product].price,
sales[day, prod_name].price,      sales[day, location].price.
sales[day, product].amount,       □
```

At this point, we have to verify whether each query of the workload can be effectively expressed on the conceptual schema in Fig. 1. The traditional methodology of the DFM [5] leads to a manual mapping of each query on the graphical representation of the schema. Clearly, this methodology can be very expensive for designers when dealing with complex schemas and numerous queries included in the workload. Furthermore, this can generate human errors and can lead to time wasting. On the other hand, in our approach, metadata can be automatically generated from a conceptual schema designed by a CASE tool. The following metadata are the description of *shipments* cube according to the predicates defined in Table 1. The metadata of the *sales* cube can be obtained in an analogous way.

```
cube(shipments).
measure(cost, shipments).
dimension(clients_dim).
dimension(orders_dim).
dimension(geo_dim).
hierarchy(orders_hier, orders_dim).
hierarchy(clients_hier, clients_dim).
hierarchy(geo_hier, geo_dim).
level(orders,1,orders_hier).
level(clients,1,clients_hier).
level(locations,1, geo_hier).
attribute(clients, client, id).
attribute(clients, client_name, desc).
attribute(orders, order, id).
attribute(locations, location, id).
attribute(locations, loc_name, desc).
cube_dim(shipments, location).
cube_dim(shipments, order).
cube_dim(shipments, client). □
```

Then, each query of the workload is translated into a specific goal by the compiler and the set of goals represents the input of the logical program we use to validate the conceptual schema. The translation made by the compiler follows.

```
goal(sales, [day, product], price),
goal(sales, [day, prod_name], price),
goal(sales, [day, product],amount),
goal(orders, [day, product],price),
goal(sales, [day, location],price). □
```

In detail, each goal is a test on the conceptual schema and the schema is validated only in the case the IE reports no errors at all. Thus, using the produced multidimensional metadata and goals, the IE executes the five tests, and, for each of them, we report the output with reference to the given goal.

Test 1: *goal(sales, [day, product], price).*
Sales is a cube, price is a measure, day is a dimensional attribute, product is a dimensional attribute, there is a valid aggregation path to day from sales, there is a valid aggregation path to product from sales.

Test 2: *goal(sales, [day, prod_name], price).*
Sales is a cube, price is a measure, day is a dimensional attribute, prod_name is not a dimensional attribute.

Test 3: *goal(sales, [day, product], amount).*
Sales is a cube, amount is not a measure.

Test 4: *goal(orders, [day, product], price).*
Orders is not a cube.

Test 5: *goal(sales, [day, location], price).*
Sales is a cube, price is a measure, day is a dimensional attribute, location is a dimensional attribute, there is a valid aggregation path to day from sales, there is no valid aggregation path to location from sales.

In Test 1, no error message is reported. This means that the schema is able to correctly provide an answer to this query. In Test 2, there is the evidence that *prod_name* is not a dimensional attribute. In fact, it has been defined as a descriptive attribute. Then, the program ends reporting an error and, as a consequence, this states that the schema is not valid in reference to the given workload. At this point, the designer can choose whether to correct the query (*product* is the correct one) or to modify the schema, by introducing a further dimensional attribute (in the case that no dimensional attribute exists for the *products* dimension). Let us assume to continue the validation process. In Test 3, the error reported is that *amount* is not a valid measure of the *sales* cube. Thus, the query could not be answered. This obliges the designer to modify the schema, by introducing the needed measure for the *sales* cube. In Test 4, the error reports that *orders* is not a cube and the designer has to introduce an *ad-hoc* cube in the schema in order to support this query. Finally, in Test 5, we have that location is a dimensional attribute but there is no aggregation path from *sales*. In fact, this dimensional attribute is part of a hierarchy that belongs to the *shipments* cube. So, the designer must (re)model the schema by adding the *location* dimension to the *sales* cube.

In conclusion, we highlight that the traditional overload in the conceptual phase, due to the manual check of the schema, is usually bypassed or ignored by designers. As a consequence, DW designers may obtain a logical schema that does not satisfy user requirements. On the other hand, our approach can be very useful to designers in order to avoid human errors and to obtain time saving, as the logical program is able to detect whether the conceptual schema supports all the queries of the workload in a unified and fast way. This leads to a high level of automation in the design process, especially in the case where the metadata generation process is integrated in the CASE tool utilized by the designer.

7 Conclusions

We have presented a novel methodology able to validate a DW conceptual schema, according to the preliminary workload defined from user requirements. This validation is executed in automatic way via a logic program, based on the Predicate Calculus. In fact, the inferential engine is a logic program that validates the conceptual schema automatically via the inferential process using a set of multidimensional metadata. If the conceptual schema is validated, then the DW designer can safely translate it into a logical one. If it is rejected, then the designer can choose whether to correct the schema or to modify the workload.

Currently, the rules of the inferential engine deal with some basic issues of the validation process, since the rationale behind this work is to test the efficacy of adopting automatic techniques in this step of the data warehouse design lifecycle. Thus, future work consists of extending the logical program in order to manage different kinds of hierarchies and to process also the selection clause of queries.

References

1. Negash, S., Gray, P.: Business Intelligence. In: Handbook on Decision Support Systems 2. International Handbooks on Information Systems, pp. 175–193. Springer, Heidelberg (2008)
2. Jarke, M., Vassiliou, M.: Foundations of Data Warehouse Quality: an Overview of the DWQ Project. In: 2nd International Conference on Information Quality, Cambridge, Mass. (1997)
3. Kimball, R.: The Data Warehouse Lifecycle Toolkit. In: Practical Techniques for Building Data Warehouse and Business Intelligence Systems, 2nd edn. John Wiley & Sons, Chichester (2008)
4. Golfarelli, M., Rizzi, S.: A Methodological Framework for Data Warehouse Design. In: 1st ACM International Workshop on Data Warehousing and OLAP, pp. 3–9. ACM, Washington (1998)
5. Golfarelli, M., Maio, D., Rizzi, S.: The Dimensional Fact Model: a Conceptual Model for Data Warehouses. Int. J. Cooperative Information Systems 7, 215–247 (1998)
6. Chaudhuri, S., Dayal, U.: An Overview of Data Warehousing and OLAP Technology. ACM Sigmod Record 26, 65–74 (1997)
7. Tryfona, N., Busborg, F., Borch Christiansen, J.G.: starER: A Conceptual Model for Data Warehouse Design. In: 2nd ACM International Workshop on Data Warehousing and OLAP, pp. 3–8. ACM, New York (1999)
8. Serrano, M.A., Calero, C., Trujillo, J., Luján-Mora, S., Piattini, M.: Empirical Validation of Metrics for Conceptual Models of Data Warehouses. In: Persson, A., Stirna, J. (eds.) CAiSE 2004. LNCS, vol. 3084, pp. 506–520. Springer, Heidelberg (2004)
9. Bonifati, A., Cattaneo, F., Ceri, S., Fuggetta, A., Paraboschi, S.: Designing Data Marts for Data Warehouses. ACM Transactions on Software Engineering and Methodology 10, 452–483 (2001)
10. Mazón, J.-N., Trujillo, J., Lechtenbörger, J.: Reconciling Requirement-Driven Data Warehouses with Data Sources via Multidimensional Normal Forms. Data & Knowledge Engineering 63, 725–751 (2007)

11. Phipps, C., Davis, K.C.: Automating Data Warehouse Conceptual Schema Design and Evaluation. In: 4th International Workshop on Design and Management of Data Warehouses, pp. 23–32 (2002)
12. Romero, O., Abelló, A.: Multidimensional Design by Examples. In: Tjoa, A.M., Trujillo, J. (eds.) DaWaK 2006. LNCS, vol. 4081, pp. 85–94. Springer, Heidelberg (2006)
13. dell'Aquila, C., Di Tria, F., Lefons, E., Tangorra, F.: Dimensional Fact Model Extension via Predicate Calculus. In: 24th International Symposium on Computer and Information Sciences, pp. 211–216. IEEE Press, Los Alamitos (2009)
14. Ballard, C., Herreman, D., Schau, D., Bell, R., Kim, E., Valencic, A.: Data Modeling Techniques for Data Warehousing. IBM Redbooks, Riverton (1998)
15. Halpin, T.A.: Conceptual Schema and Relational Database Design, 2nd edn. Prentice Hall, Australia (1995)
16. Husemann, B., Lechtenborger, J., Vossen, G.: Conceptual Data Warehouse Design. In: International Workshop on Design and Management of DataWarehouses, Stockholm, Sweden, pp. 6-1–6-11 (2000)
17. Halpin, T.A., McCormack, J.I.: Automated Validation of Conceptual Schema Constraints. In: Loucopoulos, P. (ed.) CAiSE 1992. LNCS, vol. 593, pp. 364–377. Springer, Heidelberg (1992)
18. Proper, H.: Generating Significant Examples for Conceptual Schema Validation. In: Interactive Query Formulation using Query By Navigation. Asymetrix Research Laboratory, University of Queensland, Australia (1994)
19. Costal, D., Teniente, E., Urpí, T., Farré, C.: Handling Conceptual Model Validation by Planning. In: Constantopoulos, P., Vassiliou, Y., Mylopoulos, J. (eds.) CAiSE 1996. LNCS, vol. 1080, pp. 255–271. Springer, Heidelberg (1996)
20. Huynh, T.N., Mangisengi, O., Min Tjoa, A.: Metadata for Object-Relational Data Warehouse. In: International Workshop on Design and Management of Data Warehouses, Stockholm, Sweden, pp. 3-1–3-9 (2000)
21. Sen, A.: Metadata Management: Past, Present and Future. Decision Support Systems 37, 151–173 (2004)
22. Object Management Group, Common Warehouse Metamodel Specification, vers. 1.1, vol. 1. OMG, Needham (2003)
23. Dijkstra, E.W., Scholten, C.S.: Predicate Calculus and Program Semantics. Springer, New York (1990)
24. Sterling, L., Shapiro, E.: The Art of Prolog: Advanced Programming Techniques, 2nd edn. MIT Press, Cambridge (1994)
25. Donnelly, C., Stallman, R.: Bison Version 2.1 (2005), http://www.gnu.org/software/bison
26. Paxson, V., Estes, V., Millaway, J.: Flex: the Fast Lexical Analyzer Manual Edition 2.5.35 (2007), http://flex.sourceforge.net

A Model-Driven Heuristic Approach for Detecting Multidimensional Facts in Relational Data Sources

Andrea Carmè[1], Jose-Norberto Mazón[2], and Stefano Rizzi[3]

[1] Iconsulting, Italy
a.carme@iconsulting.biz
[2] Lucentia Research Group
Dept. of Software and Computing Systems
University of Alicante, Spain
jnmazon@dlsi.ua.es
[3] DEIS - University of Bologna, Italy
stefano.rizzi@unibo.it

Abstract. Facts are multidimensional concepts of primary interests for knowledge workers because they are related to events occurring dynamically in an organization. Normally, these concepts are modeled in operational data sources as tables. Thus, one of the main steps in conceptual design of a data warehouse is to detect the tables that model facts. However, this task may require a high level of expertise in the application domain, and is often tedious and time-consuming for designers. To overcome these problems, a comprehensive model-driven approach is presented in this paper to support designers in: (1) obtaining a CWM model of business-related relational tables, (2) determining which elements of this model can be considered as facts, and (3) deriving their counterparts in a multidimensional schema. Several heuristics –based on structural information derived from data sources– have been defined to this end and included in a set of Query/View/Transformation model transformations.

1 Introduction

The development of data warehouses is based on detecting multidimensional elements from a detailed analysis of data sources. Among multidimensional elements, facts are those of highest importance since they represent events of interests for knowledge workers. Therefore, several techniques, such as guidelines or glossaries, have been developed so far to support designers in detecting multidimensional roles of elements in a relational schema (including facts). For example, in a retail domain, a table called Sales is likely to cover the role of a fact. However, these techniques may become tedious and time-consuming when the application domain is complex (in a medical domain, is a table called FertilityCycle a fact?) or, even worse, when table names are meaningless (what is the multidimensional counterpart of a table called SP_CCCM?).

T.B. Pedersen, M.K. Mohania, and A M. Tjoa (Eds.): DaWaK 2010, LNCS 6263, pp. 13–24, 2010.

Other approaches arose to support designers in tackling this task in a more automated manner [1,2,3]. However, these are focused on automatically detecting other multidimensional concepts (such as dimension hierarchies) rather than facts, so discovering facts still relies on informal techniques. Furthermore, most approaches assume that data sources are well-documented or documentation can be easily obtained; unfortunately, this is not generally true [4], and even if some documentation exists, it is likely to be out-of-date with respect to the actual data sources.

To overcome these drawbacks, in this paper we present an approach for formalizing fact detection from relational data sources without requiring additional documentation. Our approach is based on a set of heuristics, elicited from some real-world case studies we are working on. These heuristics use some syntactical information derived from the data sources, thus guiding designers in the detection of facts independently of their knowledge about the application domain. We have formalized these heuristics by means of QVT (Query/View/Transformation) transformations in a model-driven perspective, in such a way that the final multidimensional schemata are derived with a high degree of automation, thus saving time and costs. Basically, our approach consists of three tasks (see Fig. 1): (1) detect clusters of business-related tables within data sources and derive their relational CWM model, (2) support designers in properly determining which elements of this model can be considered as facts by means of a set of heuristics-based QVT model transformations, and (3) model facts, together with their dimensions and measures, in a multidimensional schema.

The remainder of this paper is structured as follows. Section 2 briefly describes the current approaches for discovering multidimensional facts. Section 3 describes our heuristics and the definition of model transformations for detecting facts. Section 4 presents an implementation of our approach and draws the conclusions.

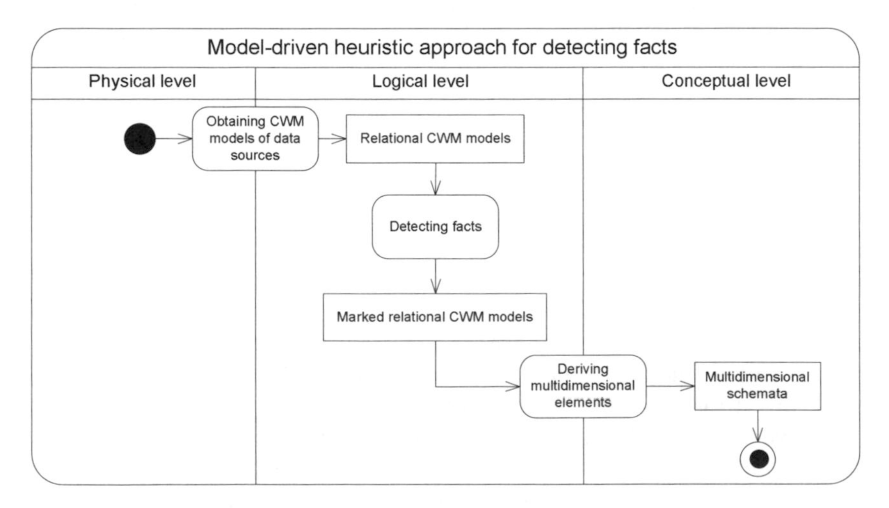

Fig. 1. Overview of our approach for detecting facts

2 Related Work

Most approaches for deriving multidimensional schemata from relational data sources (e.g., [5,6,7,8]) propose informal mechanisms (such as guidelines or glossaries) to support designers. In order to increase the level of automation of this task, other approaches use heuristics to determine which tables are good candidates to become facts. Phipps and Davis [1] propose to consider every entity in an Entity-Relationship schema that contains numerical attributes as a fact, which may be unfeasible since (1) most entities in a schema would be selected, and (2) it is assumed that an up-to-date conceptual schema of data sources is available. Jensen et al. [2] consider not only the presence of measures, but also table cardinality to identify facts; though this approach builds on a reverse-engineering stage in which relational metadata is obtained from data sources, its success highly depends on the skill of domain experts.

Two automated approaches for detecting facts are presented in [3] and [9]. Song et al. [3] propose structural heuristics to detect facts from an Entity-Relationship schema: all entities with a high number of many-to-one relationships are candidates to become facts. Not realistically, they assume that a conceptual schema is always available. Romero and Abelló [9] detect facts by expressing multidimensional SQL queries over relational data sources, and assume that those aggregated attributes in the SELECT clause which are not included in the GROUP BY clause belong to a table that is a potential fact. However, this approach depends on the ability of the users to express their own information requirements as SQL queries.

Our work is inspired by [10], that considers relational data sources as legacy systems whose documentation either is not available, or cannot be obtained, or is too complex to be easily understood through a manual analysis. To overcome these problems, they consider the development of a data warehouse as a modernization scenario which addresses the analysis of the available data sources aimed at discovering multidimensional structures. These structures are then used to derive a data-driven multidimensional schema or reconcile a requirement-driven multidimensional schema with data sources. However, the heuristics for detecting facts presented in that work are rather simplistic and deliver a single solution, which may hide the analysis potential of data sources.

3 Model-Driven Heuristic Approach for Detecting Facts

Our model-driven approach aims to support designers in marking tables from relational data sources as facts. Each table can be differently marked, thus suggesting several possibilities to designers. A set of heuristics for determining which tables are good candidates for being facts, mainly based on an analysis of functional dependencies, have been developed and formalized by using QVT (Query/View/Transformation) [11] model transformations. Our approach assumes that all database constraints (primary and foreign keys) are known, which is perfectly reasonable since these constraints can be nimbly derived [12].

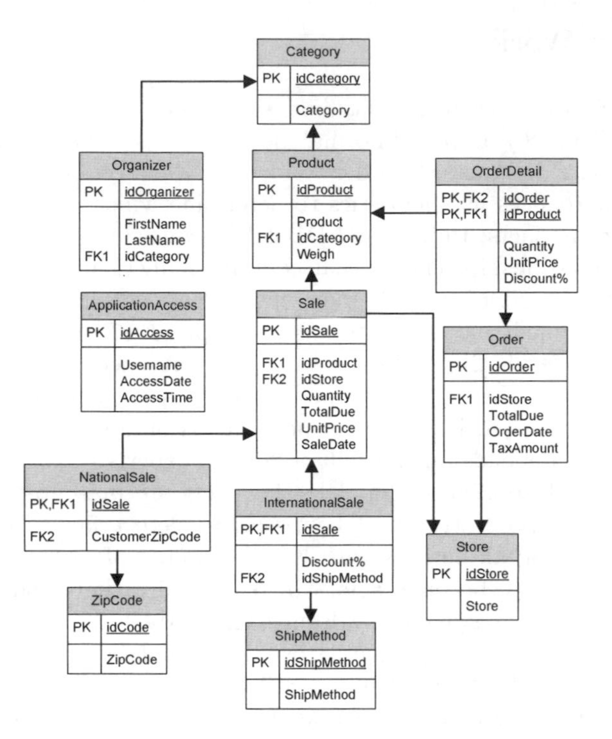

Fig. 2. Relational schema for the running example

The example we will use throughout the paper is based on the retail domain (see Fig. 2) and summarizes situations we have detected in a real case study we are working on at the Spanish fertility institute *TAHE Fertilidad*[1], which we cannot show due to confidentiality issues. Data related to sales and orders are stored, as well as stores, products, etc. Sales are specialized into national and international ones. The OrderDetail relation allows to include several products in each order.

3.1 Obtaining CWM Models of Data Sources

This phase concerns the extraction of relational elements (tables, columns, and constraints) from data sources by querying the DBMS data dictionary. It consists of two steps: (1) delimiting the relational elements related to the application domain, and (2) creating their models based on CWM.

The rationale behind the first step is that, in real-world scenarios, data sources not only store interesting data for analysis but also data about instance feeding applications, security, audit, and so on, that should be ignored when facts are being detected. The benefits of this pre-processing step are twofold: on the one hand, useless elements are not considered; on the other, heuristics will be

[1] `http://www.tahefertilidad.es`

more reliable because the required measures will be calculated by considering only interesting relational elements. Relational elements are first grouped into clusters, using a graph theory algorithm that computes connected graph components [13]. The output is a set of directed, connected graphs whose nodes and edges represent relations and functional dependencies, respectively. Then the designer, in collaboration with domain experts, manually determines which clusters are useful for analysis. In our running example, the cluster containing table ApplicationAccess is not considered, since it is supposed to be unrelated to the business domain.

During the second step, a *relational CWM* (rCWM) model is created for each selected cluster. *Common Warehouse Metamodel* (CWM) [14] consists of a set of metamodels for representing data warehouse and business intelligence metadata, including a relational metamodel that allows relational elements to be easily represented. The next phases of our approach are applied separately to each rCWM model created. Fig. 3 shows part of the rCWM model for our running example.

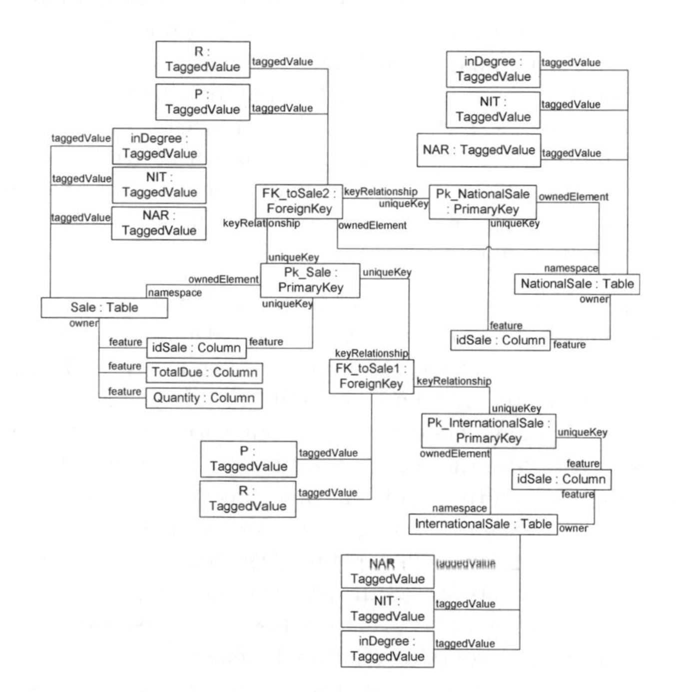

Fig. 3. Part of the relational CWM model for the running example

3.2 Detecting Facts

The fact detection process (Fig. 4) consists of several steps aimed at (1) marking relationship cardinalities, (2) calculating the in-degree of tables, (3) marking facts, (4) marking dimensions and measures, and (5) spawning analysis contexts. Note that several marks can be applied to each relational element, by adding

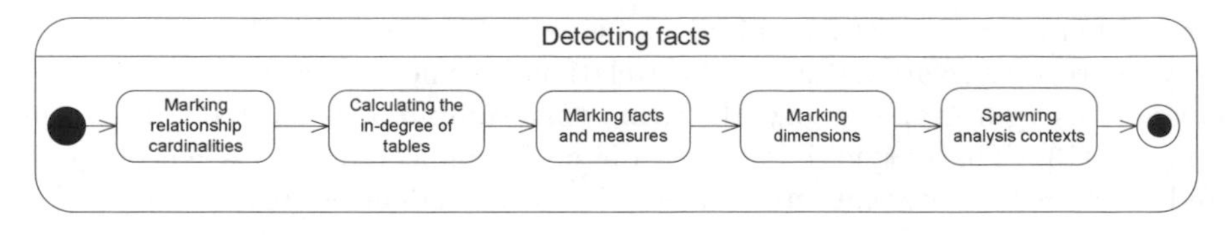

Fig. 4. Fact detection process

values to the *description attributes* provided by CWM. Before explaining the process steps, we describe the heuristics they rely on.

Heuristics. Our heuristics are based on a set of measures calculated from the tables of the rCWM model.

1. The first heuristics states that a table may be a fact if it contains a higher number of instances (NIT) than most other tables. The rationale is that a large table is frequently updated because it stores data related to dynamic events of a business process. The NIT value is retrieved querying data sources through a simple SQL query.
2. The second heuristics states that a table may be a fact if it has a large ratio of numerical attributes: $NAR = NNA/NTA$, where NNA is the number of numeric attributes and NTA is the total number of attributes of a table.
3. The third heuristics states that a table may be a fact if it has a low in-degree, i.e., few or no incoming foreign keys (an incoming foreign key for table T is a foreign key referencing the primary key of T).

To quantify qualitative terms such as "high" and "few", we computed three thresholds. Thresholds for NIT and NAR are calculated using the statistical percentile concept [15]. We have chosen the upper quartile (75-th percentile) as the NIT threshold and the lower quartile (25-th percentile) as the NAR threshold because this gave good results in our case study. Of course, further tests will be needed to find the best percentile to be used in general cases. The in-degree threshold is fixed to 1, which means considering as potential facts only tables with one or no incoming foreign key. We use 1 instead of 0 to consider some specific patterns that we will explain in the following subsections.

Each heuristic measure is stored in a CWM tagged value connected to the related table, as shown in Fig. 3. Thresholds are stored using tagged values linked to the package that contains relational elements.

Marking relationship cardinalities. The relational model has a limited expressiveness. Specifically, one-to-one relationships, that have an ad-hoc representation in the Entity-Relationship model, are not explicitly modeled in a relational schema. Indeed, the existence of a foreign key between two tables does not explain if the relationship between these tables is many-to-one or one-to-one. Since this knowledge is necessary for our approach, we use two transformations to single out two kinds of one-to-one relationships that we will call, respectively, *strong* and *weak*.

- Strong one-to-one relationships are *schema-based* since they are derived and validated within the schema structure. Precisely, a strong one-to-one relationship between two tables T and S is detected when the primary key of T is a foreign key referencing S. A QVT transformation checks this pattern inside rCWM models and marks the foreign keys involved as one-to-one.
- Weak one-to-one relationships are *instance-based*, since they are elicited from data sources instances. A weak one-to-one relationship between T and S is detected when T includes a foreign key (different from its primary key) referencing S, and at most one tuple of T has the value of the primary key of each tuple of S. In this case, no explicit schema constraint assured the correctness of this cardinality assumption; however, considering that data warehouse systems are typically fed by data sources populated with a huge amount of data –hence, instances are representative of the application domain–, we can reasonably take it as true. A specific QVT transformation has been developed for detecting this pattern by integrating the algorithm proposed in [4]. Precisely, two queries are performed over T to count the number of non-null values of its foreign key with and without duplicates; the QVT transformation stores the results, compares them, and marks the foreign key as one-to-one if they are equal.

Foreign keys not marked as one-to-one are marked as many-to-one. In our running example, the foreign keys that link NationalSale and InternationalSale to Sale are marked as (strong) one-to-one, as well as the (weak) one that connects Organizer to Category. The other foreign keys are marked as many-to-one.

Calculating the in-degree of tables. A QVT transformation rule has been defined to calculate in-degree of tables. Note that a foreign key that has already been marked as one-to-one is not taken into account here, due to the possibility to navigate these relationships in both ways. Indeed, two tables marked as facts can be linked by a foreign key expressing a one-to-one relationship.

In our running example, table Order has in-degree 1, while Sale has in-degree 0 even if it has two incoming foreign keys (from NationalSale and InternationalSale, respectively), because these were marked as one-to-one.

Marking facts and measures. A table is marked as a fact if (1) its NIT and NAR are greater or equal to the thresholds, and (2) its in-degree is 0 or 1. The comparison is made by the QVT transformation presented in Fig. 5. Then, all numerical attributes of each table T marked as fact (excluding those belonging to the primary key of T) are marked as potential measures. In our example, Sale, Order, OrderDetail, and Product meet the first constraint, so they can be marked as facts. However, Product is not marked as a fact because its in-degree is 2 (i.e. the second constraint is not fulfilled).

Marking dimensions. For each table T marked as fact, its dimensions and the related hierarchies can be derived by following many-to-one relationships as normally done in current approaches (e.g., [5,1]).

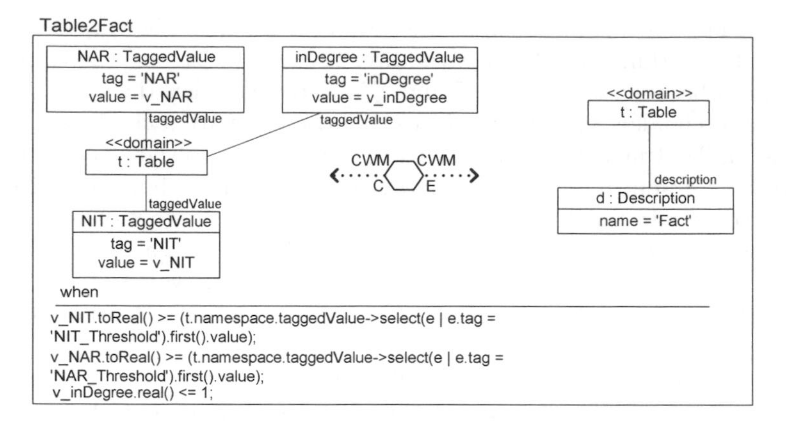

Fig. 5. QVT transformation for marking facts

Spawning analysis contexts. The aim of this phase is to create a set of models, each related to a possible analysis context, so as to generate every multidimensional solution implicitly contained in the relational data sources. This is done in two situations:

1. *Fact-dimension conflicts.* After the marking process, the marked rCWM model may present some configurations of marks that lead to inconsistencies in the multidimensional schema. These conflicts must be handled before creating the multidimensional representation of elements. Precisely, a marked rCWM model contains a conflict when a table is marked both as a fact and as a dimension. In our example, there is a conflict in the Order table. To overcome this problem, for each table T that has a conflict two rCWM models, corresponding to two different analysis contexts, are spawned: one where T is marked as dimension, one where it is marked as fact.
2. *Specialization.* When a table T marked as fact has a one-to-one foreign key referencing table S, we spawn two rCWM models: only S is marked as fact in the first one; S and T are marked as facts in the second one. For example, InternationalSale and NationalSale are both linked with one-to-one relationships to Sale. This leads to creating three rCWM models where: (1) only Sale is marked as fact, (2) Sale and InternationalSale are marked as facts, and (3) Sale and NationalSale are marked as facts.

In the end, the total number of rCWM models spawned depends on the number of conflicts and specializations in the original marked rCWM model. Precisely, the total number of rCWM models is $MN = (CN * 2) * \prod_{i=1}^{SN} SNT_i$ where CN is the number of fact-dimension conflicts, SN the number of specializations, and SNT_i the number of tables involved in the i-th specialization.

It is worth noting that an exponential number of rCWM models is obtained this way. In order to manage these high amount of models, our proposal can be easily integrated in the model-driven approach for data warehouse development proposed in [16,17], where the rCWM models can be reconciled with a conceptual schema previously defined from the information requirements of decision makers. A single multidimensional schema, that at the same time fits data sources and fulfills user requirements, is obtained this way. Due to space constraints, this reconciliation phase is not discussed in this paper.

3.3 Deriving Multidimensional Elements

The spawning phase creates one or more rCWM models. Two special patterns have been developed for handling special situations that can arise afterwards, namely (1) *skip* and (2) *merge*. Both share the same starting situation, i.e., two tables T and S marked as facts and such that T references S via a foreign key. The patterns are distinguished depending on the the mark applied to this foreign key.

1. When the foreign key is marked as many-to-one, a skip pattern is detected. In this case, T and its dimensions are not included in the multidimensional schema, so as to focus on the right granularity in each case. For example, the OrderDetail fact-marked table is skipped and Order is considered as fact. We recall that OrderDetail will be considered as fact in one or more other solutions.
2. A merge pattern is detected when the foreign key is marked as one-to-one. In this case, a fact is created whose dimensions and measures are the union of those belonging to T and S. For instance, Sale can be merged with NationalSale or InternationalSale to create facts for national and international analysis purposes, respectively.

These patterns are applied using QVT transformations, one of which (`Table2Merge`) is shown in Fig. 6b. In this merge transformation, an input pattern consisting of a table T marked as fact that refers S by means of a foreign key fk marked as one-to-one, leads to create a fact f (previously created from table S by means of the `Table2MDFact` transformation as shown in Fig. 6a). Importantly, according to the QVT transformations called in the WHERE clause,

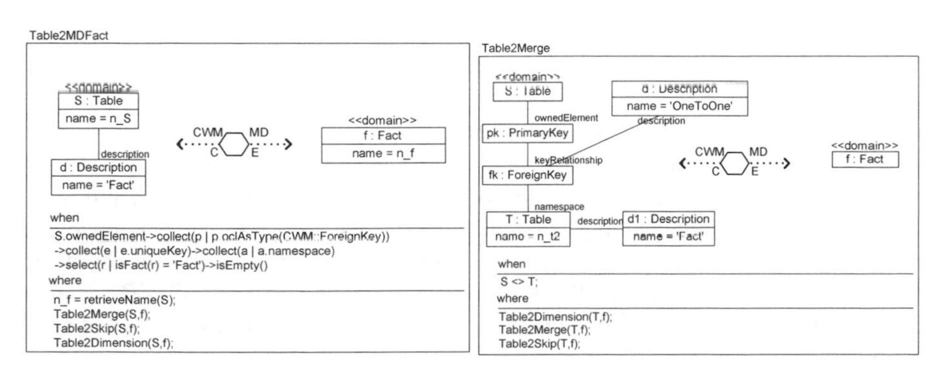

(a) Obtaining facts (b) Merging facts

Fig. 6. QVT from a marked rCWM model to a multidimensional schema

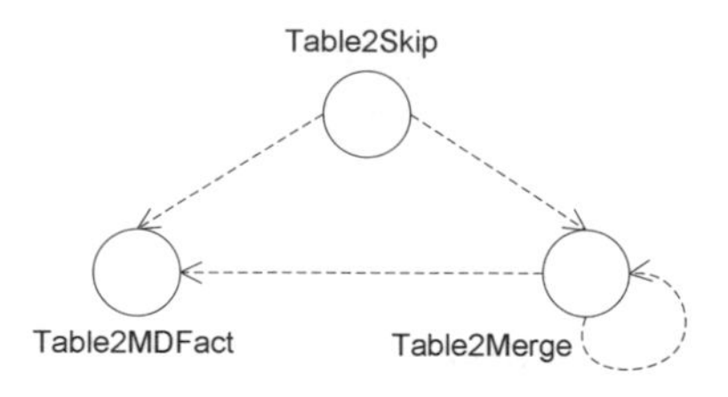

Fig. 7. Transformation execution order

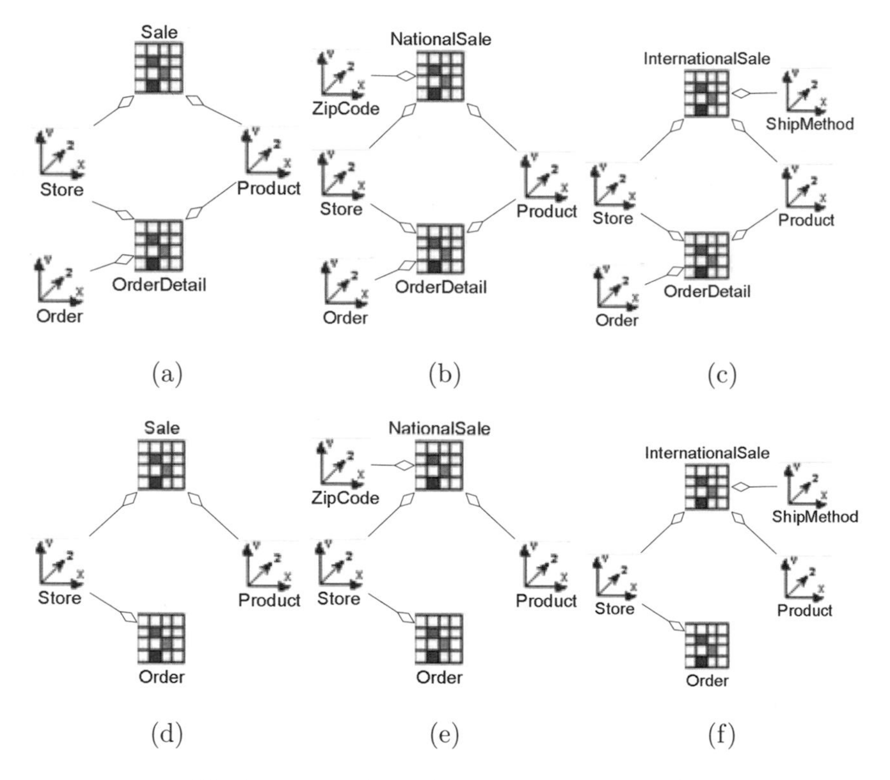

Fig. 8. Approach results over running example

the multidimensional counterparts of all the tables related to T will be related to f. Besides, when merge transformations are applied, the name of the table analyzed in the last merge transformation called is chosen as the fact name.

As to the order for applying transformations, the `Table2MDFact` transformation is executed first to create all facts, then special patterns are detected and applied by means of the QVT transformations called in the WHERE clause. The transformation flow is graphically represented in Fig. 7 using the approach defined in [18].

In Fig. 8 we present the solutions derived by applying our approach to the running example (measures and time dimensions are not shown for simplicity). The solutions in Fig. 8a, 8b, and 8c consider as facts OrderDetail and Sale in a general, national, and international analysis context respectively. The solutions in Fig. 8d, 8e, and 8f consider as facts Order rather than OrderDetail. As a whole, these solutions bring to light the full multidimensional potential of data sources; designers can then select the solution that best matches user requirements.

4 Conclusions and Future Work

Current approaches for data-driven conceptual design do not give designers a comprehensive and formal approach to detect facts. To fill this gap, in this paper we presented a model-driven approach for formalizing fact discovery in relational data sources by means of QVT transformations. Our approach is based on a set of heuristics relying on syntactical information derived from the data sources, thus guiding designers in the detection of multidimensional facts independently of their knowledge about the application domain. Remarkably, our approach has low computational complexity; the total processing time for the largest relational source schema we used for testing (about 130 tables and 140 foreign key constraints) is about 20 seconds.

The proposed model transformations have been implemented in the ECLIPSE[2] development platform. ECLIPSE is an open source project which has been conceived as a modular platform that can be extended by means of plugins in order to add more features and new functionalities. In that way, we have designed a set of modules encapsulated in a single plugin that provides ECLIPSE with capabilities for supporting our approach:

Relational module. It implements the relational metamodel contained in CWM.

Multidimensional module. The profiling mechanism of the *Unified Modeling Language* (UML) has been used to create multidimensional models.

Transformation module. It uses *mediniQVT*[3], a QVT transformation engine, in order to code and execute the mapping patterns.

Acknowledgments. This work has been supported by the QUASIMODO (PAC08-0157-0668) project from the Castilla-La Mancha Ministry of Education and Science (Spain). We would like to express our gratitude to personnel at *TAHE Fertilidad* (`http://www.tahefertilidad.es`) for their support during the development of this work.

References

1. Phipps, C., Davis, K.C.: Automating data warehouse conceptual schema design and evaluation. In: Proc. DMDW, pp. 23–32 (2002)
2. Jensen, M.R., Holmgren, T., Pedersen, T.B.: Discovering multidimensional structure in relational data. In: Kambayashi, Y., Mohania, M., Wöß, W. (eds.) DaWaK 2004. LNCS, vol. 3181, pp. 138–148. Springer, Heidelberg (2004)

[2] `http://www.eclipse.org`

[3] `http://projects.ikv.de/qvt`

3. Song, I.Y., Khare, R., Dai, B.: SAMSTAR: a semi-automated lexical method for generating star schemas from an entity-relationship diagram. In: Proc. DOLAP, pp. 9–16 (2007)
4. Alhajj, R.: Extracting the extended entity-relationship model from a legacy relational database. Inf. Syst. 28(6), 597–618 (2003)
5. Golfarelli, M., Maio, D., Rizzi, S.: The Dimensional Fact Model: A conceptual model for data warehouses. Int. J. Cooperative Inf. Syst. 7(2-3), 215–247 (1998)
6. Hüsemann, B., Lechtenbörger, J., Vossen, G.: Conceptual data warehouse modeling. In: Proc. DMDW, p. 6 (2000)
7. Böhnlein, M., von Ende, A.U.: Deriving initial data warehouse structures from the conceptual data models of the underlying operational information systems. In: Proc. DOLAP, pp. 15–21 (1999)
8. Moody, D.L., Kortink, M.A.R.: From enterprise models to dimensional models: a methodology for data warehouse and data mart design. In: Proc. DMDW, p. 5 (2000)
9. Romero, O., Abelló, A.: Multidimensional design by examples. In: Tjoa, A.M., Trujillo, J. (eds.) DaWaK 2006. LNCS, vol. 4081, pp. 85–94. Springer, Heidelberg (2006)
10. Mazón, J.N., Trujillo, J.: A model driven modernization approach for automatically deriving multidimensional models in data warehouses. In: Proc. ER, pp. 56–71 (2007)
11. Object Management Group: MOF 2.0 Query/View/Transformation, `http://www.omg.org/cgi-bin/doc?ptc/2005-11-01`
12. Soutou, C.: Relational database reverse engineering: Algorithms to extract cardinality constraints. Data Knowl. Eng. 28(2), 161–207 (1998)
13. Hopcroft, J.E., Tarjan, R.E.: Efficient algorithms for graph manipulation [h] (algorithm 447). ACM Commun. 16(6), 372–378 (1973)
14. Object Management Group: Common Warehouse Metamodel Specification 1.1, `http://www.omg.org/cgi-bin/doc?formal/03-03-02`
15. SAS Institute: Base SAS 9.1.3 Procedures Guide. Second edn. (2006)
16. Mazón, J.N., Trujillo, J., Lechtenbörger, J.: Reconciling requirement-driven data warehouses with data sources via multidimensional normal forms. Data Knowl. Eng. 63(3), 725–751 (2007)
17. Mazón, J.N., Trujillo, J.: A hybrid model driven development framework for the multidimensional modeling of data warehouses. SIGMOD Record 38(2), 12–17 (2009)
18. Meliá, S., Kraus, A., Koch, N.: MDA transformations applied to web application development. In: Lowe, D.G., Gaedke, M. (eds.) ICWE 2005. LNCS, vol. 3579, pp. 465–471. Springer, Heidelberg (2005)

Physical Design and Implementation of Spatial Data Warehouses Supporting Continuous Fields

Leticia Gómez[1], Alejandro Vaisman[2], and Esteban Zimányi[3]

[1] Instituto Tecnológico de Buenos Aires
lgomez@itba.edu.ar
[2] Universidad de Buenos Aires
avaisman@dc.uba.ar
[3] Université Libre de Bruxelles
ezimanyi@ulb.ac.be

Abstract. Although many proposals exist for extending Geographic Information Systems (GIS) with OLAP and data warehousing capabilities (a topic denoted SOLAP), only recently the importance of supporting continuous fields (i.e., phenomena that are perceived as having a value at each point in space and/or time) has been acknowledged. Examples of such phenomena include temperature, altitude, or land use. In this paper we discuss physical design issues arising when a spatial data warehouse includes a combination of spatial and non-spatial dimensions and measures, and spatio-temporal dimensions representing continuous fields. We give the syntax and semantics of the data types (and their operators) needed to support fields in SOLAP environments, and present an implementation of these types, on top of spatial-SQL. We also show how queries using the spatio-temporal operators for fields are written, parsed, and executed.

1 Introduction

In the last few years, efforts have been carried out to integrate Geographic Information Systems (GIS) [1] and OLAP (On-Line Analytical Processing) [2]. This integration, called SOLAP (standing for Spatial OLAP), aims at exploring spatial data by drilling on maps, in the same way as OLAP operates over tables and charts. This concept was introduced by Rivest *et al.* [3], who also describe the desirable features and operators a SOLAP system should have. A survey on the topic can be found in [4]. The need for sophisticated GIS-based decision support systems, for the analysis of organizational data with respect to geographic information, is encouraging OLAP and GIS vendors to integrate their products.

Advances in data analysis technologies raise new challenges. One of them is the need to handle *continuous fields*, which describe physical phenomena that change continuously in time and/or space. Examples of such phenomena are temperature, pressure, and land elevation. Besides physical geography, continuous fields (from now on, fields), like land use and population density, are used in human geography as an aid in spatial decision-making process. Formally, a field

T.B. Pedersen, M.K. Mohania, and A M. Tjoa (Eds.): DaWaK 2010, LNCS 6263, pp. 25–39, 2010.

is defined as composed of [5]: (a) a domain $\mathcal{D}$ which is a continuous set; (b) a range of values $\mathcal{R}$; and (c) a mapping function f from $\mathcal{D}$ to $\mathcal{R}$.

Although some work has been done to support querying fields in GIS, spatial multidimensional analysis of continuous data is still in its infancy. Existing multidimensional models dealing with discrete data are not adequate for the analysis of continuous phenomena. Multidimensional models and associated query languages are thus needed, to support continuous data. Recently, Vaisman and Zimányi [6] presented a conceptual model for SOLAP that supports dimensions and measures representing continuous fields, and characterized multidimensional queries over fields. They defined a *field* data type, a set of associated operations, and a multidimensional calculus supporting this data type. In this paper we go a step further, and study the translation of this conceptual data model to physical structures based on the well-known star-schema [2]. We also introduce two new data types, field and tempfield, define a semantics for the operators associated to these types, and present an implementation for them. Finally, we define an SQL-like query language over the physical structures and operators mentioned above, and provide a preliminary implementation of the language.

This paper is organized as follows. Section 2 provides an overview of related work dealing with fields. Section 3 presents the conceptual model, and introduces the field data type and its associated operators. In Section 4 we discuss the physical warehouse design to implement the conceptual model and we introduce the SQL-like language to support fields. Section 5 presents the operators of the field data type, whose implementation is shown in Section 7. Section 6 presents the query language, and Section 8 sketches how a query in this language is implemented. We conclude in Section 9.

2 Related Work

In his pioneering work on defining *algebra for fields*, Tomlin [7] proposed a so-called map algebra, based on the notion that a map is used to represent a continuous variable (e.g., temperature). There are three types of functions in Map algebra: *local*, *focal*, and *zonal*. Local functions compute a value at a certain location as a function of the value(s) at this location in other map layer(s). Focal functions compute each location's value as a function of existing values in the neighboring locations of existing layers (i.e., they are characterized by the topological predicate *touches*). Zonal functions (characterized by the topological predicate *inside*), compute a location's new value from one layer (containing the values for a variable), associated to the zone (in another map) containing the location. Câmara *et al.* [8] and Cordeiro *et al.* [9] formalized and extended these functions, supporting more topological predicates. We base our proposal on this work, and on the proposal of Mennis *et al.* [10], where map algebra operators are extended to query time-varying fields. The model and query language we present here cover those proposals, and extend them to the multidimensional setting.

Paolino *et al.* [5] introduced *Phenomena*, a visual language for querying continuous fields, based on a conceptual model where users view the world as consisting of both continuous fields and discrete objects, and are able to manipulate them in a uniform manner. Phenomena uses an extension to Spatial SQL that supports continuous fields, proposed by Laurini *et al.* [11]. GeoRaster[1] is a feature of Oracle Spatial that allows storing, indexing, querying, analyzing, and delivering raster data, and its associated metadata. GeoRaster provides specialized data types and associated operators, as well as an object relational schema, which can be used to store and manipulate multidimensional raster layers. None of these tools and languages were devised for a SOLAP setting.

Regarding *fields and multidimensional models*, the joint contribution of the GIS and OLAP communities to this problem has been limited. Shanmugasundaram *et al.* [12] proposed a data cube representation that deals with continuous dimensions. This works focuses on using the known data density to calculate aggregate queries without accessing the data. The representation reduces the storage requirements, but continuity is addressed in a limited way. Ahmed *et al.* use interpolation methods to estimate (continuous) values for dimension levels and measures, based on existing sample data values [13]. Continuous cube cells are computed on-the-fly, producing a continuous representation of the discrete cube. These proposals are based on a data model devised for OLAP, not for *spatial* OLAP, which goes against a comprehensive representation of spatial dimensions and measures. Opposite to this, our approach is based on a conceptual multidimensional model designed with spatial data in mind. Thus, continuous fields are introduced as a natural extension to this model. In order to support fields, Vaisman and Zimányi presented a conceptual model for spatio-temporal OLAP supporting fields, and a calculus to query such data. In this paper, we build on that work to propose a user-friendly SQL-like version of the calculus making use of two new data types that support spatio-temporal fields.

3 Preliminaries

We now briefly describe the conceptual model proposed in [6], extending the *MultiDim* model [14] to support fields. For this we use the example in Figure 1, which represents information about crops produced at land plots. We use this model also as our running example. There is information in vector format describing the location of land plots in provinces. Further, there are raster maps of elevation, soil type, temperature, and precipitation.

A *multidimensional schema* is a finite set of dimensions and fact relationships. A *dimension* comprises at least one *hierarchy*, which contains at least one level. A hierarchy with only one level is called a *basic hierarchy*. Levels in a hierarchy (e.g., the one formed by LandPlot and Province) are related to each other through a binary relationship that defines a partial order $\preceq$ between them. Given two consecutive related levels l_i, l_j, if $l_i \preceq l_j$ then l_i is called *child* and l_j is called *parent*. When levels in a hierarchy are spatial, they are related by a topological relationship. For example, the ⬬ pictogram in the LandPlot hierarchy indicates that a land plot is covered by its parent (a province).

[1] http://download.oracle.com/docs/html/B10827_01/geor_intro.htm

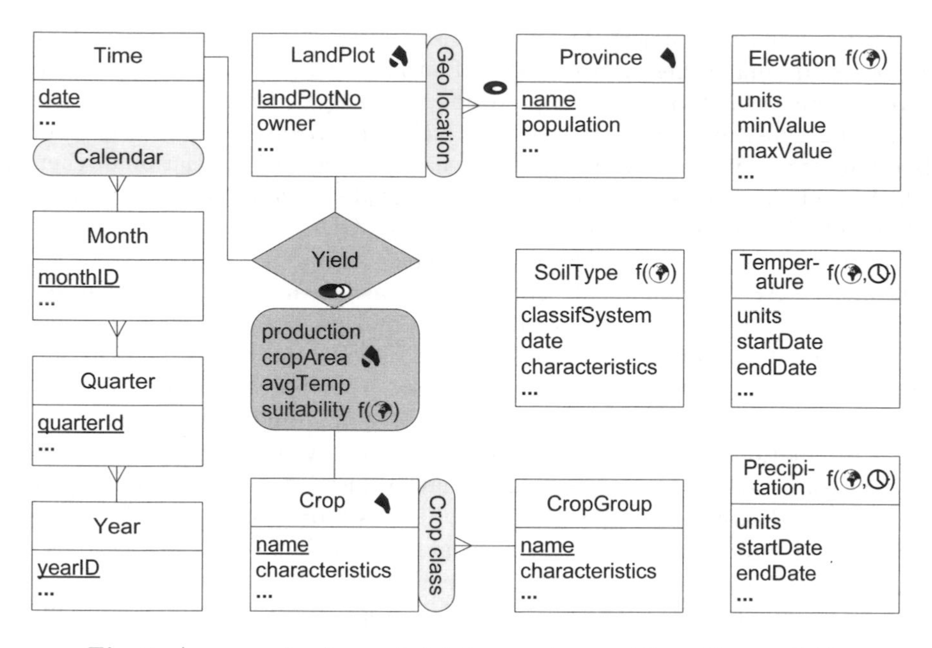

Fig. 1. An example of a spatial data warehouse with continuous fields

A level representing the least detailed data in a hierarchy is called a *leaf level* (e.g., LandPlot), and is related to at least one *fact relationship* (e.g., Yield). The latter represents an n-ary relationship between two or more leaf levels. If these levels are spatial, the relationship may also be topological and requires a spatial predicate. For example, the pictogram in Yields indicates an intersection between the spatial dimensions LandPlot and Crop. A fact relationship contains measures, which may be thematic or spatial. The former (e.g., production) are the usual alphanumeric measures in standard OLAP, and may be calculated using spatial operators, such as distance or area. The latter are represented by a geometry or a field. An example is the cropArea measure, which is computed as the intersection of land plots and crop areas. Dimension levels are composed of key attributes and property attributes. A *key attribute* of a parent level (e.g., name in Province) determines how child members are grouped for applying aggregation functions to measures. A *property attribute* contains additional features of a level; it can be spatial (represented by a geometry or field) or thematic (alphanumeric data types).

To support fields, we include the notion of field dimensions and field measures. *Non-temporal* field levels and measures are identified by the f( ) pictogram, while *temporal* ones are identified by the f(,) pictogram. A *field dimension* is a dimension containing at least one level that is a field. In our example, the field dimensions are Elevation, SoilType, Temperature, and Precipitation where the latter two are temporal field dimensions. A *field measure* is a measure represented by a continuous field. For example, suitability is a field measure computed in terms of elements in the model, e.g., the suitability at a certain point can be a function of the kind of crop, temperature, precipitation, and elevation, at that point or its vicinity. Finally, a *field hierarchy* is a set of related field levels; it allows a field to be seen at different granularities. Although not shown in our example, SoilType can define a hierarchy for soil classification, e.g., the USDA Soil Taxonomy.

Notice that in our approach field dimensions deserve particular treatment. In traditional multidimensional models, every dimension is connected to at least one fact relationship. The same approach has been followed in models introducing fields in spatial data warehouses (e.g., [13]), where dimension instances are values in the underlying domain (that may be obtained through on-the-fly interpolation). Due to the nature of continuous fields, there may be an infinite number of instances, each one corresponding to one possible value of the domain. We chose a different approach: we define a field dimension containing only one instance (corresponding to the function), and the attributes of the field dimension correspond to metadata describing it, like the units at which the values are recorded (e.g., Celsius or Farenheit for temperature). Consequently, field dimensions are part of the model, but are not tied to any particular fact table. The physical model we propose below shows the viability of this approach, and reveals the drawbacks of on-the-fly interpolation.

4 A Physical Model for DWs with Continuous Fields

We discuss next a physical data model supporting the conceptual model introduced in Section 3. Most conceptual models for spatial data warehouses proposed so far are based on the star/snowflake schema, but do not follow such schema when it comes to implementation issues. In these models it is not clear how a field (e.g., implemented as a raster grid) can fit into the standard star/snowflake schema. As mentioned before, we propose a different approach where field dimensions are not linked to the fact table. Although field dimensions are part of the model, and are considered as elements in the query language, there is no natural key to tie fields to a fact table. However, aggregations over fields can be included as pre-computed field measures in the fact table.

We represent fields using raster structures describing regular square grids. Our implementation is based on the OpenGIS specification for coverage geometry and interpolation functions of the OGC [15]. Other implementations of fields using irregular tesselations of space, such as triangulated irregular networks (TIN) or Voronoi diagrams, are possible and are left for future work. Nontemporal fields are stored in a table containing a spatial attribute (denoted geom) representing the geometry of the cell, and an alphanumeric attribute that stores its value. Temporal field tables have two additional attributes representing, respectively, the start and end instants of the value's validity interval.

Dimension tables represent both spatial and nonspatial dimensions. There can be either one table per dimension level or one table per dimension, depending on whether the dimension is normalized or not (i.e., either a star or a snowflake schema is used). *Spatial dimension tables* have an additional spatial attribute (denoted by geom) containing the geometry of the object. *Field dimension tables* have, in addition to the attributes of field tables, an attribute containing the metadata of the field (which can be implemented, for example, as an XML document). Non-field dimension tables have a surrogate identifier denoted id. This identifier allows dimension tables (spatial or not) to be linked to fact tables through a foreign key relationship.

Fact tables include references to spatial and nonspatial dimension tables. We consider four kinds of measures: (1) numeric, as in standard OLAP; (2) spatial measures; (3) field measures; and (4) field aggregations. An example of a measure of type (2) is cropArea in Figure 1. This measure, of spatial type region, represents the intersection (or any other valid spatial operation) of *all* the spatial dimensions, while taking into account the other nonspatial dimensions. In our example, at a given day, a member of the LandPlot level (say, L1) may intersect a member of the Crop level (e.g., wheat). The intersection of the geometries of both members (e.g., multipolygons according to the OGC data types) results in another geometry that is recorded in the fact table as the spatial measure cropArea. An example of measure of type (3) is suitability in Figure 1. This measure is actually a field, which can be precomputed, at each point, as a function of geographic characteristics of related spatial dimensions (LandPlot and Crop), other related dimensions (Time), other fields (Elevation, ...), and other parameters. One possible way of implementing such a measure is to have a field value associated to each instance of the fact table, in our case, to each combination of land plot, crop, and time. Measures of type (4) are aggregations of measures of type (3), in a way that resembles map algebra operations. In Figure 1, measure avgTemp indicates the average value of the temperature field at the finest granularity of the fact table, that is, at the combination of land plot, crop, and day. In other words, a tuple in the fact table Yield will have an attribute that represents this temperature.

5 The Field and Temporal Field Data Types

We define next two new data types, denoted field and tempfield, and their corresponding operations, along the lines of Güting *et al.* [16].

Field types capture the variation in space of base types. They are obtained by applying a constructor field($\cdot$). Hence, a value of type field(real) (e.g., representing altitude) is a continuous function f : point $\rightarrow$ real. We describe next some of the operations of field types.

A set of operations realize the *projection into the domain and range*. The defspace operator receives a field, and returns the geometry defining it; rangevalues receives a field, and returns the set of values that the function takes.

Another set of operators allow the *interaction with domain and range*. Operations atpoint, atpoints, atline, and atregion restrict the function to a given subset of the space defined by a spatial value. That means that the operators receive a field and a geometry, and return a field restricted to such geometry. Operations at and atrange restrict the function to a point or to a point set in the range of the function. Predicates atmin and atmax reduce the function to the points in space when its value is minimal or maximal.

Rate of change operators compute how a field changes across space. Functions partialder_x and partialder_y give, respectively, the partial derivative of the function defining the field with respect to the one of the axis x and y.

Aggregation operators take a field as argument and produce a scalar value. Operations fmin and fmax give, respectively, the minimum and maximum value taken by the function. Three field aggregation operators take as argument a field over numeric values (int or real) and return a real value. These are volume, area, surface with their standard meaning. From these basic operators, other derived operators are defined, namely favg, fvariance, and fstdev.

All operations on base or spatial types are generalized for field types. An operation is *lifted* (following [16]) to allow any of the argument types to be replaced by the respective field type and also return a corresponding field type. Intuitively, the semantics of such lifted operations is that the result is computed at each point using the non-lifted operation. *Aggregation* operators are also uplifted in the same way. For instance, an uplifted avg operator combines several fields, yielding a new field where the average is computed at each point in space. These uplifted aggregation operations correspond to Tomlin's *local functions* [7].

Focal, zonal, and global operators can be derived from the above operators. *Focal* (or *neighborhood*) operators compute a new field in which the output value at a point is a function of the values of the input field in the neighborhood "around" that point. Neighborhoods can be defined by different sizes and geometries. Different arithmetic and statistical functions can be applied to summarize neighborhood values. For example, a focalmax that computes at each point p the maximum value of the neighborhood around that point at a distance d can be defined as follows

$$\mathsf{focalmax}(f, p, d) \stackrel{\text{def}}{=} \mathsf{fmax}(\mathsf{atregion}(f, \mathsf{buffer}(p, d))).$$

Here, the buffer operator creates a surface of radius d around point p, the atregion operator restricts the field f to that surface, and the fmax operator takes the maximum value among all the values of the resulting field.

Zonal operators take as input two fields, f_1 defining the input values and f_2 defining a set of zones, and compute an output field where the value at each point is computed from all values of the input field that belong to the zone associated with that point. For example, a zonalmax that computes at each point p the maximum value of the zone to which p belongs can be defined as follows

$$\mathsf{zonalmax}(f_1, f_2, p) \stackrel{\text{def}}{=} \mathsf{fmax}(\mathsf{atregion}(f_1, \mathsf{defspace}(\mathsf{at}(f_2, \mathsf{val}(\mathsf{atpoint}(f_2, p)))))).$$

Here, atpoint restricts the field f_2 defining the zones to the point p, val takes the value v of the field at that point, at restricts f_2 to the points that have value v, defspace obtains the underlying space where f_2 takes value v, atregion restricts the input field f_1 to that space, and the fmax operator takes the maximum value among all the values of the resulting field.

Finally, *global functions* compute a field in which the value at a point is computed from potentially all the points of the underlying space. An example is the Euclidean distance which, given a set of "sources" defining objects of interest such as schools, hospitals, or roads, computes for each point p of the underlying space the distance to the closest source. If the sources are defined by a geometry g (of one of the four spatial types) such a function can be defined as follows

$$\mathsf{globaldistance}(p, g) \stackrel{\text{def}}{=} \mathsf{distance}(p, g),$$

where the distance function [16] determines the minimum Euclidean distance between the closest pair of points from the first and second arguments.

Temporal fields model phenomena whose value change along time and space. (e.g., temperature). The work in [6] defines temporal fields based on the moving types in [16]. Moving (or temporal) types are obtained by applying the constructor moving(·). Hence, moving(real) (e.g., representing the temperature at a specific point) is a continuous function f : instant $\rightarrow$ real. Temporal fields are obtained by applying a constructor tempfield(·) which is an abbreviation of moving(field(·)). We describe next some of the operators of moving types.

A set of operations realize the *projection into the domain and range*. Operations deftime and rangevalues return, respectively, the projection of a moving type into its domain and range. In other words, given a temporal field, deftime returns the intervals in which it is defined, and rangevalues returns a set with the values in its range.

Another set of operators allow the *interaction with domain and range*. Operations atinstant and atperiod restrict the function to a given time or set of time intervals. That means, given a field and a time instant (period), returns the field(s) valid at that time(s). Operations initial and final return, respectively, the (instant,value) pairs for the first and last instant of the definition time. Operation at restricts the function to a point or to a point set (a range) in the range of the function. Predicates atmin and atmax reduce the function to the times when it was minimal or maximal, respectively. The present predicate allows checking whether the temporal function is defined at an instant of time, or is ever defined during a given set of intervals. Analogously, predicate passes checks whether the function ever assumed one of the values from the range given as second argument.

Finally, as was the case for field types, all operations on nontemporal types are generalized (or lifted) for moving types. As an example, the = operator has lifted versions where one or both of its arguments can be moving types and the result is a moving Boolean. Intuitively, the semantics of such lifted operations is that the result is computed at each time instant using the non-lifted operation.

6 A SOLAP Language That Supports Fields

We now present an SQL-like query language for the model of Section 3. This model requires a language that supports different kinds of objects, namely dimensions, fact tables (spatial and non-spatial), and fields. Vaisman and Zimányi [6] proposed a query language based on the tuple relational calculus extended with aggregate functions and variable definitions proposed by Klug [17]), and showed that extending this calculus with *field* types is enough to express multidimensional queries over fields. We base our language on this calculus.

We start with a simple example that does not include fields: "For land plots located in the province of Limburg and crops in the cereals group give the maximum production by month".

```
SELECT l.landPlotNo, t.month, max(y.production)
FROM LandPlot l, Crop c, Time t, Yield y
WHERE l.province.name="Limburg" AND c.group.name="Cereal"
GROUP BY l.landPlotNo, t.month
```

Like in typical OLAP languages, we hide the structure of the dimensions, which is stored as metadata. Also, metadata allows determining which type of objects are the ones in the FROM clause (e.g., dimension tables – spatial or not –, fact tables, or fields). This query can be trivially translated to SQL as:

```
SELECT l.landPlotNo, m.month, max(y.production)
FROM LandPlot l, Province p, Crop c, Group g, Time t, Month m, Yield y
WHERE y.landPlot=l.id AND l.province= p.id AND p.name="Limburg"
AND y.crop=c.id AND c.group=g.id AND g.name="Cereal"
AND y.time=t.id AND t.month=m.id
GROUP BY l.landPlotNo, m.month
```

We next introduce fields in the language. Let us start with a simple query, not involving a fact table: "Total area at sea level in the province of Antwerp".

```
SELECT area(intersection(defspace(at(e.geom,0)),l.province.geom))
FROM Elevation e, LandPlot l
WHERE l.province.name="Antwerp"
```

Function at restricts the elevation field to the points in space that have the value 0, and defspace yields the region containing such points, which is then intersected with the province of Antwerp. The area operator is finally applied.

The next query includes a fact table: "For land plots having at least 30% of their surface at the sea level, give the average suitability value for wheat on February 1st, 2009."

```
SELECT l.LandPlotNo, favg(y.suitability)
FROM Elevation e, LandPlot l, Yield y, Crop c, Time t
WHERE area(defspace(atregion(at(e,0),l.geom)))/area(l.geom) > 0.3
AND c.name="Wheat" AND t.date="02/01/2009"
```

Here, the elevation field is restricted to the value 0 by means of function at, and the resulting field is restricted to the geometry of the land plot with function atregion. The operator defspace obtains the geometry of the restricted field, the area of this geometry is computed, and this is finally divided by the total area of the land plot. Then, the average suitability is computed using the field aggregation operation favg applied to the field measure suitability.

We now show a spatio-temporal query including fields: "Land plots at the sea level in Limburg with average temperature greater than 10 °C in March 2009 and suitability (at every point of the land plot) for a wheat crop at June 1st, 2009 greater than 1.4."

```
SELECT l.landPlotNo
FROM LandPlot l, Crop c, Time t, Temperature temp, Yield y
WHERE l.province.name="Limburg" AND
favg(avg(atperiods(atregion(temp,l.geom),["03/01/09","03/31/09"])))>10
AND intersects(defspace(at(e,0)),l.geom)
AND t.date= "1/6/2009" AND c.name="Wheat"
AND defspace(atrange(y.suitability,[1.4,-]))=l.geom
```

The temperature field, restricted to the geometry of the land plot and to March 2009, is aggregated with the avg operator (a local cubic operation). Then, favg is applied to obtain the average at the land plot, which is then compared to 10. The topological predicate intersects verifies that the land plot overlaps the region defined by the elevation field restricted to the sea level. After obtaining the instance of the fact relationship relating the land plot, the date, and the wheat crop, the suitability field for this instance is restricted to the points that have a value greater than 1.4, the region containing those points is obtained with function defspace, and it is verified that this region equals the geometry of the land plot, ensuring that every point satisfies the condition.

Finally, we show an example of a query returning a field: "Restrict the precipitation field to December, 2009, to the areas with an altitude greater than 150m, and an average production of wheat greater than one thousand tons."

```
SELECT atregion(atregion(atperiod(p,["12/1/2009","12/31/2009"]),
      defspace(atrange(e,[150,-]))),
   (SELECT l.geom
   FROM Yield y, LandPlot l, Crop c
   WHERE c.name="Wheat"
   GROUP BY l.geom
   HAVING AVG(y.production) > 1000))
FROM Elevation e, Precipitation p
```

The atperiod function restricts the precipitation field to December, 2009 and the result is restricted (inner atregion) to the space defined (defspace) by the restriction of the elevation field to values greater than 150 (atrange). The outer atregion function restricts this resulting field to the result of the inner query which returns the set of geometries for land plots having an average production of wheat greater than one thousand.

7 Implementing the Operators

We show now how the operators over fields are implemented. We designed the following experimental scenario, according to the conceptual model of Figure 1.

We downloaded field data from the WorldClim site[2], which provides layers with raster information at different resolutions. For our region of interest (a portion of Belgium), we used elevation data with a resolution of 5 arc-minutes, obtaining 655 cells, and temperature and precipitation data with a resolution of 10 arc-minutes, obtaining 185 cells. Raster data was downloaded in a generic grid format exported to ESRI Shape file format[3], an later imported to a PostgreSQL database with the PostGIS plugin[4] . This generates polygons with associated values. The units for elevation, precipitation, and temperature are meters, milimeters, and Celsius * 10, respectively. Both, precipitation and temperature data correspond to monthly values. We created synthetically dimension and fact data (e.g., land plots, crops). As we explained in Section 4, fields are stored in tables with attributes 'geom' and 'value'. In addition, temporal fields have attributes 'startDate' and 'endDate' representing the validity interval of the field.

We now show how the defspace and atregion operators are implemented. The other ones are implemented analogously. Since the actual Java code is self-descriptive, we have chosen to show this code instead of pseudo-code listings.

```
(1) Geometry defspace(String tempFieldTable) throws SQLException {
(2) String sqlDML;
(3) sqlDML= String.format("SELECT geom FROM %s", tempFieldTable);
(4) PreparedStatement pstmt = dbConn.prepareStatement(sqlDML);
(5) ResultSet rs = pstmt.executeQuery();
(6) Collection⟨Geometry⟩ geomCollection = new ArrayList⟨Geometry⟩();
(7) while (rs.next()){
(8)    Geometry aGeom = GeometryReader.getGeometry(rs.getObject(1));
(9)    geomCollection.add(aGeom);}
(10) pstmt.close();
(11) return unionAll(geomCollection);}
```

Fig. 2. A Java function to compute defspace

Figure 2 shows a Java function implementing the defspace operator. It receives as parameter the name of the table representing a field and returns the geometry over which the field is defined (i.e., the union of all the polygons that the field contains). The SQL statement in Line (3) retrieves the spatial element in the field table. The loop in Line (7) creates a collection of these geometries.

Note that Figure 3 shows two Java functions that implement the atregion operator. For implementation reasons we need to define two different functions that differ in the type of second parameter. The first atregion function receives a field and a geometry as parameters, and returns a field restricted to the boundaries of the geometry. If the geometry is empty, the field is not updated. The SQL statement in Line (5) deletes the tuples of the field that have no intersection with the geometry. The statement in Line (9) updates the spatial attribute

[2] http://www.worldclim.org/current

[3] http://www.esri.com/

[4] http://www.postgresql.org/; http://www.postgis.org/.

```
void atregion(String tempFieldTable, Geometry geom) throws SQLException {
(1) if (geom.isEmpty())
(2)   return;
(3) String sqlDML;
(4) PreparedStatement pstmt;
(5) sqlDML= String.format( "DELETE FROM %s WHERE NOT
                        INTERSECTS(geom, %s)", tempFieldTable, geom);
(6) pstmt = dbConn.prepareStatement(sqlDML);
(7) pstmt.execute();
(8) pstmt.close();
(9) sqlDML= String.format( "UPDATE %s SET geom=INTERSECTION(geom, %s)",
                        tempFieldTable, geom);
(10) pstmt = dbConn.prepareStatement(sqlDML);
(11) pstmt.execute();
(12) pstmt.close();}

void atregion(String tempFieldTable, Collection⟨Geometry⟩ geomCollection)
      throws SQLException {
(1) if (geomCollection.isEmpty())
(2)   return;
(3) atregion(tempFieldTable, unionAll(geomCollection)); }

Geometry unionAll(Collection⟨Geometry⟩ geomCollection){
(1) Geometry[] geomArray= new Geometry[geomCollection.size()];
(2) int i=0;
(3) for(Iterator⟨Geometry⟩ iter = geomCollection.iterator(); iter.hasNext(); i++) {
(4)   geomArray[i]= iter.next(); }
(5) GeometryFactory geometryFactory = new GeometryFactory();
(6) GeometryCollection polygonCollection=
      geometryFactory.createGeometryCollection(geomArray);
(7) Geometry union = polygonCollection.union();
(8) return union;}
```

Fig. 3. Java functions implementing the atregion operator

of the remaining tuples with the intersection between the field and the geometry. Since the underlying language does not provide a 'Union' operator that recursively computes the union of a set of geometries, we implemented a second version of atregion. which first computes the union of all geometries in the second parameter by invoking function unionAll. Its result is used in a call to the first atregion function explained above. Line (7) in function unionAll computes a union of geometries. Lines (5) and (6) are only for type conversion.

8 Implementing the Language

In this section we show how the last query in Section 6 is translated and executed. Figure 4 shows part of the computation of this query. The upper part of Figure 4 shows the sequence of function calls starting from the inner operator of

```
(1) String fieldTempTableNameElev= initField("Elevation");
(2) atrange(fieldTempTableNameElev, 150, Double.MAX_VALUE);
(3) Geometry unionField = defspace(fieldTempTableNameElev);
(4) String fieldTempTableNamePrec = initField("Precipitation");
(5) atperiod(fieldTempTableNamePrec, "12/1/2009", "12/31/2009");
(6) atregion(fieldTempTableNamePrec, unionField);
(7) lastPhase(fieldTempTableNamePrec);

public void lastPhase(String fieldTempTableName) throws SQLException {
(1) String sqlDML=
(2) "SELECT l.geom" +
(3) "FROM Yield y, LandPlot l, Crop c" +
(4) "WHERE c.name=\"Wheat\" " +
(5) "AND y.landPlot=l.landplotNo AND y.cropId =c.id " +
(6) "GROUP BY l.geom" +
(7) "HAVING AVG(y.production) > 1000";
(8) PreparedStatement pstmt = dbConn.prepareStatement(sqlDML);
(9) ResultSet rs = pstmt.executeQuery();
(10) Collection⟨Geometry⟩ geomCollection = new ArrayList⟨Geometry⟩();
(11) while (rs.next()){
(12)    Geometry aGeom = GeometryReader.getGeometry(rs.getObject(1));
(13)    geomCollection.add(aGeom); }
(14) pstmt.close();
(15) atregion(fieldTempTableName, geomCollection);
(16) spatialDump(fieldTempTableName, "_A");}
```

Fig. 4. Query evaluation

the SELECT clause of the query (which, remember, returns a field). Since we do not assume that field data fit in main memory, we use a temporary table that is updated by sequentially applying the functions explained in Section 7. Let us be more concrete. We have shown in Section 7 that the atregion operator updates geometries and deletes tuples. Thus, the function initField(nameOfFieldTable) (Line (1) in Figure 4) generates a temporary table containing the data in the original field table (in this case, Elevation). This table is the one that changes during the execution of the query, preserving the original field. Then, in Line (2) atrange is applied over the field returned in the previous step to delete the tuples that do not satisfy the condition (elevation > 150). A unique geometry is then generated over the result from the previous step using defspace (Line (3)). Then, a temporary table is created for the precipitation temporal field (Line (4)), atperiod is applied to the precipitation table for restricting the time frame of the field in Line (5), and atregion is applied to the field obtained in the previous step for restricting it with the geometry returned in Line (3). Finally, the function lastPhase is called. This function computes the collection of geometries corresponding to the inner query in the FROM clause. In Line (9) of lastPhase (shown in the lower part of Figure 4), the *translated* inner query is executed (where all the implicit joins are written in Line (5)), returning the land plots

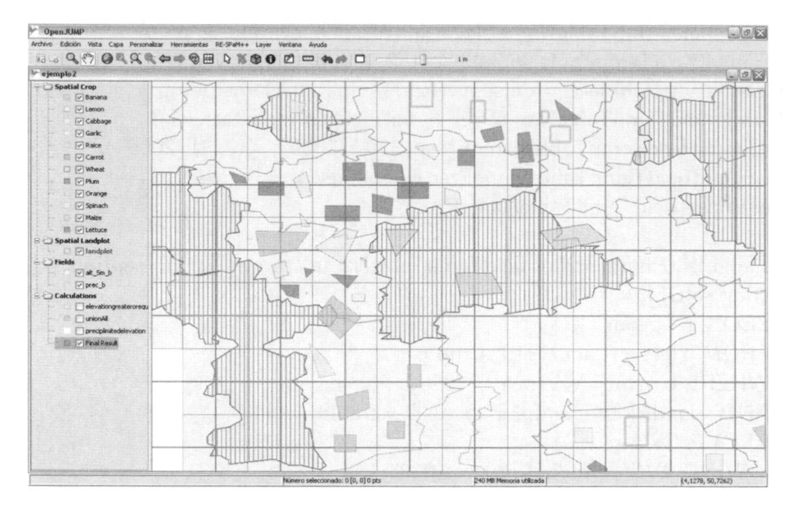

Fig. 5. The field resulting from the query

satisfying the query as a set of geometries collected in the loop in Line (11). In Line (15) atregion is invoked, and in Line (16) the result is returned.

Figure 5 shows the result of the query execution. There are two grids of different precision, one for elevation and one for precipitation. The zones with vertical bars indicate the resultant field, i.e, a precipitation field in regions with the desired altitude, and only one kind of crop.

9 Conclusion and Future Work

We have presented a physical model for spatio-temporal data warehouses that supports continuous fields. This model is based on two new data types, namely field and tempfield. These data types have a collection of operators, which we discussed. A relevant contribution of the present paper is the implementation of these operators and an associated SQL-like language that allows expressing SOLAP queries over continuous fields. The main goal of this implementation consists in showing the viability of our approach.

As future work, we will perform extensive testing of the operators and the language proposed here. Since spatio-temporal data warehouses contain huge amounts of data, optimization issues are extremely important. They include issues such as appropriate index structures, pre-aggregation, and efficient query optimization, among others. With respect to the latter, our example queries can be expressed in several ways, exploiting either the fact relationship or the geometries of the dimension levels with spatial and topological operators. Although these alternative queries yield the same result, the evaluation time of them may vary significatively, depending on the actual population of the data warehouse. Finally, we will consider other possible implementations of fields such as triangulated irregular networks (or TINs) and Voronoi diagrams.

References

1. Worboys, M.F., Duckham, M.: GIS: A Computing Perspective, 2nd edn. CRC Press, Second edn (2004)
2. Kimball, R.: The Data Warehouse Toolkit. J. Wiley and Sons, Inc., Chichester (1996)
3. Rivest, S., Bédard, Y., Marchand, P.: Toward better suppport for spatial decision making: Defining the characteristics of spatial on-line analytical processing (SOLAP). Geomatica 55, 539–555 (2001)
4. Bédard, Y., Rivest, S., Proulx, M.: Spatial online analytical processing (SOLAP): Concepts, architectures, and solutions from a geomatics engineering perspective. In: Wrembel-Koncilia (ed.) Data Warehouses and OLAP: Concepts, Architectures and Solutions, pp. 298–319. IRM Press (2007)
5. Paolino, L., Tortora, G., Sebillo, M., Vitiello, G., Laurini, R.: Phenomena: a visual query language for continuous fields. In: Proc. of ACM-GIS, pp. 147–153 (2003)
6. Vaisman, A.A., Zimányi, E.: A multidimensional model representing continuous fields in spatial data warehouses. In: Proc. of ACM-GIS, pp. 168–177 (2009)
7. Tomlin, D.: Geographic Information Systems and Cartographic Modelling. Prentice-Hall, Englewood Cliffs (1990)
8. Câmara, G., Palomo, D., de Souza, R.C.M., de Oliveira, D.: Towards a generalized map algebra: Principles and data types. In: Proc. of GeoInfo, pp. 66–81 (2005)
9. Cordeiro, J.P., Câmara, G., Moura, U.F., Barbosa, C.C., Almeida, F.: Algebraic formalism over maps. In: Proc. of GeoInfo., pp. 49–65 (2005)
10. Mennis, J., Viger, R., Tomlin, C.: Cubic map algebra functions for spatio-temporal analysis. Cartography and Geographic Information Science 32, 17–32 (2005)
11. Laurini, R., Paolino, L., Sebillo, M., Tortora, G., Vitiello, G.: A spatial SQL extension for continuous field querying. In: Proc. of COMPSAC Workshops, pp. 78–81 (2004)
12. Shanmugasundaram, J., Fayyad, U.M., Bradley, P.S.: Compressed data cubes for OLAP aggregate query approximation on continuous dimensions. In: Proc. of KDD, pp. 223–232 (1999)
13. Ahmed, T.O., Miquel, M.: Multidimensional structures dedicated to continuous spatiotemporal phenomena. In: Jackson, M., Nelson, D., Stirk, S. (eds.) BNCOD 2005. LNCS, vol. 3567, pp. 29–40. Springer, Heidelberg (2005)
14. Malinowski, E., Zimányi, E.: Advanced Data Warehouse Design: From Conventional to Spatial and Temporal Applications. Springer, Heidelberg (2008)
15. Open Geospatial Consortium Inc.: OpenGIS Abstract Specification: Topic 6: The Coverage Type and its Subtypes. OGC 07-011, Version 4 (2007)
16. Güting, R.H., Schneider, M.: Moving Objects Databases. Morgan Kaufmann, San Francisco (2005)
17. Klug, A.: Equivalence of relational algebra and relational calculus query languages having aggregate functions. Journal of the ACM 29, 699–717 (1982)

Benchmarking Spatial Data Warehouses

Thiago Luís Lopes Siqueira[1,2], Ricardo Rodrigues Ciferri[2], Valéria Cesário Times[3], and Cristina Dutra de Aguiar Ciferri[4]

[1] São Paulo Federal Institute of Education, Science and Technology, IFSP, Salto Campus, 13.320-271, Salto, SP, Brazil
[2] Computer Science Department, Federal University of São Carlos, UFSCar, 13.565-905, São Carlos, SP, Brazil
[3] Informatics Center, Federal University of Pernambuco, UFPE, 50.670-901, Recife, PE, Brazil
[4] Computer Science Department, University of São Paulo at São Carlos, USP 13.560-970, São Carlos, SP, Brazil

prof.thiago@cefetsp.br, ricardo@dc.ufscar.br, vct@cin.ufpe.br, cdac@icmc.usp.br

Abstract. Spatial data warehouses (SDW) enable analytical multidimensional queries together with spatial analysis. Mainly, three operations are related to SDW query processing performance: (i) joining large fact tables and large spatial and non-spatial dimension tables; (ii) computing one or more costly spatial predicates based on spatial ad hoc query windows; and (iii) aggregating data according to different spatial granularity levels. Several techniques to improve the query processing performance over SDW have been proposed in the literature. However, we identified the lack of a benchmark to carry out a controlled experimental evaluation of such techniques and, principally, to effectively measure the costs of the aforementioned three complex operations. In this paper, we propose a novel spatial data warehouse benchmark, called Spadawan, to provide performance evaluation environments for SDW and enable a further investigation on spatial data redundancy. The Spadawan benchmark is available at http://gbd.dc.ufscar.br/spadawan.

Keywords: spatial data warehouse, benchmarking, performance evaluation, drill-down and roll-up operations.

1 Introduction

Spatial data warehouses (SDW) enable analytical multidimensional queries together with spatial analysis. A relational SDW inherits several components of conventional data warehouses, such as fact and dimension tables, numeric measures and hierarchies that aggregate these measures according to distinct granularity levels [1]. Additionally, the SDW has spatial attributes that store vector geometries and define spatially-enabled components, such as spatial dimension tables, spatial measures and spatial hierarchies [2][3][4]. Typically, a spatial hierarchy is a *predefined* 1:N association among higher and lower granularity spatial attributes that is determined by a spatial relationship, e.g. containment, such as *(city)* $\preceq$ *(address)*. As a result, spatial OLAP

T.B. Pedersen, M.K. Mohania, and A M. Tjoa (Eds.): DaWaK 2010, LNCS 6263, pp. 40–51, 2010.

(SOLAP) operations are common roll-up and drill-down extended to hold spatial predicates [5]. Also, the well-known star and snowflake schemas may be adequately adapted to support the inclusion of spatial attributes, which introduce new storage costs and might impair query processing performance [3][6].

Mainly, three operations are related to SDW query processing performance: (i) joining large fact tables and large spatial and non-spatial dimension tables; (ii) computing one or more costly spatial predicates based on spatial ad hoc query windows; and (iii) aggregating data according to different spatial granularity levels. An example of a spatial and multidimensional query is "find out the total revenue earned by suppliers whose addresses are inside a rectangular window". This query mentions a topological relationship and a spatial *ad hoc* query window that was not previously stored in dimension tables. Another query may be issued to roll-up to the city granularity level by using a larger window that intersects the cities where the suppliers are located, for instance.

Indices and materialized views are used to provide efficient query processing over SDW, and the requirements to evaluate their efficiency are datasets with different characteristics of data volume, data distribution and data types, as well as diverse types of query concerning their selectivity. The literature mentions benchmarks for decision support and data warehouses [7][8][9], and for spatial databases [10][11], synthetic spatial datasets generators [12] and real spatial datasets (e.g. Tiger/Line, see http://www.census.gov/geo/www/tiger/). However, using them to evaluate SDW query processing requires several adaptations to comprise spatial roll-up and drill-down operations, for instance. Therefore, there is a lack of a SDW benchmark to carry out a controlled experimental evaluation and, principally, to effectively assess the costs of the aforementioned operations.

In this paper, we propose a novel **spa**tial **da**ta **wa**rehouse be**n**chmark, called Spadawan, to address the query processing performance on spatial roll-up and drill-down operations using *predefined* spatial hierarchies over SDW. As spatial predicates, the Spadawan benchmark focuses on intersection, containment and enclosure range queries. Furthermore, it comprises redundant and non-redundant SDW schemas based on the Star Schema Benchmark (SSB) [8]. Consequently, the Spadawan benchmark provides a further spatial data redundancy investigation and comparison with a non-redundant SDW schema.

This paper is organized as follows. Section 2 surveys related work. Section 3 describes the SDW schemas of the Spadawan benchmark, while Section 4 describes data loading operations according to each schema. Section 5 presents the queries of the Spadawan benchmark and their particularities. Section 6 briefly describes a case study and Section 7 concludes the paper.

2 Related Work

Benchmarks for spatial databases [10][11] are not aimed at assessing the efficiency of SOLAP operations, although they focus on the spatial predicate computation. Regarding data warehouses, TPC-D is an obsolete benchmark for decision support databases that does not support indices nor materialized views [7]. This fact motivated the proposal of the TPC-H [7], which provides individual queries that are not known in advance. However, its schema differs from the traditional star schema. The TPC-DS [9] suppresses this issue with a snowflake schema, but is aimed at data refreshing and its project is still under development. The SSB [8] extends the TPC-H to enable the analysis of historical trends and provides a set of predefined queries to run over its star schema. The SSB's queries refer to descriptive locations of suppliers and customers, since there is a predefined conventional hierarchy among attributes, i.e., *(region)* $\preceq$ *(nation)* $\preceq$ *(city)* $\preceq$ *(address)*. However, the SSB does not hold spatial attributes nor stores maps that would enable multidimensional queries with spatial predicates, which is the focus of the Spadawan benchmark.

We argue that the SSB can be adapted to maintain spatial data and therefore provide spatial roll-up and drill-down operations evaluation, by reusing synthetic or real spatial datasets This adaption requires maintaining the queries' semantics by adding spatial predicates and providing spatial predefined hierarchies based on the conventional existing ones. In this paper, we propose the Spadawan benchmark by extending the SSB to store a real spatial dataset and by altering the SSB's queries aiming at enabling spatial roll-up and drill-down operations evaluation.

3 The Spadawan Benchmark Schemas

We considered existing conceptual and logical models for SDW [2][3][4] in order to propose our SDW schemas, which extend the SSB schema by introducing spatial attributes that store geometries in spatial dimension tables, as shown in Fig. 1. The spatial attributes have the suffix *_geo* and are based on the SSB's conventional attributes that describe suppliers and customers locations, concerning their addresses, cities, nations and regions. We designed the *redundant* (Fig. 1a) and the *hybrid* (Fig. 1b) SDW schemas aiming at different purposes as follows.

According to Stefanovic et al. [3], *Customer* and *Supplier* should be considered as spatial-to-spatial dimension tables and must store all spatial attributes, as shown in Fig. 1a. Clearly, these spatial dimension tables maintain spatial data redundancy. For instance, the map for Europe is stored in every row whose supplier is located in Europe. Therefore, the *redundant* schema aims at investigating to what extent SOLAP queries performance is affected by spatial data redundancy.

On the other hand, Fidalgo et al. [4] state that, in SDW, spatial data must not be redundant and should be shared whenever is possible. Considering that the SSB's customers and suppliers share city, nation and region locations, but have individual addresses, we designed the *hybrid* schema (Fig. 1b) to comply with these characteristics that are not treated by the *redundant* schema. For instance, the *hybrid* schema's City spatial dimension table maintains distinct maps of cities where customers and suppliers reside. Therefore, the *hybrid* schema aims at evaluating the overhead of introducing additional joins costs to the query processing performance, as these joins are required to avoid spatial data redundancy.

The spatial data redundancy may also increase the number of tables to be scanned. Suppose that a spatial *ad hoc* query window intersects both customers and suppliers cities geometries. Then, in a SDW with a redundant schema (as shown in Figure 1a), two tables would be scanned, while in a hybrid schema SDW (as given in Figure 1b), a single table storing all geometries for cities would be searched.

Finally, our extensions preserved descriptive data as well as created two spatial hierarchies based on the SSB's original conventional hierarchies. They are valid for both the *redundant* and the *hybrid* schemas: (i) (*region_geo*) $\preceq$ (*nation_geo*) $\preceq$ (*city_geo*) $\preceq$ (*c_address_geo*); and (ii) (*region_geo*) $\preceq$ (*nation_geo*) $\preceq$ (*city_geo*) $\preceq$ (*s_address_geo*). According to Malinowski and Zimányi [13], these hierarchies can be classified as simple symmetric spatial hierarchies with the containment spatial relationship. We emphasize that the *hybrid* schema is not a snowflake schema, since the latter normalizes hierarchies.

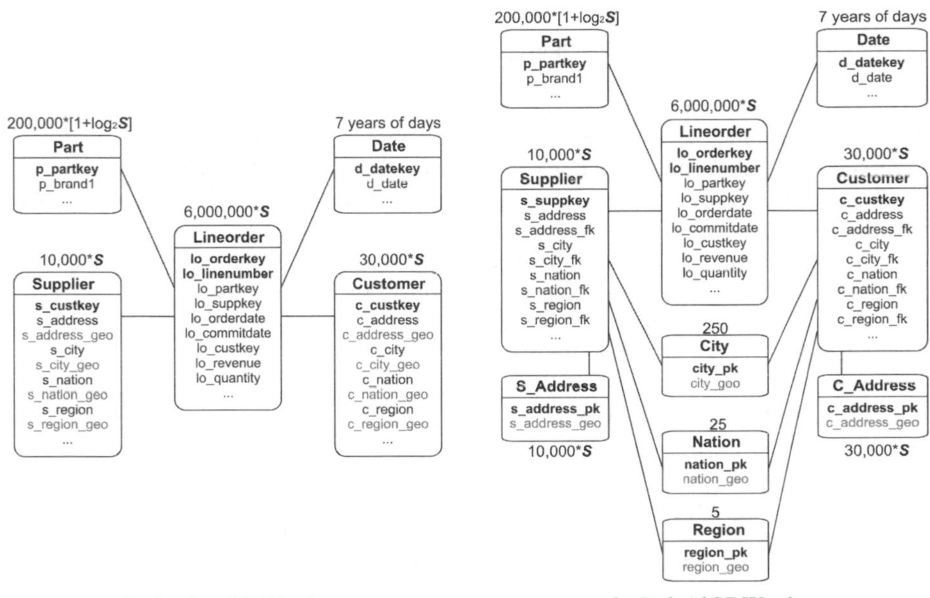

a. *Redundant* SDW schema b. *Hybrid* SDW schema

Fig. 1. The Spadawan benchmark schemas

4 Data Generation and Loading

Loading data into the SDW schemas described in Section 3 requires running the SSB data generator as well as performing other tasks depending on the selected schema. Fig. 1 shows the data cardinality of each table according to the scale factor *S* chosen to generate the SSB dataset. Regarding suppliers and customers locations, there are always 5 distinct regions, 5 nations per region and 10 cities per nation. We determined that the spatial attributes that represent cities, nations and regions are polygons, which were reused from the Tiger/Line real dataset. On the other hand, customer and supplier descriptive addresses cardinalities depend on *S*, as well as the number of customers and suppliers per city. As for geographic addresses, they are synthetic points randomly distributed inside each city polygon. We implemented a software to generate and distribute these points such that customers and suppliers have unique and distinct addresses. As a result, the spatial data volume of addresses varies according to *S*, as well as the quantity of customer and supplier addresses inside each city. Data sets that have already been used for populating the SDW redundant and hybrid schemas are available at http://gbd.dc.ufscar.br/spadawan.

The Spadawan benchmark's geometries do not suffer modifications after the data loading. Obviously, the same scale factor *S* and the same spatial dataset used for the *redundant* schema must be used for the *hybrid* schema in order to enable spatial data redundancy investigation. Section 4.1 and 4.2 describe, respectively, the data loading for the *redundant* and *hybrid* schemas, Section 4.3 discusses how to extend these schemas to increase spatial data volume and to decrease spatial predicate selectivity.

4.1 Loading the Redundant Schema

The following five steps must be performed to load the *redundant* SDW schema.

1. Load the geometries for cities, nations and regions into temporary tables.
2. Execute the SSB data generator with scale factor *S* and load its tables.
3. Run our generator of addresses, which also loads customer and supplier addresses into temporary tables.
4. Alter and update the tables *Customer* and *Supplier* to include the geometries of addresses, cities, nations and regions. Define all the constraints.
5. Discard all the temporary tables and build spatial indices supported by the DBMS (e.g. R-tree [14] or GiST [15]) on the spatial attributes.

4.2 Loading the Hybrid Schema

Loading the *hybrid* schema also requires five steps. Steps 1, 2 and 3 are similar to those described for the *redundant* schema. The remaining steps are defined as follows.

4. Alter and update the tables *Customer* and *Supplier* to include foreign keys referencing the spatial dimension tables, which are the altered temporary tables of steps 1 and 3.
5. Build spatial indices supported by the DBMS on the spatial attributes.

4.3 Increasing Data Volumes

The spatial data volumes for City, Nation and Region levels are fixed in the SSB. We argue that a fixed data volume for spatial data is unrealistic and should impose a severe drawback to Spadawan benchmark. In order to overcome this drawback, we describe the algorithm *IncreaseVolume* to enable increasing the spatial data volume and decreasing the spatial predicate selectivity. A high selectivity determines that most of the spatial objects are processed in the spatial predicate computation, while a low selectivity ensures that only a few of them is processed. The algorithm *IncreaseVolume* consists of an intermediate step between the steps 2 and 3 presented in Sections 4.1 and 4.2, and can be used to load both *redundant* and *hybrid* schemas.

The algorithm *IncreaseVolume* generates a spatial data volume *n* times larger than that built with a given scale factor *S*. Translation (line 3) is an operation that shifts a given geometry to another location, according to chosen offsets. As a result, a translation modifies all coordinates of the geometry. Specifically, the translation used in the *IncreaseVolume* algorithm must assure that: (i) geometries of the same granularity level do not overlap; and (ii) the spatial hierarchy must not be deteriorated. For instance, if *city1* is a city and was replicated and translated, the copy of *city1* must not overlap other cities. Also, if *city1* is inside *nation1*, the copy of *city1* must be inside the copy of *nation1*.

Consider that: (i) |*X*| is the cardinality of the spatial attribute *X*, i.e., the number of distinct objects that *X* can assume; (ii) *sobj* is a spatial object for the attribute *X*; (iii) *sobj.id* is the identifier for the spatial object *sobj*; and (iv) *sobj'* is a copy of the spatial object *sobj*. Then, the strategy to generate an identifier for *sobj'* is to do: *sobj'.id* ← *sobj.id* + |*X*| (line 4). Analogously, the primary key values for replicated suppliers and customers can be determined (line 6). Regarding the spatial predicate selectivity, the commented lines (lines 7 and 8) must be executed when constant selectivity is desired. Otherwise, the original selectivity will be divided by *n*. We further discuss this issue in Section 5.1.

```
    Algorithm IncreaseVolume
1   For i ← 1 To n-1
2       Replicate the initial set of geometries
3       Translate the replicated geometries to new coordinates
4       Generate new identifiers for these geometries
5       Replicate the initial dataset of the dimension tables Customer and Supplier
6       Generate new primary key values for these customers and suppliers
7       /* Replicate the initial set of spatial query windows */
8       /* Translate these windows together with the replicated geometries */
9   End-For
```

5 Queries

5.1 Ad Hoc Spatial Query Windows

Regarding the spatial query windows, they are quadratic, correlated with the spatial data, and considered *ad hoc* because their rectangles are not stored in any spatial dimension table. A spatial roll-up operation requires a set of four windows, each one associated to a granularity level (Address, City, Nation or Region) and has a specific size (as lower the granularity, smaller is the window). We defined two separate types of sets for the spatial query windows: disjoint and overlapping.

Regarding the type disjoint, consider a set of windows *d1*. Every window of *d1* has one centroid that is an address. To create *d1*'s windows, initially, one arbitrary address is chosen to be the centroid of the address window. Then, city, nation and region windows are produced subsequently by reusing the centroid of the address window, as shown in Fig. 2a. Note that the query window size is proportional to the granularity level. In order to create another set of windows *d2*, the centroid for its windows is another address, specifically chosen to assure that the windows of *d2* do not overlap any window of *d1*. As a result, all windows of different sets are disjoint, and the user can query distinct locations as previously fetched objects are not reused.

Concerning the type overlapping, consider a set of windows *o1* whose windows were created similarly to *d1*. In order to create another set of windows *o2*, any point inside the address window of *o1* is chosen to be the centroid of the new address, city, nation and region windows. As a result, all windows of different sets overlap, and the user can retrieve data related to a specific neighborhood, as shown in Fig. 2b. In fact, continuous-line windows were built using an address as centroid, while dashed-line windows had centroids obtained from any point inside the previous address window. The query window size is also proportional to the granularity level. Overlapping query windows were designed to evaluate the reuse of previously fetched objects, which is a task aided by system cache and buffers.

The Spadawan benchmark performs five roll-up/drill-down operations based on five fixed sets of disjoint query windows, as well as performs ten roll-up/drill-down operations based on ten fixed sets of overlapping query windows. Since the quantity of windows is fixed and they also have a fixed place, the number of spatial objects that satisfies the spatial predicate associated to a given window is also fixed. Therefore, replicating a set of windows together with spatial data, as described by the *IncreaseVolume* algorithm, maintains the spatial predicate selectivity constant. On the other hand, increasing only the spatial data volume by n times, divides the spatial predicate selectivity by n.

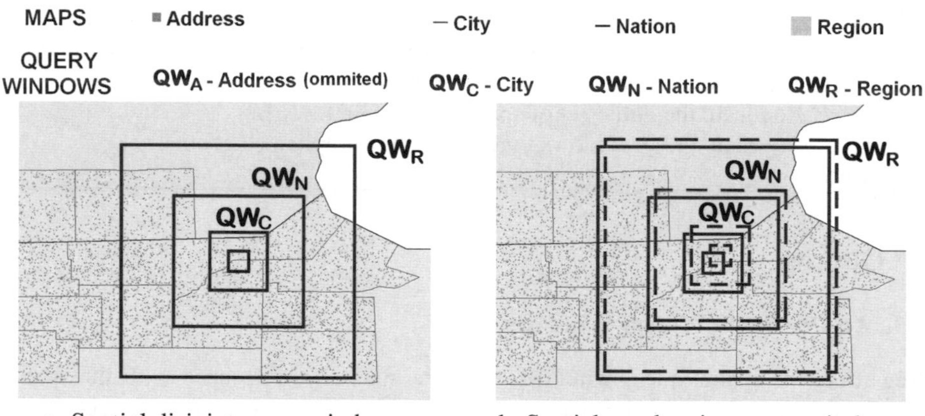

a. Spatial disjoint query windows b. Spatial overlapping query windows

Fig. 2. Spatial ad hoc query windows

5.2 Query Types 1, 2 and 3

Queries of type 1, 2 and 3 were based on query Q2.3 of the SSB and aim at evaluating the performance of: (i) at least three joins among tables, depending on the selected SDW schema; (ii) four spatial predicates computation based on *ad hoc* spatial query windows; and (iii) data aggregation according to four spatial granularity levels.

Figure 3 illustrates how a single query was transformed into a spatial roll-up operation. We replaced conventional predicates that formerly referred to nominative locations by spatial predicates involving *ad hoc* spatial query windows. Instead of asking for a single descriptive granularity level, four queries of distinct spatial granularity levels are issued subsequently, considering that: R_A is a spatial relationship to evaluate supplier addresses against the spatial query window QW_A, R_C is a spatial relationship to evaluate cities against the spatial query window QW_C, R_N is a spatial relationship to evaluate nations against the spatial query window QW_N, and R_R is a spatial relationship to evaluate regions against the spatial query window QW_R. The size of the spatial query windows QW_A, QW_C, QW_N and QW_R are distinct and decreases as the granularity level decreases. This ensures a control of the selectivity factor of the queries in different granularity levels.

As a result, Query Types 1, 2 and 3 enable data aggregation according to the four aforementioned spatial granularity levels. Query Type 1 focuses mainly on the intersection relationship (i.e. IRQ: intersection range query on the spatial predicate), while Query Type 2 focuses mainly on the containment relationship (i.e. CRQ: containment range query on the spatial predicate) and Query Type 3 focuses mainly on the enclosure relationship (i.e. ERQ: enclosure range query on the spatial predicate).

Query Type 1 is detailed in Table 1. It uses the containment spatial predicate at the Address level and the intersection predicate at City, Nation and Region levels. The QW/Extent column shows the fraction of the extent occupied by the spatial query window. For instance, the query window for Address level represents 0.001% of the extent. Table 1 lists the average number of objects that are returned per query, considering 5 roll-up operations with the sets of spatial disjoint query windows and 10 roll-up operations with the sets of spatial overlapping query windows.

Table 1 shows the selectivity factor (*SF*), which consists of the conventional *SF* multiplied by the spatial *SF*. The former is fixed and defined by the SSB as 1/1000. The later is calculated by dividing the number of returned spatial objects by the spatial attribute cardinality. For instance, at City granularity level, the spatial *SF* is 3.6/250 and therefore the query *SF* is 1/1000 * 3.6/250 (value of 0.0000144). Only one spatial *SF* was defined at Nation level to assess the efficiency when no spatial objects are returned as query answer (Table 2). This represents an extreme situation on query processing.

It is not possible to estimate the number of addresses that satisfies the spatial predicate, since the address data volume and the number of addresses inside each city depend on the scale factor *S* used to generate the SSB dataset. Therefore, we estimated the number of objects retrieved by the query as well as the *SF* for the Address level using the data generation scale factor of 1.

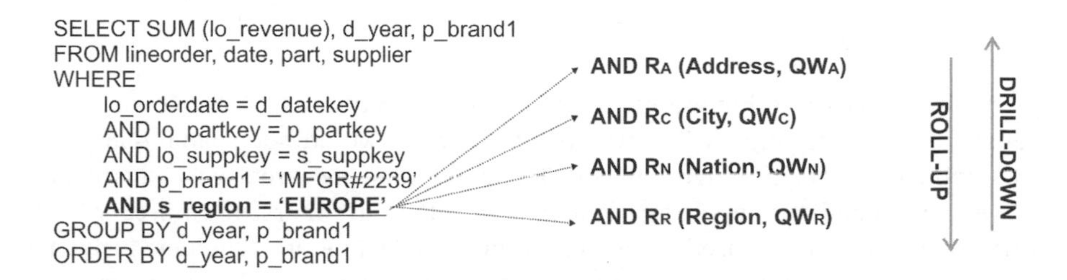

Fig. 3. The template for Query Types 1, 2 and 3.

Query Types 2 and 3 are detailed in Tables 2 and 3, respectively, and evaluate other spatial predicates using different sizes of query windows. We emphasize that all buffers and cache must be flushed at the end of each spatial roll-up operation that utilize spatial disjoint query windows. On the other hand, they must not be flushed when utilizing overlapping spatial query windows.

Table 1. Additional information for Query Type 1

			Disjoint Query Windows		Overlapping Query Windows	
Level	Predicate R	QW/Extent	Objects/query	*SF*	Objects/query	*SF*
Address	R_A = CRQ	0.001%	2.2	0.00000022	5.4	0.00000054
City	R_C = IRQ	0.05%	3.6	0.0000144	4.0	0.000016
Nation	R_N = IRQ	0.1%	1.6	0.000064	3.0	0.00012
Region	R_R = IRQ	1%	1.2	0.00024	2.0	0.0004

Table 2. Additional information for Query Type 2

			Disjoint Query Windows		Overlapping Query Windows	
Level	Predicate R	QW/Extent	Objects/query	*SF*	Objects/query	*SF*
Address	R_A = CRQ	0.01%	19.0	0.0000019	37.0	0.0000037
City	R_C = CRQ	0.1%	1.4	0.0000056	3.0	0.000012
Nation	R_N = CRQ	10%	1.2	0.000048	0.0	0.0
Region	R_R = CRQ	25%	0.4	0.00008	1.0	0.0002

Table 3. Additional information for Query Type 3

			Disjoint Query Windows		Overlapping Query Windows	
Level	Predicate R	QW/Extent	Objects/query	*SF*	Objects/query	*SF*
Address	R_A = CRQ	0.00001%	1.0	0.0000001	1.0	0.0000001
City	R_C = ERQ	0.0005%	0.8	0.0000032	1.0	0.000004
Nation	R_N = ERQ	0.001%	0.8	0.000032	1.0	0.00004
Region	R_R = ERQ	0.01%	1.0	0.0002	1.0	0.0002

5.3 Query Type 4

Query type 4, shown in Fig. 4, was based on the SSB's query Q3.3 and consists of a spatial roll-up and spatial drill-down operations with two ad hoc spatial query windows, which add an extra high join cost. Basically, this query retrieves "the revenue per year per brand for suppliers of an area x to the customers of an area y". The same granularity level is used for both customers and suppliers simultaneously. The containment spatial predicate is verified at Address level while the intersection predicate is verified at City, Nation and Region levels. Table 4 shows additional details.

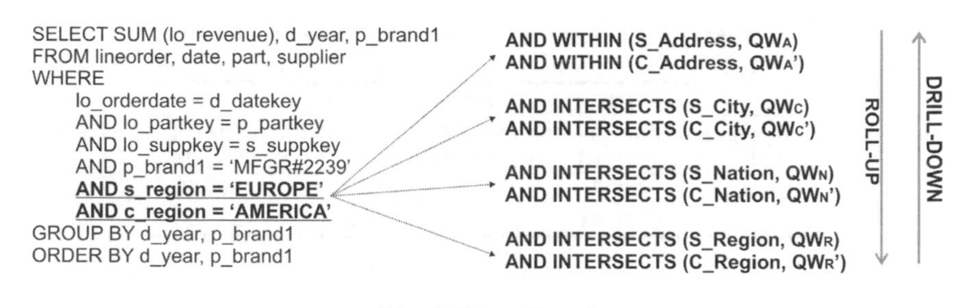

Fig. 4. Query Type 4

Table 4. Additional information for Query Type 4

			Disjoint Query Windows		Overlapping Query Windows	
Level	Predicate	QW/Extent	Objects/query	*SF*	Objects/query	*SF*
Address	CRQ	0.001%	9.1	0.00000091	11.3	0.00000114
City	IRQ	0.05%	7.2	0.0000288	9.0	0.000036
Nation	IRQ	0.1%	3.2	0.000128	5.0	0.0002
Region	IRQ	1%	2.4	0.00048	3.0	0.0006

6 Case Study

We have already used the Spadawan benchmark to investigate the impact of spatial data redundancy over SDW [6]. We loaded the following datasets: D1: the redundant schema using the scale factor $S = 10$, which occupied 150 GB; D2: the hybrid schema with $S = 10$, which occupied 15 GB; D3: the hybrid schema with $S = 6$; and D4: the hybrid schema with $S = 2$. Regarding City, Nation and Region levels, the spatial data volume remained fixed as well as the spatial predicate selectivity. The Address level data volume varied according to S.

We performed five spatial roll-up operations, using the five sets of disjoint query windows, and collected the average elapsed time at each granularity level. The GiST index was defined over the spatial attributes to enhance the spatial predicate computation. Experiments were conducted on a computer with a 2.8 GHz Pentium D processor, 2 GB of main memory, a 7200 RPM SATA 320 GB hard disk, Linux CentOS 5.2, PostgreSQL 8.2.5 and PostGIS 1.3.3.

Table 5 shows the results obtained for the datasets D1, D2, D3 and D4 for Query Type 1. It is important to observe that: (i) the spatial data redundancy drastically impaired query processing performance especially at Nation and Region levels whose cardinalities are lower; and (ii) the smaller the conventional data volume, the shorter the elapsed time to process the queries over the hybrid schema. Spatial data redundancy impaired not only the query processing performance, but also the storage requirements, since D1 occupied ten times more space than D2.

Another interesting issue was raised by evaluating Query Type 4 against the dataset D1. At Region and Nation granularity levels, we aborted the query processing after 4 days of execution, since this elapsed time was prohibitive. At City level, the query took 172,900.15 seconds (approximately 48 hours). On the other hand, the

Table 5. Elapsed times in seconds for Query Type 1

	D1	D2	D3	D4
Address	2831.23	2853.85	1803.62	594.31
City	2773.10	2758.70	1686.61	562.08
Nation	3449.76	2765.61	1694.00	545.59
Region	6200.44	2790.29	1703.31	552.94

same query issued against the dataset D2 took only 130.34 seconds, i.e., the spatial data redundancy provided an unacceptable increase of 132,900.00%.

We have developed the Spatial Bitmap Index (SB-index) [16] in order to decrease the query response time in SDW. The SB-index was also validated using the Spadawan benchmark. For further details about the performance evaluation, see [16].

7 Conclusions and Future Work

This paper proposed a novel benchmark for spatial data warehouses, called Spadawan, whose main characteristics are: (i) it generates SDW datasets composed of points and polygons in spatial attributes; (ii) it is composed of different types of SOLAP queries that enable the performance evaluation of intersection range queries, containment range queries and enclosure range queries in the spatial predicate; (iii) it enables the evaluation of spatial roll-up and drill-down operations; (iv) it provides a means of investigating spatial data redundancy in SDW by designing two distinct data schemas with spatial hierarchies and spatial dimensions; (v) it permits the adjustment of the SDW data volume and the spatial predicate selectivity; and (vi) it uses spatial query windows that may overlap each other or may be disjoint from each other. We validated the Spadawan benchmark using it to generate a performance evaluation environment to assess the impact of spatial data redundancy over SOLAP queries [6] and the efficiency of the SB-index data structure [16].

As future work, we intend to propose additional SOLAP query types to analyze drill-across operations on extended SDW schemas and to compute aggregations of geometries of spatial objects. We also plan to incorporate different spatial data, such as lines, polygons with holes and with islands, on the spatial data generation and SOLAP query processing. Another future work would be to extend the current benchmark by covering all types of classification hierarchies in addition to the predefined 1:N. The use of the Spadawan benchmark with different techniques, such as indices and materialized views, is another future work.

Acknowledgements. This work has been supported by the following Brazilian research agencies: FAPESP, CNPq, CAPES, INEP and FINEP. The first two authors thank the support of the Web-PIDE Project in the context of the Observatory of the Education of the Brazilian Government. The work carried by the third author was supported by funds from the CNPq under the Grant 479018/2009-0. The last author's work has been founded by FAPESP under the Grant 2009/06052-7.

References

1. Kimball, R., Ross, M.: The data warehouse toolkit: the complete guide to dimensional modeling. John Wiley & Sons, Inc., Chichester (2002)
2. Malinowski, E., Zimányi, E.: Advanced data warehouse design: from conventional to spatial and temporal applications (data-centric systems and applications). Springer, Heidelberg (2008)
3. Stefanovic, N., Han, J., Koperski, K.: Object-based selective materialization for efficient implementation of spatial data cubes. IEEE Trans. Knowl. Data Eng. 12(6), 938–958 (2000)
4. Fidalgo, R., Times, V.C., Silva, J., Souza, F.F.: GeoDWFrame: a framework for guiding the design of geographical dimensional schemas. In: Kambayashi, Y., Mohania, M., Wöß, W. (eds.) DaWaK 2004. LNCS, vol. 3181, pp. 26–37. Springer, Heidelberg (2004)
5. Rivest, S., Bédard, Y., Proulx, M., Nadeau, M., Hubert, F., Pastor, J.: SOLAP technology: merging business intelligence with geospatial technology for interactive spatio-temporal exploration and analysis of data. J. of Photogrammetry and Remote Sensing 60, 17–33 (2005)
6. Siqueira, T.L.L., Ciferri, C.D.A., Times, V.C., Oliveira, A.G., Ciferri, R.R.: The impact of spatial data redundancy on SOLAP query performance. J. Braz. Comp. Soc. 15(2), 19–34 (2009)
7. Poess, M., Floyd, C.: New TPC benchmarks for decision support and web commerce. SIGMOD Record 29(4), 64–71 (2000)
8. O'Neil, P., O'Neil, E., Chen, X., Revilak, S.: The star schema benchmark and augmented fact table indexing. In: TPCTC, pp. 237–252 (2009)
9. Poess, M., Smith, B., Kollar, L., Larson, P.: TPC-DS, taking decision support benchmarking to the next level. In: SIGMOD, pp. 582–587 (2002)
10. Paton, N.W., Williams, M.H., Dietrich, K., Liew, O., Dinn, A., Patrick, A.: VESPA: a benchmark for vector spatial databases. In: Jeffery, K., Lings, B. (eds.) BNCOD 2000. LNCS, vol. 1832, pp. 81–101. Springer, Heidelberg (2000)
11. Günther, O., Oria, V., Picouet, P., Saglio, J., Scholl, M.: Benchmarking spatial joins à la carte. In: SSDBM, pp. 32–41 (1998)
12. Theodoridis, Y., Silva, J.R., Nascimento, M.A.: On the generation of spatiotemporal datasets. In: SSD, pp. 147–164 (1999)
13. Malinowski, E., Zimányi, E.: Spatial hierarchies and topological relationships in the spatial MultiDimER model. In: Jackson, M., Nelson, D., Stirk, S. (eds.) BNCOD 2005. LNCS, vol. 3567, pp. 17–28. Springer, Heidelberg (2005)
14. Guttman, A.: R-trees: a dynamic index structure for spatial searching. In: SIGMOD, pp. 47–57 (1984)
15. Aoki, P.M.: "Generalizing "search" in generalized search trees". In: ICDE, pp. 380–389 (1998)
16. Siqueira, T.L.L., Ciferri, R.R., Times, V.C., Ciferri, C.D.A.: A spatial Bitmap-based index for geographical data warehouses. In: ACM SAC, pp. 1336–1342. ACM, Inc., New York (2009), http://doi.acm.org/10.1145/1529282.1529582

Discovering Community-Oriented Roles of Nodes in a Social Network

Bin-Hui Chou and Einoshin Suzuki

Department of Informatics, ISEE, Kyushu University,
Fukuoka 819-0395, Japan
chou@i.kyushu-u.ac.jp, suzuki@inf.kyushu-u.ac.jp
http://www.i.kyushu-u.ac.jp/~suzuki/slabhome.html

Abstract. We propose a new method for identifying the role of a vertex in a social network. Existing well-known metrics of node centrality such as betweenness, degree and closeness do not take the community structure within a network into consideration. Furthermore, existing proposed community-based roles are defined using cliques, and thereby it is difficult to discover vertices with only few links that bridge communities. To overcome the shortcomings, we propose three community-oriented roles, *bridges*, *gateways* and *hubs*, without knowledge on the community structure, for representing vertices that bridge communities. We believe that detecting the roles in a social network is useful because such nodes are valuable by themselves due to their intermediate roles between communities and also because the nodes are likely to provide a deeper understanding of the communities. Our method outperforms the state-of-the-art method through experiments using data of DBLP records in terms of the subjective validness of the outputs.

1 Introduction

In a social network, a vertex represents an individual and an edge between a pair of vertices represents the presence of a relationship between them. Analyzing categories of vertices and discovering social relationships between vertices are acquainted as social network analysis [9]. The methods on social network analysis not only can be examined in the field of social science but also can be applied to the field of biology, communication studies and information science [2], and thereby have received considerable attention recently.

By using network connectivity properties, social network analysis often aims to discover various categories of vertices in a network. We can find vertices of high connectivity or discover densely-connected subgroups, and subjectively assign roles to vertices based on the result of discovery. Evaluating the role of a vertex is useful in many applications such as viral marketing [1,3], epidemiology [8]. We think that the discovery of roles that connect communities is especially useful in understanding and utilizing communities as each of them is likely to play a key role in the community. For example, law enforcement agents are able to gain information of two gangs of criminals if they keep an eye on the intermediator between the two gangs.

T.B. Pedersen, M.K. Mohania, and A M. Tjoa (Eds.): DaWaK 2010, LNCS 6263, pp. 52–64, 2010.

There are many existing renowned metrics used to estimate the role of a vertex. Centrality is a measure of a vertex how it well connects to other vertices in a network. Degree, betweenness and closeness are all measures in terms of centrality [9]. However, the role of a vertex is assigned according to its rank without considering the community structure and therefore, vertices that bridge communities are not successfully detected in the measures of centrality.

Recently, [6] proposed community-based roles which are defined using the degree of a vertex and a proposed community metric. In the proposed community metric, a community is defined based on a clique, i.e, a maximal complete subgraph, and thus, a vertex that connects many cliques tends to have a large value of the degree. The proportional feature of the degree and the community metric makes it difficult to assign roles to vertices that connect communities with few links. Note that a vertex that has a small degree but connects with communities indicates that it bridges communities.

In order to overcome the shortcomings of the previous works, we propose three community-oriented roles — *bridges*, *gateways* and *hubs*. *Bridges*, *gateways* and *hubs* represent different kinds of roles, each of which is essential as a kind of relationship between communities. We believe that detecting the roles in a social network is useful because such nodes are valuable by themselves due to their intermediate roles between communities and also because the nodes are likely to provide a deeper understanding of the communities. A *bridge* is a vertex just located between two communities. A *bridge* connects communities, each of which has only one single link with it. Since each community has only one link with the *bridge*, it is controversial to cluster a bridge into any community. A *gateway* is a vertex that acts as an entrance or an exit of a community when we move from a community to another one. A *gateway* should be included in the community which most of its neighbors belong to. A *hub* is a confluent vertex, on which groups of vertices converge. Groups of vertices may be clustered into the same community or be divided into different communities, which depends on the result of community detection. When groups of vertices are divided into different communities, the *hub* should be the overlapping vertex among communities. We define the proposed roles rigorously and implement a discovery algorithm which does not require information of the community structure. The experimental results show that most of the proposed roles exist in the boundary of communities and vertices with the proposed roles are important in bridging communities.

2 Motivation and Problem Setting

Figure 1 presents a motivating example. Nodes 2, 3, 4, and 5 are members of one community and nodes 6, 7, 8, 9, 10, and 11 are members of another community in Figure 1a. Because each community in Figure 1a only has one link to node 1, we do not cluster node 1 into any community.

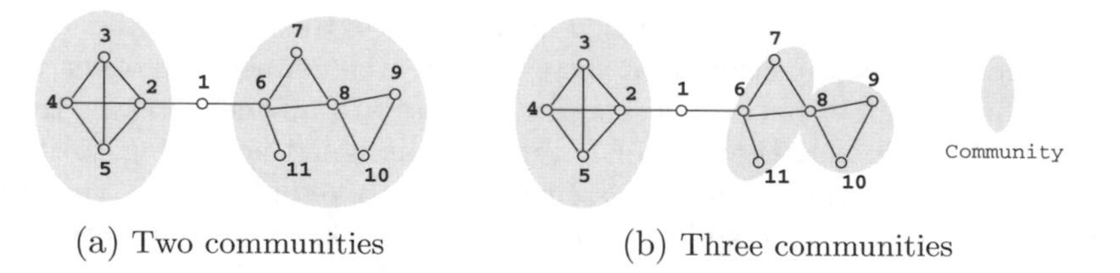

(a) Two communities (b) Three communities

Fig. 1. Motivating example

Table 1. Centrality measures (C_D, C_C and C_B are degree centrality, closeness centrality and betweenness centrality respectively. The higher the value is, the more central the vertex is.)

Node	1	2	3	4	5	6	7	8	9	10	11
C_D[1]	2	4	3	3	3	4	2	4	2	2	1
C_C[2]	0.50	0.43	0.33	0.33	0.33	0.53	0.40	0.43	0.32	0.33	0.36
C_B[3]	24	21	0	0	0	29	0	16	0	0	0

According to the rank of centrality measures shown in Table 1, node 6, followed by nodes 1, 2 and 8 in Figure 1 can be considered central or important vertices. In this example, node 1 may be viewed as the most remarkable vertex because node 1 is the important vertex that bridges two communities while the result in the centrality measures suggests node 6 to be the most important. Moreover, a centrality measures provides only information of ranking and cannot differentiate nodes 1, 2, 6 and 8 because a centrality measure does not consider the community structure.

Four roles — *ambassadors*, *bridges*, *big fish* and *loners*, are proposed in [6]. The method in [6] is designed to classify roles according to the degree of a vertex and a proposed community metric for a vertex. Nodes 2, 6 and 8 are discovered as *ambassadors* because each of them has a large degree and a large value for the proposed community metric, nodes 3, 4 and 5 are discovered as *big fish* because each of them has a large degree but a small value for the proposed community metric, and node 11 is a *loner* because node 11 has a small degree and a small value for the community metric. In the proposed community metric, a community is defined based on a clique so [6] fails to detect node 1 that connects the two communities, which are not cliques.

Considering the example in Figure 1a, we think node 1 is the most important vertex because it is located between the two communities. It becomes easy to discover these two communities if we can detect node 1. We regard nodes 2 and 6 are entrances or exits of the communities and they become important vertices

[1] $C_D(n_i) = \sum_j I(i,j)$, where $(i,j) \in E$ and I is a 0/1 indicator function.

[2] $C_C(n_i) = \frac{N-1}{\sum_{j=1}^{g} d(n_i,n_j)}$, where $d(n_i,n_j)$ is the length of the shortest path between vertices i and j and N is the number of vertices.

[3] $C_B(n_i) = \sum_{j<k} g_{jk}(n_i)$, where g_{jk} is the number of the shortest paths between vertices j and k that contain vertex i.

in bridging communities if we remove node 1 and connect nodes 2 and 6 directly. When we split the second community into two, we find that node 8 becomes a vertex connecting communities as shown in Figure 1b. This example inspires us to discover these community-oriented roles that bridge communities. We name node 1, nodes 2 and 6, and node 8 a *bridge*, *gateways*, and a *hub*, respectively.

Therefore, We tackle the problem of discovering community-oriented roles — *bridges*, *gateways* and *hubs* from a social network G. The problem setting is formalized as follows.

Input: a social network $G = \langle V, E \rangle$, where $V = \{v_1, v_2, \ldots, v_n\}$ is a non-empty finite set of vertices and E is a set of edges where an edge is binary relation of an unordered pair of distinct elements of V

Output: $V' = \{v_i \mid v_i \text{ is either a bridge, a gateway, a hub or a loner}\}$

3 Community-Oriented Roles

We examined existing methods with a motivating example in section 2. The centrality measures are used to find the most central node and [6] is designed to assign a role using the degree and the proposed community metric of a vertex. In this paper, we detect community-oriented roles by using topological information. Generally two vertices are similar if they have a link between them. We assume vertices within the same community connect with each other more densely than vertices between communities do. The more similar the vertices in the neighborhoods of a vertex are, the denser the graph formed by the vertex and its neighbors becomes. If a vertex is located between two communities, the two communities in its neighborhood are not expected to have many common vertices between them, which leads to our main idea for defining *bridges*, *gateways* and *hubs*. Thus, neighbors of a *bridge*, a *gateway* or a *hub* do not necessarily share the same common vertices. The extent to which its neighbors connect with each other decides which role a vertex belongs to and the community structure the vertex connects accordingly differs.

A property used in our definition is network transitivity or clustering [10], which is a common property in most networks. If node A links with node B and node B links with node C, nodes A and C are likely to have a connection between them. In other words, two of your friends will have a high probability of knowing each other, on account of their common acquaintance with you. This effect is quantified by the clustering coefficient C [5] [10] which implies the probability that two of one's friends are friends themselves, defined as $C = 3\Delta/\Theta$ where Δ represents the number of triangles on the graph and Θ represents the number of connected triples of vertices. Furthermore, social networks generally have a much higher value for C than the corresponding random model [5].

3.1 Bridges, Gateways and Hubs

In this paper, we focus on a simple, undirected and unweighted graph. Let $G = \langle V, E \rangle$ be a graph, where V is a set of vertices and E is a set of unordered pairs of distinct vertices. Also, a vertex possesses at most one role in this paper.

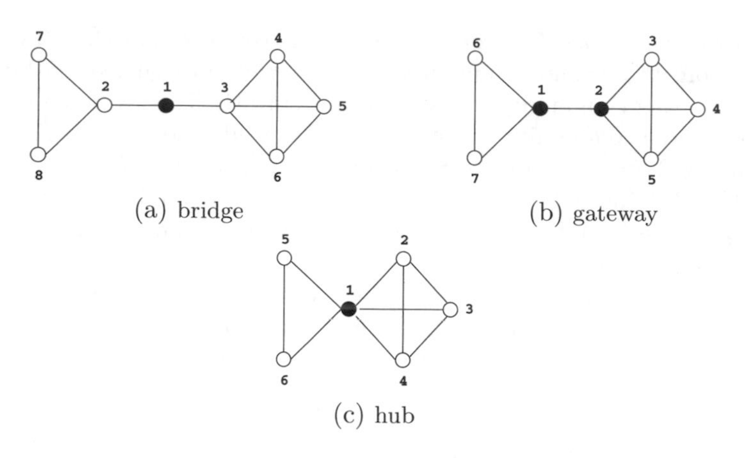

Fig. 2. Examples of a bridge (node 1 in Fig. 2a), gateways (nodes 1 and 2 in Fig. 2b), and a hub (node 1 in Fig. 2c)

Definition 1. *Let $v \in V$. v's neighborhood $N(v)$ encompasses vertices linked to v and itself.*

$$N(v) = \{u \in V \mid (v, u) \in E\} \cup \{v\}$$

Definition 2. *Let $v, u \in V, v \neq u$. v and u are connected via an intermediate node if they have only one common neighbor between them, which is denoted by $CIN(v, u)$.*

$$CIN(v, u) \Leftrightarrow |N(v) \cap N(u)| = 1$$

Note that there is no direct connection between u and v which satisfy $CIN(u, v)$ but only one common vertex between their neighborhoods. This accordingly implies that vertices u and v are likely not to belong to the same community.

Definition 3. *Let $v, u \in V, v \neq u$. v and u are strongly connected if v and u share two or more neighbors between them, which is denoted by $SC(v, u)$.*

$$SC(v, u) \Leftrightarrow |N(v) \cap N(u)| \geq 2$$

We think it is more plausible to group u and v that satisfy $SC(u, v)$ into the same community than to group two vertices that are connected via an intermediate node (CIN) because strongly connected vertices share more common vertices in their neighborhoods.

Definition 4. *A loner is a vertex v of G whose neighborhood $N(v)$ only contains itself and another vertex which has an edge to it, which is denoted by $loner(v)$.*

$$Loner(v) \Leftrightarrow |N(v)| = 2$$

A *loner* has only one association to other nodes. $\neg Loner(v)$ denotes that v is not a *loner*.

Definition 5. *A vertex $v \in V$ is called a bridge if v's neighbors are not loners and any two nodes other than v in v's neighbors have a CIN relation. $Bridge(v)$ denotes a vertex v is a bridge.*

$$Bridge(v) \Leftrightarrow \forall x, y \in N(v) - \{v\}, x \neq y : CIN(x, y) \wedge \neg loner(x) \wedge \neg loner(y)$$

We use Figure 2a as an example to explain a *bridge.* Let v, x, y in Definition 5 be 1, 2, 3, respectively. Nodes 2 and 3 that are neighbors of node 1 are not *loners* and they only have a common vertex: node 1, i.e., $CIN(2, 3)$, so node 1 in Figure 2a is a *bridge.* A *bridge* is a vertex located between two communities. A *bridge* connects communities, each of which has only one single link with it so we need to check any two neighbors of a *bridge.* Also, a *bridge*'s neighbors cannot be a *loner* since a *loner* does not form a community. As we said, we think that it is controversial to cluster a *bridge* into any community so it is a vertex independent of communities.

Definition 6. *A vertex $v \in V$ is called a gateway if it satisfies the following conditions, which is denoted by $gateway(v)$. First, it has two neighbors that are strongly connected (SC). Second, it has another neighbor that is not a loner and does not share any common neighbor except v with v's other neighbors.*

$$\begin{aligned} Gateway(v) \Leftrightarrow\ & (1) \exists x, y \in N(v) - \{v\}, x \neq y : SC(x, y) \\ & (2) \exists z \in N(v) - \{v\}, \forall u \in N(v) - \{v, z\} : \\ & \neg loner(z) \wedge CIN(z, u) \end{aligned}$$

We use Figure 2b as an example to explain a *gateway.* Let v, x, y, z, u in Definition 6 be 1, 6, 7, 2, 6 or 7, respectively. Nodes 6 and 7 satisfy $SC(6, 7)$ because nodes 6 and 7 have three common vertices (i.e., nodes 1, 6, and 7) in their neighborhoods and node 2 that is not a *loner* has only one common vertex (i.e., node 1) with other neighbors of node 1. This example shows that node 1 in Figure 2b is a *gateway.* A *gateway* acts as an entrance or an exit when we move from a community to another one. In condition (1) of the definition, x and y that are strongly connected implies the existence of a community while in condition (2), z which is not a *loner* implies that there exist one vertex which does not belong to the same community with other neighbors. These two conditions makes a *gateway* act as an entrance to a community or an exit from a community.

Definition 7. *A vertex $v \in V$ is called a hub, denoted by $hub(v)$, if there exist w, x, y, and z which are neighbors of v and which satisfy the following conditions. w and x are strongly connected, and y and z are strongly connected as well. w and y are connected via an intermediate node, and x and z are connected via an intermediate node as well.*

$$\begin{aligned} Hub(v) \Leftrightarrow & \exists w, x, y, z \in N(v) - \{v\}, w \neq x, y \neq z : \\ & SC(w, x) \wedge SC(y, z) \wedge CIN(w, y) \wedge CIN(x, z) \end{aligned}$$

We use Figure 2c as an example to explain a *hub*. Let v, w, x, y, z in Definition 7 be 1, 2, 3, 5, 6, respectively. Nodes 2 and 3 have four common vertices (i.e., nodes 1, 2, 3 and 4) and hence satisfy $SC(2,3)$, and similarly, nodes 5 and 6 have three common vertices (i.e., nodes 1, 5 and 6) and hence satisfy $SC(5,6)$. Nodes 2 and 5 have only one common vertex (i.e., node 1), satisfying $CIN(2,5)$, and similarly nodes 3 and 6 satisfy $CIN(3,6)$. Thus, node 1 in Figure 2c is a *hub*. A *hub* is a confluent vertex, on which groups of vertices converge. We regard a *hub* is a center of groups of vertices, as each pair of neighbors that are strongly connected form a group of vertices and the *CIN* conditions imply that they may belong to different groups of vertices. Groups of vertices may be clustered into the same community or be divided into different communities, which depends on the procedure of community detection explained in the next section.

3.2 Discovery Priority of Bridges, Gateways and Hubs

To detect the three kinds of roles, our algorithm checks for each vertex $v \in V$ if v satisfies the conditions in definitions 5, 6, 7. Note that each $CIN(a,b)$ and $SC(x,y)$ can be checked in $O(d^2)$ time, where d represents the degree of G, because $a, b, x, y \in N(v)$. Hence the complexity of our algorithm is $O(nd^2)$, where n represents the number of nodes in G.

In this paper, we assume that one vertex possesses at most one role. However, a vertex is possibly assigned as a *gateway* and a *hub* at the same time according to the definitions. To avoid this problem, we search roles in the following order: *bridges*, *gateways* and *hubs*. Once a role is assigned, another role cannot be assigned. The order is determined in terms of the community detection (graph clustering) which aims at maximizing the edges within the community and minimizing the edges between communities.

For clarity, we use a *gateway* and a *hub* to explain the order we determine in the context of community detection, in which an edge bridging communities of a role is deleted in each step. Suppose we have detected a *gateway* and a *hub*, each of which connects two communities. In the case of the *gateway*, there is one of its edges spanning communities so two communities will be discovered by only removing one edge. However, in the case of the *hub*, a *hub* is a vertex connecting two communities so two communities will be discovered by removing the edges which link to one of the communities. Hence, when a *gateway* and a *hub* both exist in a graph, the edge of the *gateway* which spans communities will be considered to be removed first according to the aim of community detection, which leads to our intuition for determining the order.

4 Evaluation by Experiments

We have implemented a discovery algorithm of the proposed roles in C language and evaluate them using both synthetic and real datasets. The proposed roles including *bridges*, *gateways*, *hubs*, *loners* are compared with the method named *rawComm* in [6] which proposes four kinds of roles including *ambassadors*, *bridges*, *big fish* and *loners*, and the validness of the outputs are compared subjectively.

Four kinds of roles of *rawComm* are defined using the degree of a vertex and a proposed community metric that estimates the number of communities linked to a vertex, in which a community is defined based on a clique. An *ambassador* is a vertex which has a large degree and a large value for the community metric; a *bridge* is a vertex which has a small degree but a large value for the community metric; a *big fish* is a vertex which has a large degree but a small value for the community metric; a *loner* is vertex which has a small degrees and a small value for the community metric. We follow the way in [6] to normalize the degree and the proposed community metric value of a vertex between 0 and 1, and assign roles by determining the value of a threshold r from 0 to 1 required in *rawComm*. The threshold r is used to discover roles by classifying values (i.e., the degree and the value of the proposed community metric) assigned to a vertex. *rawComm* can become more useful for communities defined by means other than clique by using probabilities values that contain community information such as the probability that two linked nodes are in the same community. We compare our proposed roles with two variations of *rawComm*, one which utilizes community information and the other which does not utilize community information.

4.1 Synthetic Data

To evaluate the proposed roles in one network, we generate a synthetic graph with 21 vertices and 37 edges as shown in Figure 3. We intentionally settled the size of the graph relatively small to clearly demonstrate the results of our method and *rawComm*. Figure 3a shows the result of our method, Figure 3b shows the result of *rawComm* without using community information, and Figures 3c and 3d are the results of *rawComm* that use community information with $r = 0.25$ and $r = 0.65$, respectively. To validate our proposed roles, we use normalized cut [7] that is a well-known clustering technique to show the community structure. Nodes circled within a gray oval in Figure 3a are clustered into one community.

As shown in Figure 3a, our method discovers node 6 as a *hub*. In Figure 3b, *rawComm* without the community information does not discover node 6 while *rawComm* discovers node 6 as an *ambassador* after importing the community information as shown in Figure 3c.

Moreover, we see nodes 1, 2, 4 and 5 are important vertices that bridge two communities from the community structure shown in Figure 3a. In our method, node 1 is distinguished as a *bridge* from other vertices, and nodes 2, 3, 4 and 5 are recognized as *gateways*. However, *rawComm* fails to discover nodes 1, 2, 4 and 5 since they are not vertices that have large degrees or connect many cliques when $r = 0.25$. Node 1 is discovered as a *bridge* when $r = 0.65$ in Figure 3d while much more vertices are assigned *loners*. From the observations, we see that vertices (i.e., nodes 1, 2, 4 and 5) that bridge communities are assigned roles in our method while *rawComm* cannot detect all vertices bridging communities and requires that the value of the threshold r is settled appropriately for discovering some of the roles.

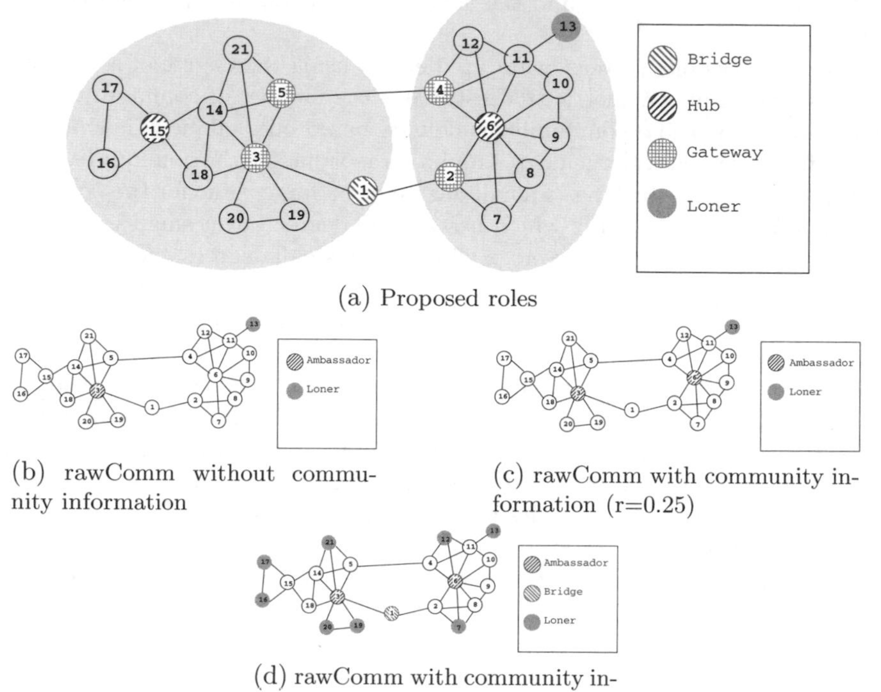

(a) Proposed roles

(b) rawComm without community information

(c) rawComm with community information (r=0.25)

(d) rawComm with community information (r=0.65)

Fig. 3. Evaluation of proposed roles using synthetic data

4.2 DBLP Data

Figures 4 and 5 show the experimental results that use the data from DBLP[1]. DBLP provides bibliographic information on major computer science journal and proceedings. For the experiment, we extracted the data of IJCAI between 2005 and 2009 from DBLP, and generated a coauthorship network, where a vertex represents an author and two authors are linked by an edge when they have coauthored at least one paper. The data of IJCAI 2005–2009 encompasses 2197 vertices and 6412 edges, and we only show a subgraph for clarity in Figure 4 because similar results are also observed in other subgraphs. Note that we only compare our method with *rawComm* with community information in this experiment.

In Figure 4a, *rawComm* discovers two kinds of roles. Nodes 1 and 3 are *ambassadors* and they are vertices which have large values for both the degree and the community metric value. Nodes 12, 14, 15, 16, 19, 20, and 31 are *loners* and they are vertices which have small degrees compared to other vertices. Similarly, we tuned the value of r in order to examine whether we can find more roles, but only one more vertex (node 17) of *ambassadors* is found while most vertices are

[1] http://www.informatik.uni-trier.de/~ley/db/

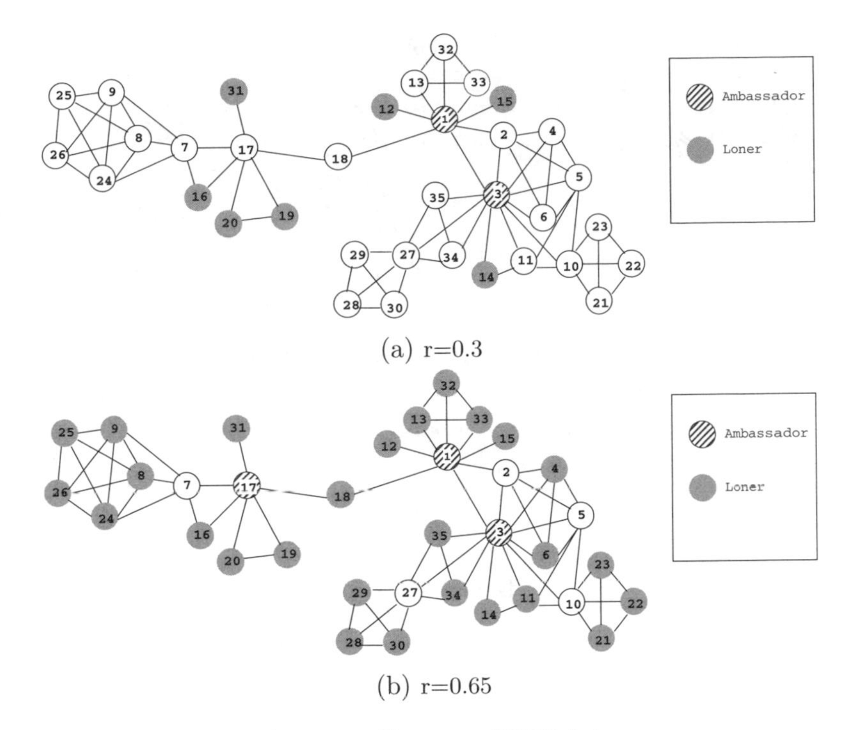

(a) r=0.3

(b) r=0.65

Fig. 4. rawComm on DBLP data

grouped into the role of *loners* when $r = 0.65$ (Figure 4b). Since the value for the community metric proposed in [6] tends to be proportional to the degree, *rawComm* appears to classify vertices into *ambassadors* and *loners*, which is also observed in [6]. Note that *rawComm* fails to distinguish node 18 although it is a node which bridges two communities when the number of communities is three (Figure 6b).

To show the relationship between the proposed community-oriented roles and the community structure, Figures 6a, 6b and 6c show clustering results when the number of given communities is 2, 3 and 4, respectively, by utilizing the normalized cut method.

By referring to Figures 5 and 6, we have the following findings of our *bridges*, *gateways* and *hubs*. Node 18 discovered as a *bridge* is an important vertex in bridging two communities, nodes 1 and 17 are entrances to communities and neighbors of node 3 are divided into different communities, which fit our intuition. Although nodes 7 and 10 are not vertices in bridging communities in the clustering results in Figure 6, they turn out to be vertices in bridging communities when the number of communities increases, which corresponds to our anticipation. From the results, we can conclude that the border between two communities are often vertices with the proposed roles such as nodes 1 and 3 when the number of communities is two and nodes 1, 3, 17, and 18 when the number of communities is three. With this regard, our method outperforms *rawComm* since

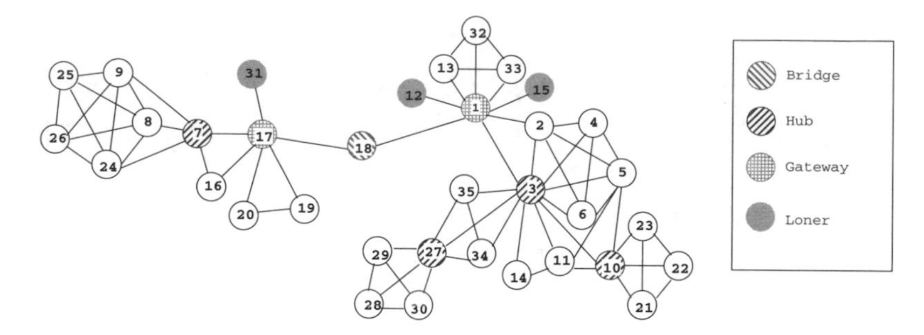

Fig. 5. Proposed roles on DBLP data

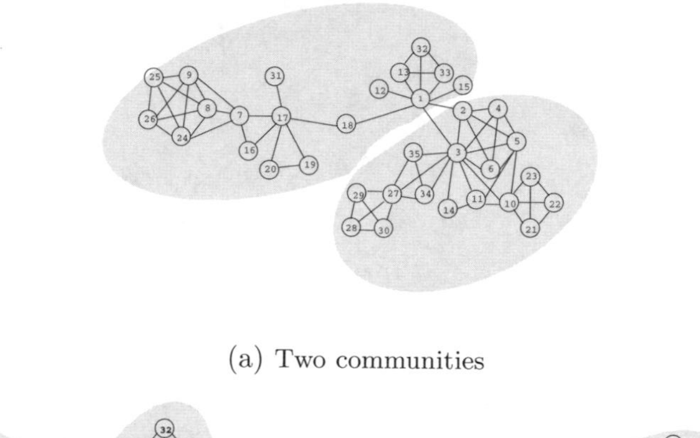

(a) Two communities

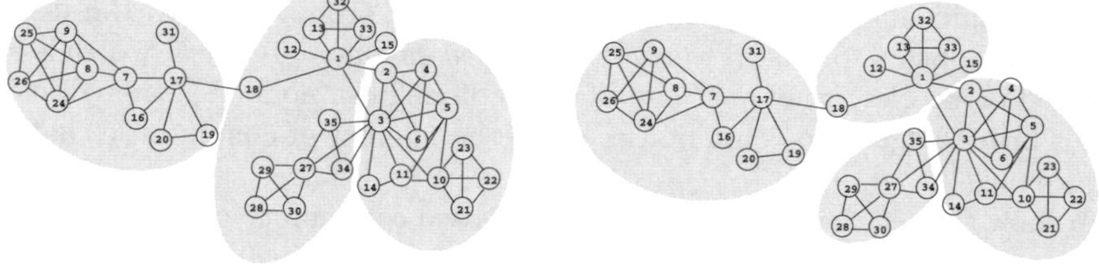

(b) Three communities (c) Four communities

Fig. 6. Comparing with clustering results that use the method of Normalized Cut

rawComm only discovers vertices with large degrees that connect relatively many cliques while our method discovers vertices that connect communities with few links such as node 18 as well.

4.3 Analysis on the Proposed Orientation of Community

To examine whether three community-oriented roles — *bridges*, *gateways* and *hubs* are important vertices in bridging communities, we further perform another experiment which shows how frequently the proposed roles appear around

Table 2. Summary of the analysis (A cell of the columns of Bridge, Gateway, Hub represents the number of correct roles/the number of discovered role for *bridges*, *gateways* and *hubs*, respectively. A cell of the column of Community represents the number of communities whose boundary contain proposed roles/the number of detected communities. A number followed by a % represents the accuracy rate.)

Confname	Bridge	Gateway	Hub	Community
KDD	0/1 (0%)	35/38 (92%)	214/269 (80%)	225/227 (99%)
IJCAI	6/12 (50%)	47/59 (80%)	180/217 (83%)	213/215 (99%)

the boundary of communities. The data for the experiment are collected from KDD2009, KDD2008-2009, KDD2007-2009, KDD2006-2009, KDD2005-2009, IJCAI2009, IJCAI2007-2009, and IJCAI2005-2009. We extracted all sub-connected graphs each of whose size is larger than ten, and performed community detection by using the clustering method introduced in [4] which measures how good the division is. The number of clusters is optimally determined by [4] and we do not have to assign the number of clusters so we use it as the method of community detection to simplify the experiment.

We check whether nodes that are discovered as community-oriented roles bridge communities and summarize the experimental result in Table 2. Here we use the result of community detection as our ground truth to compute the accuracy rate. A vertex discovered as a proposed role is judged as a correct role if its neighbors belong to different communities.

In Table 2, the number of vertices discovered as *bridges* is the smallest and the number of vertices discovered as *hubs* is the largest, which corresponds to our anticipation because there exist many cliques in social networks. *Gateways* and *hubs* have high accuracy rates while *bridges* has low accuracy rates in Table 2. For the result, we examined the graph structures and found that a part of vertices that link to *bridges* are too few to form a community, which results in low accuracy rates. From the accuracy rate for communities that amounts to 99% in Table 2, we can conclude that most of the vertices with the proposed roles are found in the boundary of communities and they are important in bridging communities.

5 Conclusions and Future Work

In this paper, we proposed three community-oriented roles, namely *bridges*, *gateways* and *hubs*, which are important roles in bridging communities. A role is assigned to a vertex based on the relationship between its neighbors. The more similar the vertices shared between its neighbors are, the denser the graph formed by the vertex and its neighbors becomes. Similarly, if the neighbors of a vertex rarely share common vertices, it implies that the vertex is likely to be the vertex bridging communities because community detection aims at maximizing edges within communities but minimizing edges between communities. Our method is validated through experiments and is shown to be able to discover vertices that bridge communities relatively accurately without knowledge on the community structure.

As discussed in the previous sections, the grouping problem of the proposed roles may be considered differently in community detection. For example, it fits our intuition not to cluster *bridges* into any community and to view *hubs* as overlapping vertices. Therefore, our future work is to develop an algorithm which detects communities highly accurately by considering roles of vertices.

References

1. Domingos, P., Richardson, M.: Mining the Network Value of Customers. In: Proceedings of the Seventh ACM SIGKDD International Conference on Knowledge Discovery and Data Mining (KDD), pp. 57–66 (2001)
2. Han, J., Kamber, M.: Data Mining: Concepts and Techniques. Morgan Kaufmann, San Francisco (2005)
3. Kempe, D., Kleinberg, J.M., Tardos, É.: Influential Nodes in a Diffusion Model for Social Networks. In: Caires, L., Italiano, G.F., Monteiro, L., Palamidessi, C., Yung, M. (eds.) ICALP 2005. LNCS, vol. 3580, pp. 1127–1138. Springer, Heidelberg (2005)
4. Newman, M.E.J.: Finding Community Structure in Networks Using the Eigenvectors of Matrices. Physical Review E (Statistical, Nonlinear, and Soft Matter Physics) 74(3), 036104 (2006)
5. Newman, M., Park, J.: Why Social Networks are Different from Other Types of Networks. Physical Review E 68(3), 36122 (2003)
6. Scripps, J., Tan, P.N., Esfahanian, A.H.: Node Roles and Community Structure in Networks. In: Proceedings of the Ninth WebKDD and the First SNA-KDD 2007 Workshop on Web Mining and Social Network Analysis (WebKDD/SNA-KDD), pp. 26–35 (2007)
7. Shi, J., Malik, J.: Normalized Cuts and Image Segmentation. IEEE Transactions on Pattern Analysis and Machine Intelligence 22, 888–905 (1997)
8. Wang, Y., Chakrabarti, D., Wang, C., Faloutsos, C.: Epidemic Spreading in Real Networks: An Eigenvalue Viewpoint. In: Proceedings of the 22nd International Symposium on Reliable Distributed Systems (SRDS), pp. 25–34 (2003)
9. Wasserman, S., Faust, K.: Social Network Analysis: Methods and Applications. Cambridge University Press, Cambridge (1994)
10. Watts, D.J., Strogatz, S.H.: Collective Dynamics of 'Small-World' Networks. Nature 393(6684), 440–442 (1998)

A Graph-Based Clustering Scheme for Identifying Related Tags in Folksonomies

Symeon Papadopoulos[1,2], Yiannis Kompatsiaris[1], and Athena Vakali[2]

[1] Informatics and Telematics Institute, CERTH
57001, Thessaloniki, Greece
{papadop,ikom}@iti.gr

[2] Department of Informatics, Aristotle University,
54124, Thessaloniki, Greece
avakali@csd.auth.gr

Abstract. The paper presents a novel scheme for graph-based clustering with the goal of identifying groups of related tags in folksonomies. The proposed scheme searches for core sets, i.e. groups of nodes that are densely connected to each other by efficiently exploring the two-dimensional core parameter space, and successively expands the identified cores by maximizing a local subgraph quality measure. We evaluate this scheme on three real-world tag networks by assessing the relatedness of same-cluster tags and by using tag clusters for tag recommendation. In addition, we compare our results to the ones derived from a baseline graph-based clustering method and from a popular modularity maximization clustering method.

Keywords: graph-based clustering, community detection, folksonomies, tag recommendation.

1 Introduction

Collaborative (or Social) Tagging is nowadays a common feature of content sharing web applications that enables users to: (a) upload new, or bookmark existing content and, (b) annotate it by means of free-text keywords (tags). Such applications, examples of which are delicious[1], flickr[2] and Bibsonomy[3], are commonly referred to as Social Tagging Systems (STS). Currently, STS attract huge amounts of traffic, which results in the emergence of massive grassroots content annotation and organization schemes, referred to as folksonomies [1,2]. Folksonomies comprise three types of entities, namely users, resources and tags, as well as the associations among them [3,4].

Folksonomies constitute a direct encoding of the views of a large number of users on how content items should be organized through a flexible annotation scheme (tagging). By analyzing the structure and content of folksonomies, one

[1] http://delicious.com/
[2] http://www.flickr.com/
[3] http://bibsonomy.org/

T.B. Pedersen, M.K. Mohania, and A M. Tjoa (Eds.): DaWaK 2010, LNCS 6263, pp. 65–76, 2010.

can expect to gain valuable insights into the topic and vocabulary structure of the system. To this end, *tag clustering* has lately attracted significant research interest due to its value in several Information Retrieval (IR) use case scenarios [5,6,7,8,9,10,11]. Tag clustering is commonly understood as a process that groups the tags of an STS in a way such that members of the same tag cluster are perceived by users as *related* to each other. Despite the subjective element in judging the degree of relatedness between tags, tag clusters are expected to correspond to meaningful topic areas, which can be useful in a series of tasks, such as information exploration and navigation [5,6], automatic content annotation [8], user profiling [9], content clustering [10,11] and tag recommendation [12,13].

To date, tag clustering has been dealt with either by conventional clustering algorithms, such as K-means [10] and Hierarchical Agglomerative Clustering [8,9], or, more recently, by use of *community detection* methods [5,6,7]. Conventional clustering schemes are frequently troubled by two shortcomings: (a) the need for providing the number of clusters as input to the algorithm, and (b) their computational complexity. Community detection methods address both of these needs, since they do not require the number of clusters (communities) to be known a priori and they are typically more efficient in terms of computations. However, *modularity maximization* methods [14], which constitute the bulk of community detection methods, are troubled by the so-called "super-community" problem, i.e. they produce few communities with very large sizes and numerous communities with small sizes. Having tag clusters of such highly skewed size distribution can be detrimental to the aforementioned IR tasks.

For that reason, we introduce in this paper a hybrid graph-based tag clustering scheme, referred to in short as HGC, which attempts to address the aforementioned constraints. HGC is based on the notion of (μ, ϵ)-cores [15], groups of nodes that have a large number of common neighbors to each other. HGC conducts an efficient search over the (μ, ϵ) parameter space and identifies the associated core sets. Subsequently, a core set expansion step is conducted based on a local modularity measure [16]. This expansion enables the resulting clusters to overlap with each other, which is particularly important for the problem of tag clustering, since tags are typically used in multiple contexts and senses.

The rest of the paper is structured as follows. Section 2 discusses existing work on the topic of tag clustering and its applications. Section 3 presents HGC, the proposed hybrid graph-based solution to the problem of tag clustering. HGC is evaluated and compared against existing clustering schemes in Section 4. The paper concludes in Section 5.

2 Related Work

The problem of tag clustering has recently attracted increasing research interest since it is a challenging task from a data mining perspective, but at the same time it also holds the potential for benefiting a variety of IR applications. For instance, tag clustering is considered important for eliciting a topic hierarchy for a tagging system and improving content retrieval and browsing [8]. Similar conclusions are reached by [5] who point that the use of raw tag information limits content exploration and discovery, thus creating the need for an additional level of organization through tag clustering. In [9], tag clusters are used as a nexus between users and their interests. Using tag clusters instead of plain tags for profiling user interests proved beneficial for personalized content ranking. An additional application of tag clustering is presented in [7]. There, the tag clusters were used as a means of identifying the different contexts of use for a given tag, i.e. for sense disambiguation. It was shown that using the tag clusters results in improved results compared to the use of external resources such as WordNet.

The methods used for performing the tag clustering largely fall under one of two approaches: (a) conventional clustering techniques, such as Hierarchical Agglomerative Clustering (HAC) [8,9] and (b) community detection methods [5,6,7]. HAC suffers from high complexity (quadratic to the number of tags to be clustered) and the need to set ad-hoc parameters (e.g. three parameters need to be set in the clustering scheme used in [9]). Community detection methods largely address the shortcomings of HAC since efficient implementations exist with a complexity of $O(Nlog(N))$ for finding the optimal grouping of N tags into communities. Furthermore, community detection methods rely on the measure of modularity [14] as a means to assess the quality of the derived cluster structure. Thus, modularity maximization methods do not require any user-defined parameters. However, a problem of modularity maximization methods pointed in [6] and confirmed by our experiments is their tendency to produce clusters with a highly skewed size distribution, which makes them unsuitable for the problem of tag clustering.

3 Description of HGC

The proposed scheme builds upon the notion of (μ, ϵ)-cores introduced in [15] and briefly described in subsection 3.1. The original algorithm, referred to as SCAN [15], suffers from two problems. First, it needs two parameters, namely μ and ϵ, to be provided as input. Second, it leaves a substantial number of nodes unassigned to clusters. As a result, its utility is limited in IR tasks such as tag recommendation. For that reason, our scheme conducts an efficient iterative search over the parameter space (μ, ϵ) in order to discover cores for multiple values of the parameters (subsection 3.2). Finally, the identified cores are expanded, as described in subsection 3.3, by maximizing a local measure of modularity [16] in order to increase the number of nodes that are assigned to communities and to enable overlap among communities.

3.1 Core Set Discovery

The definition of (μ, ϵ)-cores is based on the concepts of *structural similarity*, *ϵ-neighborhood* and *direct structure reachability*.

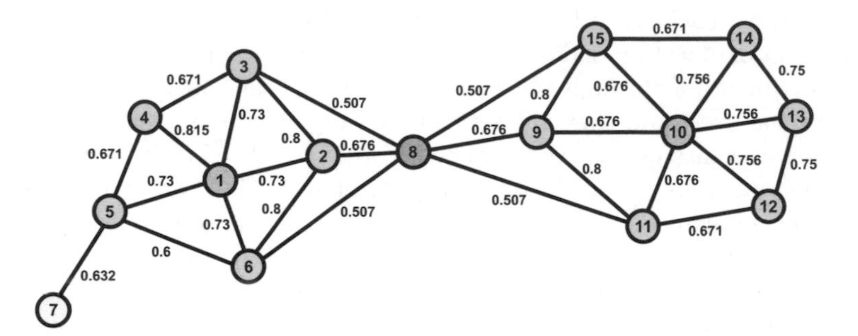

Fig. 1. Example of community structure in an artificial network. Nodes are labeled with successive numbers and edges are labeled with the structural similarity value between the nodes that they connect. Nodes 1 and 10 are (μ, ϵ)-cores with $\mu = 5$ and $\epsilon = 0.65$. Nodes 2-6 are structure reachable from node 1 and nodes 9, 11-15 are structure reachable from node 10. Thus, two community seed sets have been identified: the first consisting of nodes 1-6 and the second consisting of nodes 9-15.

Definition 1. *The* ***structural similarity*** σ *between two nodes* v *and* w *of a graph* $G = \{V, E\}$ *is defined as:*

$$\sigma(v, w) = \frac{|\Gamma(v) \cap \Gamma(w)|}{\sqrt{|\Gamma(v)| \cdot |\Gamma(w)|}} \tag{1}$$

where $\Gamma(v)$ *is the* structure *of node* v: $\Gamma(v) = \{w \in V | (v, w) \in E\} \cup \{v\}$.

Definition 2. *The* ϵ***-neighborhood*** *of a node is the subset of its structure containing only the nodes that are at least* ϵ*-similar with the node; in math notation:*

$$N_\epsilon(v) = \{w \in \Gamma(v) | \sigma(v, w) \geq \epsilon\} \tag{2}$$

Definition 3. *A vertex* v *is called a* (μ, ϵ)***-core*** *if its* ϵ*-neighborhood contains at least* μ *vertices:* $CORE_{\mu,\epsilon}(v) \Leftrightarrow |N_\epsilon(v)| \geq \mu$.

Definition 4. *A node is directly* ***structure reachable*** *from a* (μ, ϵ)*-core if it is at least* ϵ*-similar to it:* $DirReach_{\mu,\epsilon}(v, w) \Leftrightarrow CORE_{\mu,\epsilon}(v) \wedge w \in N_\epsilon(v)$.

Once the (μ, ϵ)-cores of a network have been identified, it is possible to start attaching adjacent nodes to them provided that they are reachable through a chain of nodes which are directly structure reachable from each other. We call the resulting set of nodes as a *community seed set*. The rest of the nodes are considered to be *hubs* or *outliers* depending on whether they are adjacent to more than one community core sets or not. An example of computing structural similarity values for the edges of a network and then identifying the underlying (μ, ϵ)-cores, hubs and outliers of the network is illustrated in Figure 1. This technique for collecting community seed sets is computationally efficient since its complexity is $O(\overline{k} \cdot n)$ for a network of n nodes and average degree $\overline{k}$. Computing the structural similarity values of the m network edges introduces an additional $O(\overline{k} \cdot m)$ complexity in the community detection.

3.2 Parameter Space Exploration

One issue that is not addressed in [15] pertains to the selection of parameters μ and ϵ. Setting a high value for ϵ (the maximum possible value for ϵ is 1.0) will render the core detection step very eclectic, i.e. few (μ, ϵ)-cores will be detected. Moreover, higher values for μ will also result in the detection of fewer cores (for instance, all nodes with degree lower than μ will be excluded from the core selection process). For that reason, we employ an iterative scheme, in which the community seed set selection operation is carried out multiple times with different values of μ and ϵ so that a meaningful subspace of these two parameters is thoroughly explored and the respective (μ, ϵ)-cores are detected.

The exploration of the (μ, ϵ) parameter space is carried out as depicted in Figure 2. We start by a very high value for both parameters. Since the maximum possible values for μ and ϵ are k_{max} (maximum degree on the graph) and 1.0 respectively, we start the parameter exploration by two values dependent on them (for instance, we could select $\mu_0 = 0.5 \cdot k_{max}$ and $\epsilon_0 = 0.9$; the results of the algorithm are not very sensitive to this choice). We identify the respective (μ, ϵ) cores and associated core sets and then relax the parameters in the following way. First, we reduce μ; if it falls below a certain threshold (e.g. $\mu_{min} = 4$), we then reduce ϵ by a small step (e.g. 0.05) and we reset $\mu = \mu_0$. When both μ and ϵ reach a small value ($\mu = \mu_{min}$ and $\epsilon = \epsilon_{min}$), we terminate the community seed set detection step. This exploration path ensures that first high quality communities will be discovered and subsequently less profound ones will also be detected. In order to speed up the parameter exploration process, we employ a logarithmic sampling strategy when moving along the μ parameter axis. The computational complexity of the proposed parameter scheme is a multiple of the original SCAN. The multiplicative factor is $C = s_\epsilon \cdot s_\mu$, where s_ϵ is the number of samples along the ϵ axis ($\simeq 10$) and s_μ is the number of samples along the μ axis ($\simeq \log k_{max}$).

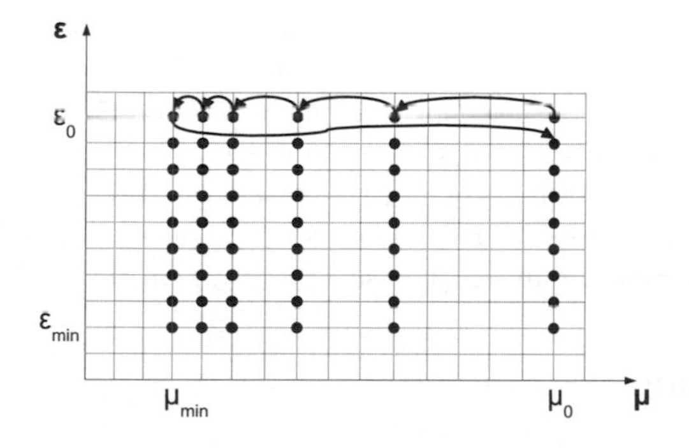

Fig. 2. Depiction of the (μ, ϵ) parameter space exploration path. The upper values μ_0 and ϵ_0 are set in relation to their maximum possible ones ($\mu_{max} = k_{max}$ and $\epsilon_{max} = 1.0$). The lower values are set to $\mu_{min} = 4$ and $\epsilon_{min} = 0.4$ since cores with lower values than these are of inconsistent quality.

3.3 Core Set Expansion

Starting from a community seed set S, the second step in the proposed community detection method involves an expansion process, which aims at attaching additional nodes, which are relevant, to the initial community seed set. The expansion step is essential for deriving higher quality communities since the community seed sets produced by the previous step may fail to include in the communities nodes that are of importance for them. In the case of tag communities, this would lead to tag communities that would miss some important keywords and would thus be less representative of their topic. In addition, it is due to this expansion step that overlap among communities is possible since the previous step produces non-overlapping community seed sets.

The community expansion step is based on the maximization of a local measure of community quality, namely *subgraph modularity* introduced in [16]. The modularity of a subgraph $S \in V$ is defined as the ratio of the number of intra-community edges (edges connecting nodes within S) over the number of edges sticking out of S (Equation 3). Obviously, the larger such a value is, the more well separated the subgraph is from the rest of the graph. In the extreme case of a disconnected subgraph, its modularity value tends to infinity:

$$M(S) = \frac{ind(S)}{outd(S)} = \frac{|\{(v, w) \in E | v, w \in S\}|}{|\{(v, w) \in E | v \in S \wedge w \in V - S\}|} \tag{3}$$

The proposed expansion step is based on a greedy maximization scheme, i.e. it successively attaches nodes to community S as long as their addition increases the subgraph modularity $M(S)$ of the community. The set of nodes that are considered as candidates for attachment to S are pooled from the "community frontier", i.e. the set of all nodes that are adjacent to at least one node of the community. Each candidate node is tentatively attached to the community and the new value of its modularity is computed. This computation can be performed very efficiently in an incremental fashion based on the values of $ind(S)$ and $outd(S)$ before the tentative attachment of the candidate node to the community.

Nodes with very high degree[4] are not considered in this process for two reasons: (a) to reduce the computational complexity of the expansion step, (b) to prevent the expansion process from creating a "gigantic" community. The node resulting in the maximum increase of modularity for the community is considered a member of the community and the process is repeated for the rest of the candidate nodes (it is possible that there is no increase of modularity by adding a node to the community, in which case no expansion takes place).

4 Evaluation

In order to gain insights into the behavior of community detection in real-world tagging systems, we conduct an evaluation study comparing the performance

[4] We create a degree-ordered list of nodes for the whole graph and consider as high-degree nodes the top 10% of them.

Table 1. Folksonomy datasets used for evaluation

(a) Basic folksonomy statistics

Dataset	#triplets	U	R	T
BIBSONOMY-200K	234,403	1,185	64,119	12,216
FLICKR-1M	927,473	5,463	123,585	27,969
DELICIOUS-7M	7,501,032	112,950	1,332,796	251,352

(b) Tag graph statistics (for large component)

Dataset	$\|V\|$	$\|E\|$	$\overline{k}$	$\overline{cc}$
BIBSONOMY-200K	11,949	236,791	39.63	0.6689
FLICKR-1M	27,521	693,412	50.39	0.8512
DELICIOUS-7M	216,844	3,443,367	31.76	0.8018

of our method (HGC) against two competing community detection methods on three datasets coming from different tagging applications, namely BibSonomy, Flickr and Delicious. The first of the two community detection methods under study is the well-known greedy modularity maximization scheme presented by Clauset, Newman and Moore (CNM) [18][5] and the second is the SCAN algorithm of [15], which is extended by HGC. The three datasets used for our study are described below and basic information on their size is presented in the upper part of Table 1.

BIBSONOMY-200K: BibSonomy is a social bookmarking and publication sharing application. The BibSonomy dataset was made available through the ECML PKDD Discovery Challenge 2009[6]. We used the "Post-Core" version of the dataset, which consists of a little more than 200,000 tag assignments (triplets) and hence the label "200K" was used to form the dataset name.

FLICKR-1M: Flickr is a popular online photo sharing and organizing application. For our experiments, we used a focused subset of Flickr comprising approximately 120,000 images that were located within the city of Barcelona (by use of a geo-query). In total, the number of tag assignments for this dataset approaches one million.

DELICIOUS-7M: Delicious is a popular social bookmarking service for managing and sharing bookmark collections. We used a snapshot of the Delicious bookmark collection corresponding to January 2006, comprising seven million tag assignments. This dataset is a subset of the collection studied in [19].

Starting from each dataset, we built a tag graph, considering an edge between any two tags that co-occur in the context of some resource. The raw graph contained a large component and several very small components and isolated nodes. For the experiments we used only the large component of each graph,

[5] We used the publicly available implementation of this algorithm, which we downloaded from `http://www.cs.unm.edu/~aaron/research/fastmodularity.htm`

[6] `http://www.kde.cs.uni-kassel.de/ws/dc09`

which accounts for more than 99% of the size of the raw graph for all three datasets. Some basic statistics of the analyzed large components are presented in the lower part of Table 1. The nodes of the three tag graphs appear to have a high clustering coefficient on average, which indicates the existence of community structure in them. We applied the three competing clustering schemes, CNM, SCAN and HGC, on the tag graphs and proceeded with the analysis of the derived communities. Since SCAN is parameter-dependent, we performed the clustering multiple times for many (μ, ϵ) combinations and selected the best solution.

Our first observation concerns the community structure produced by CNM. When considering the applications of tag clustering, it is hard to imagine that the highly imbalanced cluster structure produced by CNM can be of much benefit. For instance, knowing that two tags belong to the same huge cluster is not very informative of their semantic relation; in fact, there are many pairs of tags within such huge clusters that are not actually related to each other. Table 2 presents several such examples of unrelated tags which were placed in the same cluster. Having these tags in the same cluster is not only uninformative but it is actually misleading and thus potentially harmful for use within some IR task.

Table 2. Examples of unrelated tags that were assigned by CNM to the same community. Examples from the three largest communities of each dataset are presented.

Dataset	Examples of unrelated tags in the same community
BIBSONOMY-200K	hannover, nutritional, ebusiness, bishop, vivaldi, sunsets, skyscapes, recycle, antiracist, patentbibliometrics
	informationretrieval, magnetic, robotics, kolmogorov, wordnet, socialinformatics, thermodynamics, metaphysics, ...
	webdesign, windows, torrent, puzzle, vmware, geotagging, mov, techcrunch, cpplib, baseballplayers
FLICKR-1M	spanien, common chimpanzee, star wars, renault, restaurant, prostitution, olympicstadium, large windows, infrared
	barcelona, watermelon, photon awards, birthday, mediterranean, palm tree, fine arts, volkswagen, building, logistics
	roma, double bass, crowd surfing, environment, lomography, flickr babes, sombrero, basketball, bruce springsteen
DELICIOUS-7M	geekiness, telepathy, scifihorror, britneyspears, theflintstones, sportculture, environmentalhealth, uspatent, argentina, ...
	education, capetown, flashwebsites, businessanalyst, newjournalism, adventuretravel, musicnetwork, scienceastrophysics, ...
	food, island, bike, jersey, federal, climate, ghosts, athletics, enviroment, imperialism

In contrast, Table 3 presents several examples of interesting tag clusters discovered by HGC. Close examination of the tags contained in them reveals their close semantic and contextual association. In the case of CNM these clusters are contained in the aforementioned gigantic communities together with numerous unrelated tags, thus their utility is limited. On the other hand, the plain SCAN method can only identify subsets of these clusters, which is expected to harm the recall performance of the IR applications making use of them.

Table 3. Examples of interesting tag communities discovered by HGC. In the case of CNM, these communities are "hidden" within the gigantic communities discovered by CNM. In contrast, in the case of SCAN, these communities are smaller since they do not include tags from the community expansion step.

Dataset	Examples of interesting HGC tag communities
BIBSONOMY-200K	mpg, tif, jpeg, mpc, ico, wma, swf, fileconversion, txt, midi, psd, wmi, ogg, avi, psp, tiff, odg, mdb, kar, divx, wmv, qcp, odp, ods, rtf, odt, jpg, mov, amv, png, flv, flac, mmf, gif, sxw, amr, ...
	israelis, middleast, terrorism, middleastpeace, peaceprocess, onevoice, palestinians, conflictresolution, extremism, hatred
	urlogic, lymphatic, neoplasms, virus, pathophysiology, microbial, hemic, physician, doctor, musculoskeletal, respiratory, student, hepatological, viral, infections, hematological, gastrointestinal
FLICKR-1M	salad, spansih gastronomy, catalan food, modena, bacallà, colmenillas, bread with tomato, marinated, gastronomy, merluzzo, ec, marinado, cod, vinegar, bacalao, foie, meatfest, duck foie, ...
	george clooney, sean connery, jude law, antonio banderas, jennifer lopez, tom cruise, penelope cruz, viggo mortensen, ...
	series, australian, federer, conde godó, open, moya, tenerife, atp, las palmas gran, garros, torneo, murray, tamarasit, roland, roddick, podcast, bernardes, sharapova, djokovic, wta, wawrinka, campeonato, canarias, usopen, enric molina, chela gran, ...
DELICIOUS-7M	apollomission, saturnrocket, spacecrew, crewflight, navylieutenant, flightcommander, colonelwhite, americanastronauts, lieutenantcolonel, edwardwhite, spacewalk, capekennedy
	herbiehancock, dextergordon, chrispotter, brianblade, grantgreen, adamrogers, donaldbyrd, theloniousmonk, leemorgan, larrygoldings, hardbop, weatherreport, marcjohnson, mainstreamjazz, artblakey, billevans, joehenderson, joshuaredman, charlieparker, ...
	danacarvey, commercialparodies, thehanukkasong, richardpryor, stevemartin, wilferrell, chrisfarley, billmurray, adamsandler, kingtut, alecbaldwin, mikemyers, churchlady, chevychase, ...

Finally, we used the derived tag clusters in the context of tag recommendation in order to quantify their effect on the IR performance of a cluster-based tag recommendation system. More specifically, we created a simple recommendation scheme, which, based on an input tag, uses the most frequent tags of its containing cluster to form the recommendation set. In case more than one tags are provided as input, the system produces one tag recommendation list (ranked by tag frequency) for each tag and then aggregates the ranked list by summing the tag frequencies when of tags belonging to more than one list. Although this recommendation implementation is very simple, it is suitable for benchmarking the utility of cluster structure since it is directly based on it.

The evaluation process was conducted as follows: We divided the available tag assignments for each dataset into two sets, one used for training and the other used for testing. Based on the training set, we built the corresponding tag graph and produced the tag clusters based on the three competing methods. Then, by using the tag assignments of the test set, we quantified the extent to which the cluster structure found by use of the training set could help predict the tagging activities of users on the test set. For each test resource tagged with L tags, $K < L$ tags were used as input to the tag recommendation algorithm and the rest $L - K$ were predicted. In that way, both the number of correctly predicted tags and the one of missed tags is known. In addition, a filtering step was applied on the tag assignments of the test set. Out of the test tag assignments, we removed the tags that (a) did not appear in the training set, since it would be impossible to recommend them and (b) were among the top 5% of the most frequent tags, since in that case recommending trivial tags (i.e. the most frequent within the dataset) would be enough to achieve high performance.

Table 4 presents a comparison between the IR performance of tag recommendation when using the CNM, SCAN and HGC tag clusters. According to it, using the HGC tag clusters results in far better tag recommendations than by use of CNM across all three datasets. For instance, in the FLICKR-1M dataset, the HGC-based recommendation achieves six times higher precision than the CNM-based one (22.98% compared to 3.73%). A large part of the CNM-based recommendation failure can be attributed to the few gigantic communities that dominate its community structure. Compared to the best run of SCAN, HGC performs better in terms of number of unique correct suggestions, recall and $P@1$, but worse in terms of precision. In terms of F-measure, SCAN performs slightly better in two out of the three datasets, but HGC performs better in the third dataset. Given the fact that SCAN requires parameter tuning to achieve this performance and that HGC provides more correct unique suggestions, we conclude that the HGC tag cluster structure is more valuable in the context of tag recommendation. Since HGC extends SCAN in two steps (multiple iterations of SCAN and expansion of communities), we also ran tests to establish the relation of performance change to each of these steps: the multiple SCAN iteration step was responsible for a small part of the drop in precision and a measurable part of the increase in recall, while the expansion step was the main reason behind the increase in recall and the largest part of the drop in precision.

Table 4. IR performance of CNM, SCAN and HGC community structures in tag recommendation. The following notation is used: R_T denotes the number of correct tags according to the ground truth, R_{out} the number of tag suggestions made by the recommender, R_{TP} the number of correct suggestions, U_{TP} the number of unique correct suggestions, P, R, and F stand for precision, recall and F-measure respectively, and P@1, P@5 denote precision at one and five recommendations respectively.

	BIBSONOMY-200K			FLICKR-1M			DELICIOUS-7M		
	CNM	SCAN	HGC	CNM	SCAN	HGC	CNM	SCAN	HGC
R_T		15,216			55,875			56,893	
R_{out}	15,056	4,958	11,814	55,605	22,463	49,851	56,166	13,974	33,107
R_{TP}	272	1,120	**1,406**	2,074	10,419	**11,454**	1,022	3,624	**6,258**
U_{TP}	189	717	**837**	305	1,399	**1,666**	459	1,506	**2,628**
P (%)	1.81	**22.59**	11.90	3.73	**46.38**	22.98	1.82	**25,93**	18.90
R (%)	1.79	7.36	**9.24**	3.71	18.65	**20.50**	1.80	6.37	**11.00**
F (%)	1.80	**11.10**	10.40	3.72	**26.60**	21.67	1.81	10.23	**13.91**
P@1 (%)	1.68	3.96	**5.09**	1.95	8.02	**9.85**	1.64	2.78	**7.95**
P@5 (%)	2.18	**29.06**	17.27	3.41	**46.84**	21.27	2.35	**36.91**	29.49

5 Conclusions

We presented a parameter-free graph-based clustering scheme that is particularly suited to the task of tag clustering. The proposed scheme is based on the discovery of (μ, ϵ)-cores for multiple sets of (μ, ϵ) values and a subsequent expansion based on a local measure of cluster quality. We evaluated the proposed scheme on three real-world datasets and compared its performance against a modularity maximization clustering algorithm (CNM) and the basic (μ, ϵ)-core detection scheme (SCAN), which our proposal extends. We demonstrated that the tag clusters produced by our method are of significantly higher quality than the ones derived by CNM and achieve higher performance when used in the context of tag recommendation. Compared to SCAN, our method produces clusters with higher coverage (i.e. containing more related tags to the cluster topic). In the task of tag recommendation, the HGC clusters resulted in higher recall, but lower precision compared to SCAN. In addition, they led to a higher number of unique correct recommendations. Given also the fact that SCAN needs parameter tuning, we consider our clustering scheme as more suitable for identifying groups of related tags in folksonomies.

Acknowledgments. This work was supported by the WeKnowIt and GLOCAL projects, partially funded by the European Commission, under contract numbers FP7-215453 and FP7-248984 respectively.

References

1. Mathes, A.: Folksonomies - Cooperative Classification and Communication Through Shared Metadata (2004), http://www.adammathes.com/academic/computer-mediated-communication/folksonomies.html

2. Vander Wal, T.: Folksonomy Coinage and Definition (2007), http://www.vanderwal.net/folksonomy.html
3. Mika, P.: Ontologies are us: A unified model of social networks and semantics. In: Gil, Y., Motta, E., Benjamins, V.R., Musen, M.A. (eds.) ISWC 2005. LNCS, vol. 3729, pp. 522–536. Springer, Heidelberg (2005)
4. Hotho, A., Jäschke, R., Schmitz, C., Stumme, G.: Information Retrieval in Folksonomies: Search and Ranking. In: Sure, Y., Domingue, J. (eds.) ESWC 2006. LNCS, vol. 4011, pp. 411–426. Springer, Heidelberg (2006)
5. Begelman, G., Keller, P., Smadja, F.: Automated Tag Clustering: Improving search and exploration in the tag space (2006), http://www.pui.ch/phred/automated_tag_clustering
6. Simpson, E.: Clustering Tags in Enterprise and Web Folksonomies. Technical Report HPL-2008-18 (2008)
7. Au Yeung, C.M., Gibbins, N., Shadbolt, N.: Contextualising Tags in Collaborative Tagging Systems. In: Proceedings of 20th ACM Conference on Hypertext and Hypermedia, Turin, Italy, June 29-July 1, pp. 251–260. ACM, New York (2009)
8. Brooks, C.H., Montanez, N.: Improved annotation of the blogosphere via autotagging and hierarchical clustering. In: Proceedings of WWW 2006: 15th International Conference on World Wide Web, pp. 625–632. ACM, New York (2006)
9. Gemmell, J., Shepitsen, A., Mobasher, B., Burke, R.: Personalizing Navigation in Folksonomies Using Hierarchical Tag Clustering. In: Song, I.-Y., Eder, J., Nguyen, T.M. (eds.) DaWaK 2008. LNCS, vol. 5182, pp. 196–205. Springer, Heidelberg (2008)
10. Giannakidou, E., Koutsonikola, V.A., Vakali, A., Kompatsiaris, Y.: Co-Clustering Tags and Social Data Sources. In: Proceedings of WAIM 2008: 9th International Conference on Web-Age Information Management, pp. 317–324. IEEE, Los Alamitos (2008)
11. Java, A., Joshi, A., Finin, T.: Detecting Commmunities via Simultaneous Clustering of Graphs and Folksonomies. In: Proceedings of WebKDD 2008: KDD Workshop on Web Mining and Web Usage Analysis (2008)
12. Sigurbjörnsson, B., van Zwol, R.: Flickr tag recommendation based on collective knowledge. In: Proceedings of WWW 2008: 17th International Conference on World Wide Web, pp. 327–336. ACM, New York (2008)
13. Li, X., Snoek, C.G.M., Worring, M.: Learning Social Tag Relevance by Neighbor Voting. IEEE Transactions on Multimedia 11(7), 1310–1322 (2009)
14. Newman, M.E.J., Girvan, M.: Finding and evaluating community structure in networks. Physical Review E 69, 026113 (2004)
15. Xu, X., Yuruk, N., Feng, Z., Schweiger, T.A.: SCAN: A Structural Clustering Algorithm for Networks. In: Proceedings of KDD 2007: 13th International Conference on Knowledge Discovery and Data Mining, pp. 824–833. ACM, New York (2007)
16. Luo, F., Wang, J.Z., Promislow, E.: Exploring Local Community Structures in Large Networks. In: Proceedings of the 2006 IEEE/WIC/ACM International Conference on Web Intelligence, pp. 233–239. IEEE Computer Society, Los Alamitos (2006)
17. Papadopoulos, S., Kompatsiaris, Y., Vakali, A.: Leveraging Collective Intelligence through Community Detection in Tag Networks. In: Proceedings of CKCaR 2009 Workshop in K-CAP 2009 Conference, Redondo Beach, California, USA (2009)
18. Clauset, A., Newman, M.E.J., Moore, C.: Finding community structure in very large networks. Physical Review E 70, 066111 (2004)
19. Wetzker, R., Zimmermann, C., Bauckhage, C.: Analyzing social bookmarking systems: A del.icio.us cookbook. In: Proceedings of ECAI 2008 Workshop on Mining Social Data (MSoDa), Patras, Greece, pp. 26–30 (July 2008)

Frequent Sub-graph Mining on Edge Weighted Graphs

Chuntao Jiang, Frans Coenen, and Michele Zito

The University of Liverpool
Ashton Building, Ashton Street
Liverpool, L69 3BX, United Kingdom
{c.jiang,coenen,michele}@liv.ac.uk

Abstract. Frequent sub-graph mining entails two significant overheads. The first is concerned with candidate set generation. The second with isomorphism checking. These are also issues with respect to other forms of frequent pattern mining but are exacerbated in the context of frequent sub-graph mining. To reduced the search space, and address these twin overheads, a weighted approach to sub-graph mining is proposed. However, a significant issue in weighted sub-graph mining is that the *anti-monotone property*, typically used to control candidate set generation, no longer holds. This paper examines a number of edge weighting schemes; and suggests three strategies for controlling candidate set generation. The three strategies have been incorporated into weighted variations of gSpan: ATW-gSpan, AW-gSpan and UBW-gSpan respectively. A complete evaluation of all three approaches is presented.

Keywords: Weighted Transaction Graph Mining, Weighted Frequent Sub-graph Mining, Weighting Schemes.

1 Introduction

Graph mining is concerned with the identification of patterns within graph data of various forms. One form of graph mining is frequent sub-graph mining which aims to identify frequently occurring patterns (sub-graphs) across a collection of "small" graphs or within one "large" graph. This paper concentrates on the first (also sometimes referred to as *transaction graph mining*).

Frequent sub-graph mining techniques [3, 5, 6, 8, 11, 12] have parallels with more established frequent pattern mining techniques such as those used in, for example, Association Rule Mining (ARM). Thus, in common with other forms of frequent pattern mining, frequent sub-graph mining entails two significant overheads: candidate set generation and isomorphism checking. However, these overheads are exacerbated because of the nature of graph data. In the case of candidate set generation the potential number of size $K+1$ sub-graphs that can be generated from size K graphs is exponentially greater than in the case of more standard forms of frequent pattern mining. With respect to isomorphism checking, the process of comparing a candidate pattern with the input data to

T.B. Pedersen, M.K. Mohania, and A M. Tjoa (Eds.): DaWaK 2010, LNCS 6263, pp. 77–88, 2010.

determine the support (frequency) of the candidate is significantly more complex in the case of frequent sub-graph mining than in more standard forms of frequent pattern mining such as ARM.

The overheads associated with frequent sub-graph mining are compounded when the support threshold is low. The solution advocated in this paper is based on the observation that, for many applications, some edges (nodes) in the input graph set can be considered to be more significant than others. Therefore, sub-graph patterns that include edges (nodes) with high weight values should be considered more important than those with low weight values if they both satisfied the support threshold. This concept is illustrated in this paper by considering a social network mining scenario.

Weighted frequent sub-graph mining advocates the use of *weighted support counts* to identify weighted frequent sub-graphs. Hence, the "computational burden" of sub-graph mining can be considerably alleviated by generating a set of weighted frequent sub-graphs. The concept of edge weightings can be encapsulated in a number of ways (for reasons of clarity only edge weighted graphs are considered in this paper although much of the discussion is equally applicable to node, or node and edge, weighted graphs).

Regardless of whether edge or node weighting is adopted, a significant issue encountered in weighted sub-graph mining is that the *anti-monotone property*, whereby if a K size sub-graph is not frequent none of its $K+1$ super-graphs will be frequent, typically used to restrict the size of the search space in standard pattern mining, no longer holds if weightings are applied in a naive manner. Thus any proposed weighted sub-graph mining mechanism must either be defined in such a way that the property continues to hold, or an alternative pruning strategy must be adopted.

Three edge weighting schemes are considered in this paper: (i) *Average Total Weighting* (ATW), (ii) *Affinity Weighting* (AW) and (iii) *Utility Based Weighting* (UBW). The three approaches have been incorporated into three weighted variations of the gSpan algorithm (ATW-gSpan, AW-gSpan, and UBW-gSpan).

The rest of this paper is organised as follows. A problem definition overview is presented in Section 2. The proposed edge weighting mechanisms are considered in Section 3. Experiments to evaluate the proposed techniques, and the ensuing results, are presented in Section 4. Some conclusions are presented in Section 5.

2 Problem Definition

This section introduces the necessary graph-theoretic and mining definitions. In the context of this paper a graph is defined as a finite structure G formed by a set of nodes $V = \{v_1, v_2, \ldots\}$, a set of edges $E = \{e_1, e_2, \ldots\}$, a set of vertex and edge labels $\mathcal{L}$, and a mapping $\phi_{v/e} : \mathcal{L} \rightarrow V/E$. With respect to the work described here the edge labels are assumed to be numeric so that they can be used in the calculation of relative weightings. Depending on the particular application, edges will be either *undirected* pairs over V, or *directed* (ordered) pairs.

Let $T = \{G_1, G_2, \cdots, G_t\}$ be a collection of (transaction) graphs. The support set of g is defined as $\delta_T(g) = \{t | g \subseteq G_t\}$, i.e. the set of transaction graphs where g is a sub-graph of G_t. The cardinality of the support set, $|\delta_T(g)|$ then defines the support of g with respect to T.

Definition 1. *Given a database T, a graph g, and a minimum support $\tau \in (0, 1]$, the graph g is said to be* frequent *(in T) if $|\delta_T(g)| \geq \tau \times t$. The* frequent sub-graph mining problem *is thus to find all the frequent sub-graphs in T.*

The focus of this paper is on edge weighted graphs. Therefore, the graphs in T are assumed to have weights associated with their edges. Let W_T be a weighting function that assigns a weight to any sub-graph g. The *weighted support* of g with respect to T, $wsup_T(g)$, is then:

$$wsup_T(g) = W_T(g) \times |\delta_T(g)|. \tag{1}$$

Note that the function of $W_T(g)$ needn't be a number between zero and one. By defining the weighting function, $W_T(g)$, in an appropriate manner it is possible to ensure that the anti-monotone property holds; otherwise other method, such as some heuristic based pruning technique, is required to limit the search space.

3 Graph Weighting Mechanisms

Most research work in frequent sub-graph mining [6,8,5,11] assumes each discovered frequent sub-graph is equally important. A lot of redundant and repetitive frequent patterns may therefore exist in the final result. If the size of the graph set is substantial and the minimum support threshold is very low, a typical frequent sub-graph mining task can often not be completed within a fixed period of time due to the exponential complexity of the search space. If we put emphasis on differentiating each discovered frequent sub-graph according to its importance, either as definded by the user or derived from the application domain, the computational complexity can be reduced without compromising the effectiveness of the frequent pattern discovery process. However, when a weighting scheme is integrated into the process of graph mining in a naive manner, the well-known *anti-monotone property*, which is used frequently to reduce the search space, may no longer be satisfied. Two strategies can be identified to address this dilemma: (a) adopt an interestingness measure which does satisfy the property; (b) ignore the property and adopt some alternative heuristic to reduce the computational overhead incurred by not satisfying the property.

In the context of weighted frequent sub-graph mining, weightings associated with a sub-graph pattern g can be defined in a number of manners. Three approaches are introduced in this paper: (i) Average Total Weighting (ATW), (ii) Affinity Weighting (AW), (iii) Utility Based Weighting (UBW). The first two approaches satisfy the anti-monotone property while the last one adopts an alternative pruning heuristic. The last two approaches employ two parameters to control the mining result while the first one uses one parameter only. Each approach is discussed in further detail in the following three subsections.

3.1 Average Total Weighting (ATW)

In the ATW approach inspired by the work [10], the weight for a sub-graph g is calculated by dividing the sum of the average weights in graphs that contain g with the sum of the average weights across the entire data set T. Thus:

Definition 2. *Given an edge weighted graph g with edge weights $\{w_1, w_2, \cdots, w_k\}$, the average weight associated with g is defined as $W_{avg}(g) = \frac{\sum_{i=1}^{k} w_i}{k}$.*

Where w_i can be user defined or calculated by some weighting methods.

Definition 3. *Given a set of graphs $T = \{G_1, G_2, \cdots, G_t\}$, the total weight of this set of graphs is defined as $W_{sum}(T) = \sum_{i=1}^{t} W_{avg}(G_i)$.*

Definition 4. *Given an arbitrary sub-graph g with its support set $\delta_T(g)$, the weight function of g with respect to T, $W_T(g)$, is defined as*

$$W_T(g) = \frac{\sum_{G_i \in \delta_T(g)} W_{avg}(G_i)}{W_{sum}(T)} \tag{2}$$

Definition 5. *A sub-graph g is weighted frequent with respect to T, if $|\delta(g)| \times W_T(g) \geq \tau \times t$, where $0 < \tau \leq 1$ is a minimum support threshold.*

From the above it can be easily inferred that the function $W_T(g)$, as defined by Equation 2, satisfies the anti-monotone property. Therefore, if a k-candidate is not frequent, then any of its $(k+1)$-supersets can be safely pruned from this branch in the lattice of candidates during the $k+1$ candidate generation process. It should be noted, however, that the approach will tend to bias large transaction graphs over smaller transaction graphs, thus is best applied to graph sets where the individual graphs are of a similar size.

3.2 Affinity Weighting (AW)

The Affinity Weighting (AW) approach is founded on two elements to restrict the growth of the search space: (i) a graph distance measure, and (ii) a weighting ratio. For a sub-graph g to be frequent both must be greater than specified user thresholds. The graph distance measure is calculated using an appropriately defined support weighting function, $W_T(g)$. This is defined as follows. Let g be a candidate pattern for a database $T = \{G_1, G_2, \cdots, G_t\}$. In the context of AW we define:

$$W_T(g) = \frac{1}{|V(g)|} \sum_{G_i \in \delta_T(g)} \frac{|V(G_i)| - |V(g)|}{|V(G_i)|}. \tag{3}$$

Where $V(G_i)$ is the set of vertices in transaction graph G_i and $V(g)$ is the set of vertices in the sub-graph g. Observe that $W_T(g)$ satisfies:

$$W_T(g) = \frac{|\delta_T(g)|}{|V(g)|} - \sum_{G_i \in \delta_T(g)} \frac{1}{|V(G_i)|} \tag{4}$$

It should be noted that adding nodes to g can only reduce the value of the above expression because the support$(|\delta_T(g)|)$ cannot be increased; the sum contains as many terms as $|\delta_T(g)|$ and each of these cannot be larger than $1/|V(g)|$. Thus $W_T(g)$ as defined above, insures that the weighted support of g is non-increasing (i.e. anti-monotone) in $|V(g)|$.

The graph distance measure is directed at the number of nodes contained in a graph, the weighting ratio concerned with the edge weights (which are assumed to reflex numeric values). The weighting ratio of an edge-weighted graph g is a function $c(g)$ returning a value between zero and one which is decreasing in the number of edges of g. Given an edge weighted sub-graph g with edge weights $W = \{w_1, w_2, \cdots, w_k\}$ the weighting ratio function which is similar to [13], $c(g)$, is defined as follows:

$$c(g) = \frac{MIN_{w_i \in W}\{w_i\}}{MAX_{w_j \in W}\{w_j\}}. \tag{5}$$

Definition 6. *An edge-weighted graph g is a weighted frequent (i.e. weighted affinity) pattern within a data set $T = \{G_1, G_2, \cdots, G_t\}$, with respect to a support threshold $\tau > 0$ and weighting ratio threshold $\gamma \in [0, 1]$, if the following two conditions (C1 and C2) are satisfied:*

$$\textbf{(C1)}\ wsup_T(g) \geq \tau \times t, \qquad \textit{and} \qquad \textbf{(C2)}\ c(g) \geq \gamma.$$

Definition 6 leads to an alternative pruning strategy which, may be used as part of any frequent sub-graph mining algorithms. During the candidate selection phase, the mining will keep track of the weighted support and weighting ratio of all candidates and discard all those candidates that do not satisfy at least one of **(C1)** and **(C2)**.

3.3 Utility Based Weighting (UBW)

The previous two approaches both satisfy the anti-monotone property. In this section an alternative weighting scheme which does not hold the property is proposed. The Utility Based Weighting (UBW) scheme is influenced by ideas suggested in [1, 2]. As in the case of AW scheme, the UBW scheme is founded on two elements: (i) weighted support and (ii) the share (SH) of a sub-graph. Thus:

Definition 7. *Given a sub-graph g with edges $E(g) = \{e_1, e_2, \cdots, e_k\}$. For each $e_i \in E(g)$, two vertices connecting e_i are v_1 and v_2. Their associated support sets (the graphs in T where they appear) are given as $\delta_T(v_1)$ and $\delta_T(v_2)$. The Jaccard similarity coefficient between the two vertices is defined as $jC(e_i) = |\delta_T(v_1) \cap \delta_T(v_2)|/|\delta_T(v_1) \cup \delta_T(v_2)|$. The weighting function of g, $W_T(g)$, is then defined as*

$$W_T(g) = \frac{1}{\sum_{e_i \in E(g)} jC(e_i)} \tag{6}$$

From the above it is clear that $W_T(g)$ satisfies the anti-monotone property. From Section 2 the weighted support is given by $wsup_T(g) = W_T(g) \times |\delta_T(g)|$.

Definition 8. *Given an edge weighted graph set* $T = (G_1, \ldots, G_t)$ *with edge weights* $\{w_1, w_2, \cdots, w_k\}$ *for each transaction graph* G_j *and a sub-graph* g*. Let* $g \subseteq G_j$*, the weight of* g *denoted as* $W(g, G_j)$*, is the sum of the weights of the edges which occurred in* G_j*. That is,* $W(g, G_j) = \sum_{e_i \in g, g \subseteq G_j} w_i$*. The total weight of* T*, denoted as* $TW(T)$*, represents the sum of edge weights in* T*, where* $TW(T) = \sum_{G_j \in T} \sum_{e_i \in G_j} w_i$*. The total weight of* $\delta_T(g)$*, is defined as* $TW(\delta_T(g)) = \sum_{G_j \in \delta_T(g)} \sum_{e_i \in G_j} w_i$*.*

Definition 9. *The graph weight of* g *with respect to* T*, denoted as* $GW(g)$*, is the sum of the weight of the* g *in each transaction graph* $G_j \in \delta_T(g)$*. That is,* $GW(g) = \sum_{G_j \in \delta_T(g)} W(g, G_j)$*.*

Definition 10. *The share of a sub-graph* g*, denoted as* $SH(g)$*, is the ratio of the graph weight of* g *with respect to* T *to the total weight of* T*. Thus:*

$$SH(g) = \frac{GW(g)}{TW(T)} \tag{7}$$

Given a share threshold λ*, a sub-graph* g *is SH-frequent if* $SH(g) \geq \lambda$*; otherwise,* g *is SH-infrequent.*

Theorem 1. *Given a* $T = (G_1, \ldots, G_t)$*, a sub-graph* g*, and a threshold* λ*, if* $TW(\delta_T(g)) < \lambda \times TW(T)$*, all super-graphs of* g *are SH-infrequent.*

Proof. Let h be an arbitrary super-graph of g. Clearly, $GW(h) \leq TW(\delta_T(h)) \leq TW(\delta_T(g))$. If $TW(\delta_T(g)) < \lambda \times TW(T)$ holds, $GW(h) < \lambda \times TW(T)$. That is, $SH(h) = GW(h)/TW(T) < \lambda$. Therefore, h is SH-infrequent. □

By Theorem 1, if $TW(\delta_T(g)) < \lambda \times TW(T)$, all super-graphs of g and g are SH-infrequent and can be pruned; otherwise, g is a candidate sub-graph.

Definition 11. *An edge-weighted graph* g *is a weighted frequent pattern for a graph set* $T = (G_1, \ldots, G_t)$ *with respect to a support threshold* $\tau > 0$ *and share threshold* $\lambda \in (0, 1]$ *if the following two conditions are satisfied.*

(D1) $wsup_T(g) \geq \tau \times t$, *and* **(D2)** $SH(g) \geq \lambda$.

4 Experiments and Results

This section describes a sequence of experiments designed to:

(i) Demonstrate that the proposed weighting schemes can more efficiently generate frequent sub-graphs than without using weightings. In many cases, as will be demonstrated, use of the weighting schemes allows frequent sub-graphs to be identified where this would not be possible using an unweighted approach because of this computational overhead the latter would entail.

Table 1. CTS graph set statistics

	Norfolk	Cornwall	GB
# graphs	53	53	53
Max # edges	77	412	30107
Average # edges	54	262	23055
Max # nodes	99	409	23660
Average # nodes	70	284	18749
node label count	614	2195	81153
Edge label count	6	12	46

(ii) Compare and contrast the three proposed weighted sub-graph mining techniques.

The experiments were conducted using a projection of the cattle movement database in operation in Great Britain (GB). This application domain is described in Section 4.1. The original gSpan algorithm available to the authors could not process directed graphs with self cycles. Therefore an extended gSpan algorithm (extGspan), which can process directed graphs with self cycles, was implemented in order to compare the proposed weighted approaches with the un-weighted case. Results from the experiments are presented in Sub-sections 4.2 and 4.3.

4.1 The Cattle Tracking System Database

For the experiments the Cattle Tracking System (CTS) database, in operation in GB, was used. This was provided by the Department for the Environment, Food and Rural Affairs (DEFRA) from the Rapid Analysis and Detection of Animal Risk (RADAR) project[1]. The database provides a record of cattle movements. Each record includes information such as the sender and receiver location IDs, animal ID, animal breed, etc. Three distinct transaction graph datasets were extracted from the CTS database such that nodes represented cattle location (farms, markets, slaughter houses, etc) and edges the movement of cattle between locations (the edges are directed by the direction of the cattle movement). Transaction graph sets for all of Great Britain (GB), and two areas within GB (Norfolk and Cornwall) were extracted. Edges were annotated with a weighting, indicating the number of cattles moved, and a label, indicating the type of movement (e.g. farmToFarm, farmToMarket, etc). For each data set the data from 1 January 2005 to 31 December 2005 was selected and divided into 7-day "episodes" due to the 6-day movement restriction [9] that applies to farms in GB. Statistics for each of the data sets are given in Table 1. Note that the *GB* data set is significantly larger than the *Cornwall*[2], which in turn was larger than

[1] http://www.defra.gov.uk/foodfarm/farmanimal/diseases/vetsurveillance/radar/project.htm

[2] Cornwall is a county in the SW of GB known for its substantial dairy herds.

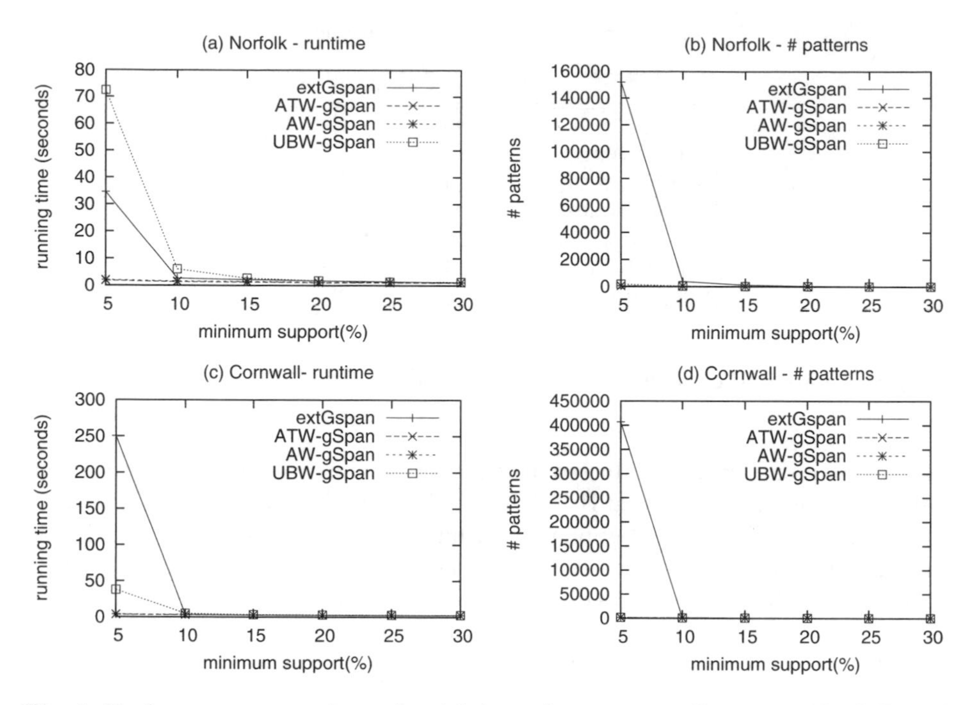

Fig. 1. Performance comparison of weighting schemes vs. extGspan on *Norfolk* and *Cornwall* data sets (using a range of support values from 5% to 30%)

the *Norfolk*[3] data set. It should also be noted that all the transaction graphs feature directed edges and self cycles.

4.2 Comparison between Weighted and Non-weighted Approaches

In this subsection the proposed weighting schemes (ATW-gSpan, AW-gSpan, and UBW-gSpan) are compared with the extended gSpan algorithm in terms of efficiency (runtime and the number of frequent sub-graphs generated). For AW-gSpan, $\gamma = 0.6$ was chosen as the weighgting ratio threshold, and $\lambda = 8\%$ was used as the share threshold for UBW-gSpan. The judstification for these γ and λ values is given in Sub-section 4.3 below.

Figure 1 shows the performance of the weighting schemes and extGspan on the *Norfolk* and *Cornwall* data sets (recall that extGspan does not make any use of weightings). It can be clearly seen from the figure that all four algorithms display a similar behaviour when the support value is between 10% to 30%, however the number of patterns generated by the extGspan algorithm increase abruptly when the support value is decreased to below 10%. From Fig. 1 it can be observed that: (i) significantly more frequent sub-graphs (at support threshold below 10%) are found using the non-weighted extGspan algorithm than using any of the weighting schemes, indicating the advantages offered using the weighted approaches, (ii)

[3] Norfolk is a county in the East of GB.

the ATW and AW schemes run faster than the UBW scheme, this is because the pruning technique adopted by UBW schem is not strong enough compared with the anti-monotone based pruning methods used by ATW and AW schemes.

Experiments (not shown) using extGspan and the *GB* data set failed to produce any results (because of memory errors) unless the support thresholod was set to 30% or above, a threshold at which only one node size sub-graph are discovered. Thus it was not possible to conduct any meaningful comparison between the weighted frequent sub-graph mining algorithms and a non-weighted approach using the *GB* data set.

4.3 Comparison of Weighting Schemes

In this subsection the three proposed weighting schemes are compared with one another using the large *GB* dataset. As above, γ was initially set to 0.6 and λ to 8% for use with AW-gSpan and UBW-gSpan algorithms. Figure 2 shows the performance of the weighting schemes on the *GB* dataset. In Fig. 2 (a), each curve depicts the number of patterns generated against the minimum support value used. From the figure it can be seen that UBW-gSpan produces the least number of patterns while AW-gSpan produces the most. Figure 2 (b) indicates the "run time" for the approaches using the same sequence of support threshold values. From the figure it can be seen that UBW-gSpan is the most "expensive", indicating that the cost of finding a minimum number of patterns is higher compared to the other two mechanisms. ATW-gSpan is the most economical.

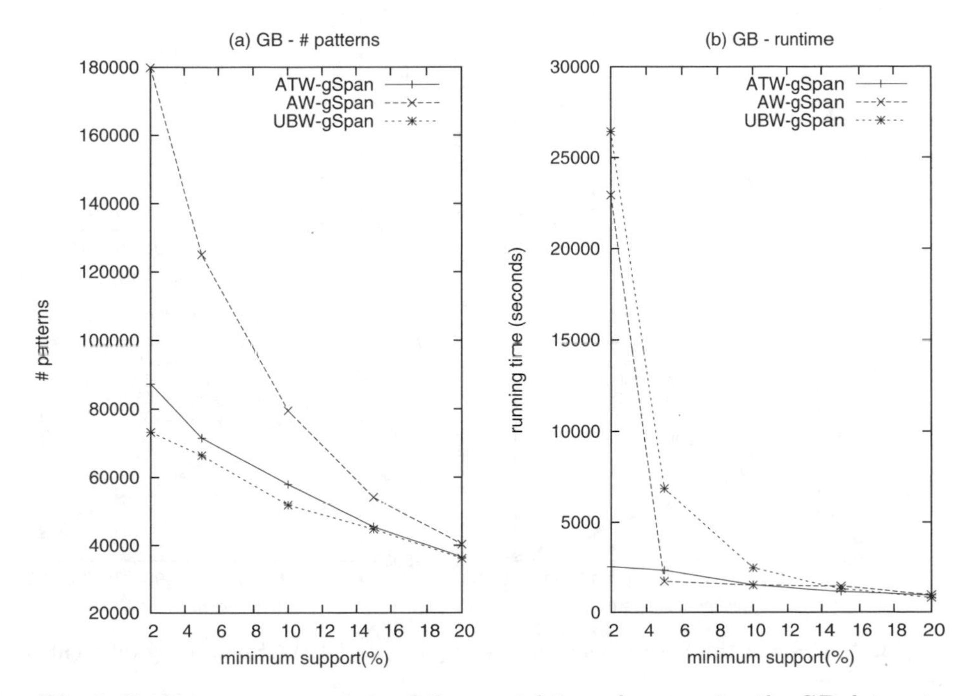

Fig. 2. Performance comparison of three weighting schemes using the *GB* data set

Reference to Fig. 1(a) and (b) confirm these results. UBW-gSpan is also expensive with respect to the *Norflok* and *Cornwall* data sets. In fact inspection of Fig. 1(a) indicates that UBW-gSpan is more expensive than applying extGspan in the case of the*Norfolk* data indicating that the cost of reducing the number of patterns is high when using UBW-gSpan. Although it should be noted that with respect to the *GB* data set extGspan was unable to process this data set at all (using realistic support thresholds). It is interesting to note in Fig. 2 (b) that as the support threshold is reduced the effect on run-time is much smaller for ATW-gSpan than the other two weighting schemes. More generally, from Fig. 2, it can be seen that (as might be expected) runtime increases significantly as the support threshold is reduced.

Figure 3 displays the effect on performance of different values for the weighting ratio threshold (γ) used in conjunction with AW-gSpan, and the share threshold (λ) used with UBW-gSpan, for a range of support threshold values from 4% to 12%. From Fig. 3 (a) and (c) it can be seen that the run time increased as the γ value is decreased, while a marginal increase in the number of patterns is witnessed. With respect to Fig. 3 (b) and (d) it can be seen that the run time increases as the λ value is decreased, while a small corresponding increase in the number of identified patterns is witnessed. However, increasing the λ value beyond 8% seems to have very little effect on the number of patterns. Overall it was found that a γ value of 0.6 and a λ value of 0.8% was the most appropriate.

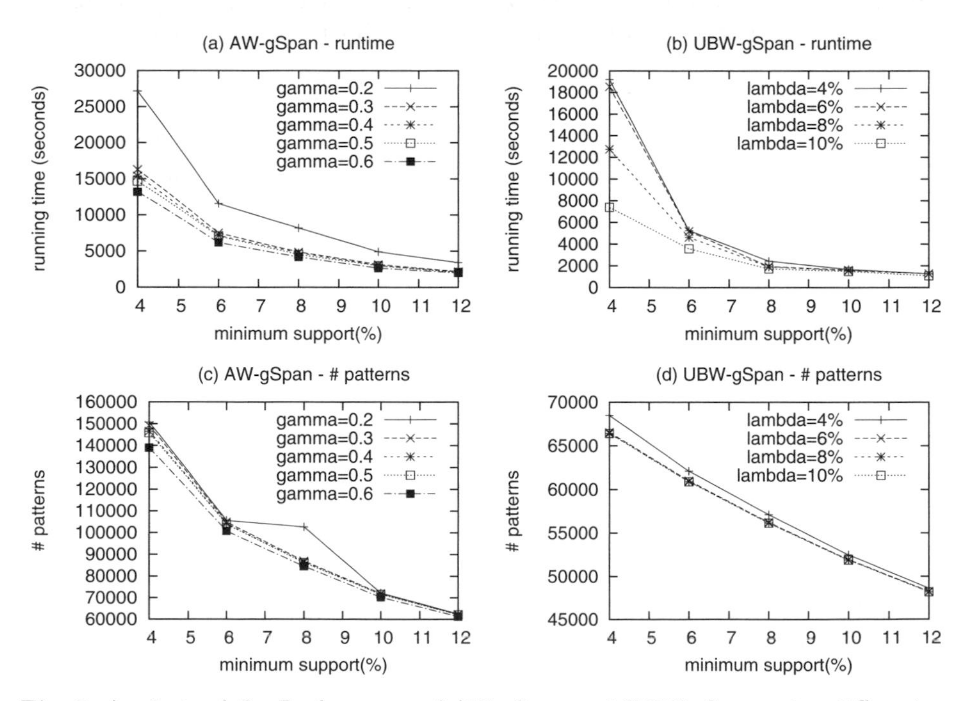

Fig. 3. Analysis of the Performance of AW-gSpan and UBW-gSpan using different γ and λ values

4.4 Quality of Results

The above experiments indicate that the proposed weighting approaches can be successfully applied so that frequent sub-graphs can be identified in large collections of graphs (such as those extracted from the CTS database) which could not otherwise be mined using more conventional graph mining approaches. The proposed weighting mechanisms operate by identifying the most "significant" edges. The question that remains is then to ask "are we finding the right frequent sub-graphs?". To answer this question the research team applied the weighting techniques to a number of classification problems. Two data sets were used, an MRI scan data set and a text mining data set where the scans and documents had been processed into a graph representation and labelled. Weighted graph mining techniques were then applied to the graph sets to produce collections of frequent sub-graphs. These sub-graphs were then interpreted as features in a feature space and used to represent the individual records using a standard feature vector representation (where each element represents a frequent sub-graph). Standard classification algorithms were then applied. The results generated were comparable with results obtained using alternative, more conventional, classification approaches thus indicating that the "right sub-graphs" had been identified. Space limitations prevent a full presentation and discussion of these results in this paper, however interested readers can refer to [4] and [7] for reports on the MRI scan and text mining experiments respectively.

5 Conclusions

This paper has proposed a solution to frequent sub-graph mining where the size of the input data is such that standard graph mining algorithms (such as gSpan) are unable to derive any appropriate results because of the computational overheads involved. Three weighting mechanisms are proposed (ATW-gSpan, AW-gSpan, and UBW-gSpan) designed to reduce to overall search space by identifying the most relevant sub-graphs. The weighting schemes assume edge weightings, but similar techniques may be applied with respect to nodes. Experiments comparing the operation of the weighting schemes to a non-weighted version of gSpan indicate that many fewer patterns are derived. The research team have established that the reduced pattern set are the "right" pattern set by applying the results using classification scenarios. The reported experiments indicate that UBW-gSpan finds the least number of patterns will requiring the largest amount of run-time. ATW-gSpan provides the best compromise, a limited number of patterns found in reasonable time (especially at low support threshold values). Experiments were also conducted with respect to the most suitable γ and λ to be used with respect to AW-gSpan and UBW-gSpan respectively. Overall it was found that a γ value of 0.6 and a λ value of 0.8% was the most appropriate.

Acknowledgements

We would like to thank the Department for the Environment, Food and Rural Affairs (DEFRA) for providing us the data. We are grateful to Dr. Christian Setzkorn from the Faculty of Veterinery Science, University of Liverpool for extracting the simplified form of the data and Mrs Puteri Nor Ellyza Nohuddin for assisting us to get the data.

References

1. Barber, B., Hamilton, H.J.: Extracting Share Frequent Itemsets with Infrequent Subsets. Journal of Data Mining and Knowledge Discovery 7, 153–185 (2003)
2. Carter, C.L., Hamilton, H.J., Cercone, N.: Share based Measures for Itemsets. In: Komorowski, J., Żytkow, J.M. (eds.) PKDD 1997. LNCS, vol. 1263, pp. 14–24. Springer, Heidelberg (1997)
3. Cook, D.J., Holder, L.B.: Substructure Discovery Using Minimum Description Length and Background Knowledge. Journal of Artificial Intelligenc Research 1, 231–255 (1994)
4. Elsayed, A., Coenen, F., Jiang, C., Garca-Fiana, M., Sluming, V.: Corpus Callosum MR Image Classification. Journal of Knowledge Based Systems (to appear 2010)
5. Huan, J., Wang, W., Prins, J.: Efficient Mining of Frequent Subgraph in the Presence of Isomorphism. In: Proceedings of the 2003 International Conference on Data Mining, ICDM 2003 (2003)
6. Inokuchi, A., Washio, T., Motoda, H.: An Apriori-based Algorithm for Mining Frequent Substructures from Graph Data. In: Proceedings of the 4th European Conference on Principles and Practice of Knowledge Discovery in Databases (2000)
7. Jiang, C., Coenen, F., Sanderson, R., Zito, M.: Text Classification using Graph Mining-Based Feature Extraction. The Journal of Knowledge Based Systems (to appear 2010)
8. Kuramochi, M., Karypis, G.: Frequent Subgraph Discovery. In: Proceedings of IEEE International Conference on Data Mining (2001)
9. Robinson, S.E., Christley, R.M.: Identifying Temporal Variation in Reported Births, Deaths and Movements of Cattle in Britain. Journal of BMC Verterinary Research (2006), doi:10.1186/1746-6148-2-11
10. Tao, F., Murtagh, F., Farid, M.: Weighted Association Rule Mining using Weighted Support and Significance Framework. In: The Ninth ACM SIGKDD International Conference on Knowledge Discovery and Data Mining (ACM SIGKDD 2003), Washington DC, USA, pp. 661–666 (2003)
11. Yan, X., Han, J.: gSpan: Graph-based Substructure Pattern Mining. In: Proceedings of 2002 International Conference on Data Mining (2002)
12. Yan, X., Han, J.: CloseGraph: Mining Closed Frequent Graph Patterns. In: Proceedings of the Ninth ACM SIGKDD International Conference on Knowledge Discovery and Data Mining, Washington DC, USA, pp. 286–295 (2003)
13. Yun, U.: WIS: Weighted Interesting Sequential Pattern Mining with a Similar Level of Support and/or Weight. ETRI Journal 29(3), 336–352 (2007)

$\mathcal{F}$&$\mathcal{A}$: A Methodology for Effectively and Efficiently Designing Parallel Relational Data Warehouses on Heterogenous Database Clusters

Ladjel Bellatreche[1], Alfredo Cuzzocrea[2], and Soumia Benkrid[3]

[1] LISI/ENSMA Poitiers University, France
bellatreche@ensma.fr
[2] ICAR-CNR and University of Calabria, Italy
cuzzocrea@si.deis.unical.it
[3] National High School for Computer Science (ESI), Algeria
s_benkrid@esi.dz

Abstract. In this paper we propose a comprehensive methodology for designing *Parallel Relational Data Warehouses* (PRDW) over *database clusters*, called *$\mathcal{F}$ragmentation&$\mathcal{A}$llocation* ($\mathcal{F}$&$\mathcal{A}$). $\mathcal{F}$&$\mathcal{A}$ assumes that cluster nodes are *heterogeneous* in *processing power* and *storage capacity*, contrary to traditional design approaches that assume that cluster nodes are instead *homogeneous*, and fragmentation and allocation phases are performed in a *simultaneous manner*, contrary to traditional design approaches that instead perform these phases in an *isolated manner*. Also, a naive replication algorithm that takes into account the heterogeneous characteristics of our reference architecture is proposed. Finally, our proposal is experimentally assessed and validated against the widely-known data warehouse benchmark *APB-1 release II*.

1 Introduction

In this paper, we focus the attention to the context of query optimization techniques over *relational Data Warehouses* (RDW) developed on top of *cluster environments* [14]. A RDW is usually modeled by means of a *star schema* consisting of a huge *fact table* and a number of *dimension tables*, similarly to the widely-known data warehouse benchmark *APB-1 release II* [4], where the fact table *Sales* is joint to the following four dimension tables: *Product*, *Customer*, *Time*, *Channel*. *Star queries* are typically executed against RDW. Star queries retrieve aggregate information (e.g., based on standard SQL aggregate operators like `SUM`, `COUNT` etc) from *measures* stored in the fact table by applying *selection conditions* on joint dimension table columns, and they are extensively used as conceptual basis for more complex *OLAP queries*, which, in turn, are exploited to extract useful summarized knowledge from RDW for decision making purposes.

Unfortunately, evaluating OLAP queries over RDW typically demands for a high-performance that is difficult to ensure over large amounts of multidimensional data, even because such queries are usually complex in nature [2]. This complexity is mainly due to the presence of joins and aggregation operations

T.B. Pedersen, M.K. Mohania, and A M. Tjoa (Eds.): DaWaK 2010, LNCS 6263, pp. 89–104, 2010.

over huge fact tables, which very often involve billions of tuples to be accessed and processed. In order to speed-up OLAP queries over RDW, several optimization approaches, mainly inherited from classical database technology, have been proposed in literature. Among others, we recall *materialized views* [12], *indexing* [20], *data partitioning* [3], *data compression* [7] etc. Despite this, it has been demonstrated that the sole use of these approaches singularly is not sufficient to gain efficiency during the evaluation of OLAP queries over RDW [21]. As a consequence, in order to overcome limitations deriving from these techniques, high-performance in database technology, including RDW [11,9], has traditionally been achieved by means of *parallel processing methodologies* [16].

Following this major trend, the most important commercial database systems vendors (e.g., Oracle, IBM, Microsoft, NCR, Sybase etc.) have recently proposed solutions able to support parallelism within the core layer of their DBMS. Unfortunately, these solutions still remain expensive for small and medium enterprises, so that *database cluster technology* represents an efficient low-cost alternative to tightly-coupled multiprocessing database systems [14]. A *database cluster* can be defined as a cluster of personal computers (PC) such that each of them runs an off-the-shelf sequential DBMS [14]. The set of DBMS relying in the cluster are then orchestrated by means of an ad-hoc middleware that implements parallel processing mechanisms and techniques, being this middleware able to support typical DBMS functionalities/services (e.g., storage, indexing, querying etc) in a transparent-for-the-user manner, just like end-users were interacting with a *singleton* DBMS. Starting from this low-cost technology solution, in our research we focus the attention on the application scenario represented by the so-called *parallel relational Data Warehouses* (PRDW) over database clusters, i.e. RDW that are developed on top of a cluster of databases that implements parallel processing mechanisms and techniques.

Similarly to the traditional context of *distributed and parallel databases* [16], the design of a PRDW on a database cluster can be achieved by means of a *general* design methodology consisting by the following steps: (*i*) *fragmenting* the input data warehouse schema; (*ii*) *allocating* the so-generated fragments; (*iii*) *replicating* fragments in order to ensure high-performance during data management and query evaluation activities. By examining the active literature, few proposals on how to design a PRDW on a database cluster exist [11,14]. These approaches can be classified into two main classes. The first class of proposals assume that data are already partitioned and allocated, and propose solutions to route OLAP queries across nodes of the database cluster in order to improve query performance [17,18]. The other class of proposals instead propose solutions to partition and allocate data across database cluster nodes [14]. Most importantly, the majority of approaches devoted to the design of a PRDW over a database cluster assume that all nodes of the cluster are *homogenous*, i.e. they have the same *processing power* and *storage capacity*. By looking at the peculiarities of the target application scenario, it is easy to understand how this assumption is not always true, as a cluster of PC with heterogeneous characteristics in terms of storage and processing capacity may exist. Therefore, it clearly follows the interest for PRDW design methodologies over database clusters characterized by *heterogeneous* nodes, in all the phases, including *data partitioning*, *fragment allocation*, and *data replication*, which is the main goal of our research.

Data fragmentation[1] is a fundamental phase of any PRDW design methodology, and can also be considered as a *pre-condition* for PRDW design [1]. Data fragmentation can be of the following two kinds [16]: (*i*) *horizontal fragmentation*, according to which table instances are decomposed into *disjoint partitions*; (*ii*) *vertical fragmentation*, according to which table instances are split into *disjoint sets of attributes*. Horizontal partitioning is the most popular solution used to design PRDW [1,21,22,11,14]. In previous PRDW design methodologies research efforts, horizontal partitioning algorithms do not control the number of generated fragments, except [1,5]. As a consequence, the number of fragments generated by the partitioning phase can be larger than the number of nodes of the database cluster. In turn, this causes flaws in the allocation and replication phases.

Allocation is the phase that places fragments generated by the partition phase across nodes of the database cluster. Allocation can be either *redundant*, i.e. with replication, or *non redundant*, i.e. without replication [16]. Some literature approaches advocate a *full replication* in order to ensure a high *intra-query parallelism* [14]. This solution demands for the availability of very large amounts of disk space, as each node must be ideally able to house the *entire* data warehouse. As a consequence, data updates become prohibitively expensive. On the basis of this main observation, we assert that replication must be *partial*, meaning that database cluster nodes house *portions* of the original data warehouse. Once fragments are placed and replicated, global OLAP queries against the target PRDW are re-written over fragments and evaluated on the *parallel machine*.

State-of-the-art PRDW design methodologies on database clusters proposals suffer from the following two main limitations. First, they focus the attention on homogenous database clusters, i.e. database clusters where nodes have the same *processing power* and *storage capacity*. Second, fragmentation and allocation phases are usually performed in an *isolated* (or *iterative*) manner, meaning that the designer first partitions his/her data warehouse using his/her favorite fragmentation algorithm and then allocates generated fragments on the parallel machine using his/her favorite allocation algorithm. This approach completely ignores the inter-dependency between fragmentation and allocation phases, which, contrary to this, can instead seriously affect the final performance of data management and OLAP query evaluation activities performed against the PRDW. Starting from these breaking evidences, in this paper we propose and experimentally assess an innovative methodology for designing PRDW on database clusters, called $\mathcal{F}ragmentation\&\mathcal{A}llocation$ ($\mathcal{F}$&$\mathcal{A}$), which overtakes the limitations above. To the best of our knowledge, our research is the first one in literature that addresses the issue of designing PRDW on heterogeneous database clusters via a combined fragmentation/allocation strategy.

[1] In this paper, we use the terms "fragmentation" and "partitioning" interchangeably.

The paper is organized as follows. Section 2 summarizes existing approaches that focus on iterative PRDW design methodologies. In Section 3, we provide a rigorous formalization of the PRDW design problem on heterogeneous database clusters, by also putting in emphasis limitations deriving from traditional iterative design methodologies. Section 4 describes our comprehensive methodology $\mathcal{F}\&\mathcal{A}$ for designing PRDW on heterogeneous database clusters, where partitioning and allocation phases are performed simultaneously. In Section 5, we provide the experimental results obtained from testing the performance of $\mathcal{F}\&\mathcal{A}$ against the widely-known data warehouse benchmark *APB-1 release II* [4]. Finally, Section 6 concludes the paper summarizing the main findings of our research, and proposing directions for future work.

2 Related Work

In this Section, we provide a brief overview on state-of-the-art approaches focusing on fragmentation and allocation techniques for supporting PRDW over database clusters [11,14,17,18].

Furtado [11] discusses partitioning strategies for *node-partitioned data warehouses*. The main suggestion coming from [11] can be synthesized in a "best-practice" recommendation stating to partition the fact table on the basis of the *larger* dimension tables (given a ranking threshold). In more detail, each larger dimension table is first partitioned by means of the *Hash mode* approach via its primary key. Then, the fact table is again partitioned by means of the Hash mode approach via foreign keys referencing the larger dimension tables. Finally, the so-generated fragments are allocated according to two alternative strategies, namely *round robin* and *random*. Smaller dimension tables are instead fully-replicated across the nodes of the target data warehouse. The fragmentation approach [11] does not take into account specific star query requirements, being such queries very often executed against data warehouses, and it does not consider the critical issues of controlling the number of generated fragments, like in [3,22].

In [14], Lima *et al.* focus the attention on data allocation issues for database clusters. Authors recognize that how to place data/fragments on the different PC of a database cluster in the dependence of a given criterion/goal (e.g., query performance) plays a critical role, hence the following two straightforward approaches can be advocated: (*i*) full replication of the target database on *all* the PC, or (*ii*) meaningful partition of data/fragments across the PC. Starting from this main intuition, authors propose an approach that combines partition and replication for OLAP-style workloads against database clusters. In more detail, the fact table is partitioned and replicated across nodes using the so-called *chained de-clustering*, while dimension tables are fully-replicated across nodes. This comprehensive approach enables the middleware layer to perform load balancing tasks among replicas, with the goal of improving query response time. Furthermore, the usage of chained de-clustering for replicating fact table partitions across nodes allows the designer not to detail the way of selecting the number of replicas to be used during the replication phase. Just like [11], [14] does not control the number of generated fact table fragments.

To summarize, the most relevant-in-literature approaches related to our research are mainly oriented towards the idea of performing the fragmentation and allocation phases over database clusters in an isolate and iterative manner.

3 Formalization of the PRDW Design Problem on Heterogeneous Database Clusters

In this Section, we introduce a rigorous formalization of the PRDW design problem on heterogeneous database clusters, which will be used as reference formalism throughout the paper. Formally, given:

- a data warehouse schema $\mathcal{DWS}$ composed by d dimension tables $\mathcal{D} = \{D_0, D_1, \ldots, D_{d-1}\}$ and one fact table $\mathcal{F}$ – as in [11,14], we suppose that all dimension tables are replicated over the nodes of the database cluster and are fully-available in main memories of cluster nodes;
- a database cluster machine $\mathcal{DBC}$ with M nodes $\mathcal{N} = \{N_0, N_1, \ldots, N_{M-1}\}$, each node N_m, with $0 \leq m \leq M-1$, having a *proper* storage S_m and *proper* processing power P_m, which is straightforwardly modeled in terms of the number of operations that N_m can process in the reference temporal unit;
- a set of star queries $\mathcal{Q} = \{Q_1, Q_2, \ldots, Q_{L-1}\}$ to be executed over $\mathcal{DBC}$, being each query Q_l, with $0 \leq l \leq L-1$, characterized by an *access frequency* f_l;
- a *maintenance constraint* $\mathcal{W} : W > M$ representing the number of fragments W that the designer considers relevant for his/her target allocation process, called *fragmentation threshold*;

the problem of designing a PRDW described by $\mathcal{DWS}$ over the heterogeneous database cluster $\mathcal{DBC}$ consists in *fragmenting the fact table $\mathcal{F}$ into N_F fragments and allocating them over different $\mathcal{DBC}$ nodes such that the total cost of executing all the queries in $\mathcal{Q}$ can be minimized while storage and processing constraints are satisfied across nodes in $\mathcal{DBC}$, under the maintenance constraint $\mathcal{W}$.*

Based on the formal statement above, it follows that our investigated problem is composed by two sub-problems, namely data partitioning and fragment allocation. Each one of these problems is known to be *NP-complete* [3,19,13]. In order to deal with the PRDW design problem over database clusters, two main classes of methodologies are possible: *iterative design* methodologies and *combined design* methodologies. Iterative design methodologies have been proposed in the context of traditional distributed and parallel database design research. The idea underlying this class of methodologies consists in first fragmenting the RDW using *any* partitioning algorithm, and then allocating the so-generated fragments by means of *any* allocation algorithm. In the most general case, each partitioning and allocation algorithm has its own cost model. The main advantage coming from these traditional methodologies is represented by the fact that they are straightforwardly applicable to a large number of even-heterogenous parallel and distributed environments (e.g., *Peer-to-Peer Databases*). Contrary

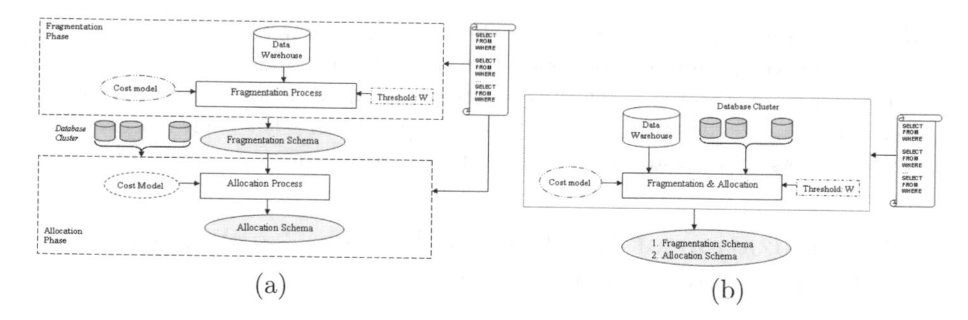

Fig. 1. Iterative PRDW Design Methodology over Heterogeneous Database Clusters (a) and Combined PRDW Design Methodology over Heterogeneous Database Clusters – The $\mathcal{F}\&\mathcal{A}$ Approach (b)

to this, their main limitation is represented by the fact that they neglect the interdependency between the data partitioning and the fragment allocation phase, respectively. Figure 1 (a) summarizes the steps of iterative design methodologies.

To overcome limitations deriving from using iterative design methodologies, the combined design methodology $\mathcal{F}\&\mathcal{A}$ we propose in our research consists in *performing the allocation phase/decision at fragmentation time, in a simultaneous manner*. Figure 1 (b) illustrates the steps of our approach. Contrary to the iterative approach that uses two cost models (i.e., one for the fragmentation phase, and one for the allocation phase), $\mathcal{F}\&\mathcal{A}$ uses only one cost model that monitors whether the *current* generated fragmentation schema is "useful" for the *actual* allocation process.

4 $\mathcal{F}\&\mathcal{A}$: A Combined PRDW Design Methodology over Heterogeneous Database Clusters

In this Section, we describe in detail our combined PRDW design methodology over heterogeneous database clusters, $\mathcal{F}\&\mathcal{A}$. We first focus the attention on the data partitioning phase, which, as stated in Section 1, is a fundamental and critical phase for any PRDW design methodology [1]. A distinctive characteristic of $\mathcal{F}\&\mathcal{A}$ is represented by the fact that, similarly to [1,5], it allows the designer to control the number of generated fragments, which should be a mandatory requirement for *any* PRDW design methodology in cluster environments (see Section 1). Then, we move the attention on data allocation issues and, finally, we provide the main algorithm implementing our proposed methodology.

4.1 Data Partitioning

In our proposed research, we make use of horizontal (data) partitioning, which can be reasonably considered as the core of $\mathcal{F}\&\mathcal{A}$. Specifically, our data partitioning approach consists in fragmenting dimension tables D_j in $\mathcal{D}$ by means of *selection predicates* of queries in $\mathcal{Q}$, and then using the so-generated *fragmentation schemes*, denoted by $\mathcal{FS}(D_j)$, to partition the fact table $\mathcal{F}$. Formally, a selection predicate is of kind: $A_k\ \theta\ V_k$, such that: (*i*) A_k models an attribute of a dimensional table D_j in $\mathcal{D}$; (*ii*) V_k models an attribute value in the *universe of instances* of $\mathcal{DWS}$; (*iii*) θ models an *equality or comparison predicate* among attributes/attribute-values, i.e. $\theta \in \{=, <, >, \leq, \geq\}$. The fact table partitioning method that derives from this approach is known-in-literature under the term "*referential partitioning*", which has recently been incorporated within the core layer of the DBMS platform *Oracle11G* [10].

Example 1. To illustrate how our proposed fragmentation process works, let us consider the *APB-1 release II* schema [4], which is characterized by the following dimensional tables: $\mathcal{D} = \{Product, Customer, Time, Channel\}$, and the following fact table: $\mathcal{F} = \{Sales\}$. Furthermore, suppose that the dimension table *Time* is partitioned into two fragments, namely $Time_{2007}$ and $Time_{2008}$, by means of the attribute *Year*, as follows: $Time_{2007} = \sigma_{Year=2007}(Time)$, $Time_{2008} = \sigma_{Year=2008}(Time)$, such that σ represents the selection predicate. As a consequence, the fact table *Sales* is fragmented on the basis of the partitioning scheme of the dimensional table *Time* into the following two fragments, namely $Sales_{2007}$ and $Sales_{2008}$, such that $Sales_{2007} = Sales \ltimes Time_{2007}$ and $Sales_{2008} = Sales \ltimes Time_{2008}$, where $\ltimes$ represents the *semi-join operator*.

Based on the data partitioning approach above, the number of fragments N_F generated from the fact table $\mathcal{F}$ is given by the following expression: $N_F = \prod_{j=0}^{d-1} \Phi_j$, such that Φ_j, with $0 \leq j \leq d-1$, denotes the number of horizontal fragments of the dimension table D_j in $\mathcal{D}$, and d denotes the number of dimension tables in $\mathcal{DWS}$. Such a decomposition of the fact table may generate a large number of fragments [21,3].

4.2 Naive Solution

In $\mathcal{F}\&\mathcal{A}$, we introduce the concept of *fragmentation scheme candidate* of a dimensional table D_j in $\mathcal{D}$, denoted by $\mathcal{FS}_C(D_j)$. Intuitively enough, a fragmentation scheme candidate is a fragmentation scheme generated during the execution of the algorithm implementing $\mathcal{F}\&\mathcal{A}$ and that *may belong* to the final solution represented by the set of N_F fact-table fragments allocated across nodes of the target database cluster.

A critical role in this respect is played by the solution used to represent-in-memory fragmentation scheme candidates, as this, in turn, impacts on the performance of the proposed algorithm. In our implementation, given a dimensional table D_j in $\mathcal{D}$, we model a fragmentation scheme candidate of D_j as a *multi-dimensional array* $\mathcal{A}_j$ such that rows in $\mathcal{A}_j$ represent so-called *fragmentation attributes* of the partitioning process (namely, attributes of D_j), and columns in $\mathcal{A}_j$ represent *domain partitions* of fragmentation attributes. Given an attribute A_k of D_j, a domain partition $\mathcal{P}_D(A_k)$ of A_k is a partitioned representation of the domain of A_k, denoted by $Dom(A_k)$, into *disjoint sub-domains* of $Dom(A_k)$, i.e. $\mathcal{P}_D(A_k) = \{dom_0(A_k), dom_1(A_k), \ldots, dom_{|\mathcal{P}_D(A_k)|-1}(A_k)\}$, such that $dom_h(A_k) \subseteq Dom(A_k)$, with $0 \leq h \leq |\mathcal{P}_D(A_k)| - 1$, denotes a sub-domain of $Dom(A_k)$, and the following property holds: $\forall\ h_p, h_q : h_p \neq h_q$, $dom_{h_p}(A_k) \bigcap dom_{h_q}(A_k) = \emptyset$. Given an attribute A_k of D_j, a number of alternatives for generating a domain partition $\mathcal{P}_D(A_k)$ of $Dom(A_k)$ exist. Among all the available solutions, $\mathcal{F}\&\mathcal{A}$ makes use of the set of queries $\mathcal{Q}$ to this end (see Section 3). Coming back to the structural definition of $\mathcal{A}_j$, each cell of $\mathcal{A}_j$, denoted by $\mathcal{A}_j[k][h]$, stores an integer value that represents the number of attribute values of A_k belonging to the sub-domain $dom_h(A_k)$ of $Dom(A_k)$. It is a matter of fact to notice that $\mathcal{A}_j[k][h] \in [0 : |\mathcal{P}_D(A_k)|]$ (see Figure 2).

Based on the multidimensional representation model for fragmentation scheme candidates above, for each dimension table D_j in $\mathcal{D}$, the final fragmentation scheme of D_j, $\mathcal{FS}(D_j)$, is generated according to the following role-based semantics:

- all cells in $\mathcal{A}_j$ of a fragmentation attribute A_k of D_j have *different* values $\mathcal{A}_j[k][h]$, then all sub-domains of $Dom(A_k)$ will be used to partition D_j;
- all cells in $\mathcal{A}_j$ of a fragmentation attribute A_k of D_j have the *same* value $\mathcal{A}_j[k][h]$, then the attribute A_k will not participate to the fragmentation process;
- a sub-set of cells in $\mathcal{A}_j$ of a fragmentation attribute A_k of D_j have the *same* value $\mathcal{A}_j[k][h]$, then the corresponding sub-domains of $Dom(A_k)$ will be merged into one sub-domain only, and then used to partition D_j.

Example 2. Consider again the *APB-1 release II* schema [4]. Suppose that the fragmentation process example is driven by the following fragmentation attributes: *Class*, *Group* and *Family*, all belonging to the dimension table *Product*. Also, suppose that the domains of these attributes are the following: *Dom(Class)* $= \{C_1, C_2, C_3\}$, *Dom(Group)* $= \{G_1, G_2, G_3\}$ and $Dom(Family) = \{F_1, F_2, F_3\}$, and that the domain of each attribute is decomposed into three distinct sub-domains, as shown in Figure 2 (a). Figure 2 (b) shows a fragmentation scheme candidate example $\mathcal{A}_{Product}$ of *Product* for the running fragmentation process example. Note that attribute $A_2 = Family$ is not concerned by the fragmentation process, as all its cells in $\mathcal{A}_{Product}[2][h]$ have the same value, with $0 \leq h \leq 2$. On the other hand, based on fragmentation scheme shown in Figure 2 (b), the dimension table *Product* will be fragmented into $3 \times 2 = 6$ horizontal fragments, hence the fact table *Sales* will be also partitioned into 6 (fact-table) fragments accordingly.

Based on the formal model of fragmentation scheme candidates above, the naive solution to the PRDW design problem over database clusters we propose, which represents a first attempt of the algorithm implementing $\mathcal{F}\&\mathcal{A}$, makes use of a *hill climbing heuristic* [8], which consists of the following two steps:

1. find an initial solution $\mathcal{I}_0$ – $\mathcal{I}_0$ may be obtained via using a *random* distribution for filling cells of fragmentation scheme candidates for each fragmentation attribute A_k of dimensional tables D_j in $\mathcal{D}$;

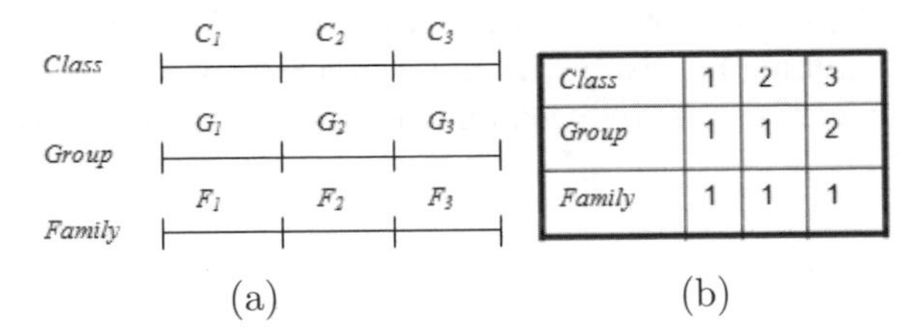

Class	1	2	3
Group	1	1	2
Family	1	1	1

(b)

Fig. 2. Attribute Domain Partitions (a) and a Fragmentation Scheme Candidate $\mathcal{A}_{Product}$ of the dimension table *Product* (b) of the Running Example

2. iteratively improve the initial solution $\mathcal{I}_0$ by using the hill climbing heuristic until no further reduction in the total query processing cost due to evaluating queries in $\mathcal{Q}$ can be achieved, and the storage and processing constraints are satisfied, under the maintenance constraint $\mathcal{W}$.

It should be noted that, since the number of fragmentation scheme candidates generated from $\mathcal{DWS}$ is finite, the hill climbing heuristic will *always* complete its execution, thus finding the final solution $\mathcal{I}_F$. This ensures the convergence of the naive solution at a theoretical level.

4.3 Improved Solution

The previous naive solution can be improved by introducing two specialized operators, namely *Merge* and *Split*, which allow us to further reduce the total query processing cost due to evaluating queries in $\mathcal{Q}$. Let us now focus on the formal definitions of these operators.

Given a fragmentation attribute A_k of a dimension table D_j in $\mathcal{D}$ having $\mathcal{FS}(D_j)$ as fragmentation scheme, *Merge* takes as input two domain partitions of A_k in $\mathcal{FS}(D_j)$, namely $\mathcal{P}_D^p(A_k)$ and $\mathcal{P}_D^q(A_k)$, an returns as output a new fragmentation scheme for D_j, denoted by $\mathcal{FS}'(D_j)$, where $\mathcal{P}_D^p(A_k)$ and $\mathcal{P}_D^q(A_k)$ are merged into a singleton domain partition of A_k, denoted by $\mathcal{P}_D^{p,q}(A_k)$. *Merge* reduces the number of fragments generated by means of the fragmentation scheme $\mathcal{FS}(D_j)$ of D_j, hence it is used when the number of generated fragments does not satisfy the maintenance constraint $\mathcal{W}$ (see Section 3). Formally, *Merge* is defined as follows:

$$Merge : \langle A_k, D_j, \mathcal{FS}(D_j), \mathcal{P}_D^p(A_k), \mathcal{P}_D^q(A_k)\rangle \rightarrow \langle A_k, D_j, \mathcal{FS}'(D_j), \mathcal{P}_D^{p,q}(A_k)\rangle \quad (1)$$

Given a fragmentation attribute A_k of a dimension table D_j in $\mathcal{D}$ having $\mathcal{FS}(D_j)$ as fragmentation scheme, *Split* takes as input a domain partition of A_k in $\mathcal{FS}(D_j)$, $\mathcal{P}_D(A_k)$, an returns as output a new fragmentation scheme for D_j, denoted by $\mathcal{FS}'(D_j)$, where $\mathcal{P}_D(A_k)$ is split into two distinct domain partitions of A_k, denoted by $\mathcal{P}_D^p(A_k)$ and $\mathcal{P}_D^q(A_k)$, respectively. *Split* increases the number of fragments generated by means of the fragmentation scheme $\mathcal{FS}(D_j)$ of D_j. Formally, *Split* is defined as follows:

$$Split : \langle A_k, D_j, \mathcal{FS}(D_j), \mathcal{P}_D(A_k)\rangle \rightarrow \langle A_k, D_j, \mathcal{FS}'(D_j), \mathcal{P}_D^p(A_k), \mathcal{P}_D^q(A_k)\rangle \quad (2)$$

On the basis of these operators running on fragmentation schemes of dimensional tables, the hill climbing heuristic still finds the final solution $\mathcal{I}_F$, while the total query processing cost can be reduced and the maintenance constraint $\mathcal{W}$ can be satisfied.

4.4 Data Allocation

The data allocation phase of $\mathcal{F}\&\mathcal{A}$ is performed simultaneously to the data fragmentation/partitioning phase. Basically, each fragmentation scheme candidate generated by the algorithm implementing $\mathcal{F}\&\mathcal{A}$ is allocated across nodes of the target database cluster, with the goal of minimizing the total query processing cost queries in $\mathcal{Q}$ over *all* nodes, while satisfying the storage and processing constraints on *each* node. In more detail, during the allocation phase the following concepts/data-structures are used:

- *Fragment Placement Matrix* (FPM) $\mathcal{M}_P$, which stores the positions of a fragment across nodes (recall that fragment replicas may exist). To this end, $\mathcal{M}_P$ rows model fragments, whereas $\mathcal{M}_P$ columns model nodes. $\mathcal{M}_P[i][m] = 1$, with $0 \leq i \leq N_F - 1$ and $0 \leq m \leq M - 1$, if the fragment F_i is allocated on the node N_m in $\mathcal{N}$, otherwise $\mathcal{M}_P[i][m] = 0$.
- *Fragment Size* $Size(F_i)$, which models the size of the fragment F_i in terms of the number of its instances across the nodes. $Size(F_i)$ is estimated by means of selection predicates. Since each node N_m in $\mathcal{N}$ has its own storage capacity S_m, the storage constraint associated to F_i across all nodes of the target database cluster can be formally expressed as follows:

$$\forall m \in [0 : M-1] : \sum_{i=0}^{N_F-1} \mathcal{M}_P[i][m] \times Size(F_i) \leq S_m \tag{3}$$

- *Fragment Usage Matrix* (FUM) [15] $\mathcal{M}_U$, which models the "usage" of fragments by queries in $\mathcal{Q}$. To this end, $\mathcal{M}_U$ rows model queries, whereas $\mathcal{M}_U$ columns model fragments. $\mathcal{M}_U[l][i] = 1$, with $0 \leq l \leq L-1$ and $0 \leq i \leq N_F - 1$, if the fragment F_i is involved by the query Q_l in $\mathcal{Q}$, otherwise $\mathcal{M}_U[l][i] = 0$. An *additional* column is added to $\mathcal{M}_U$ for representing the access frequency fr_l of each query Q_l in $\mathcal{Q}$ (see Section 3). In order to evaluate a query Q_l in $\mathcal{Q}$ on a node N_m in $\mathcal{N}$, N_m must store *relevant fragments* for Q_l. Based on our theoretical framework, a fragment F_i is relevant iff the following property holds: $\mathcal{M}_P[i][m] = 1 \wedge \mathcal{M}_U[l][i] = 1$, with $0 \leq i \leq N_F - 1$, $0 \leq m \leq M-1$ and $0 \leq l \leq L-1$.

  Example 3. Let $Q = \{Q_1, Q_2, Q_3, Q_4\}$ and $F = \{F_1, F_2, F_3, F_4, F_5, F_6, F_7, F_8\}$ be the set of queries and generated fragments, respectively. The corresponding FUM is shown in Table 1.
- *Fragment Affinity Matrix* (FAM) $\mathcal{M}_A$, which models the "affinity" between two fragments F_{i_p} and F_{i_q}. To this end, $\mathcal{M}_A$ rows and columns both model fragments, hence $\mathcal{M}_A$ is a symmetric matrix. $\mathcal{M}_A[i_p][i_q]$, with $0 \leq i_p \leq N_F - 1$ and $0 \leq i_q \leq N_F - 1$, stores the sum of access frequencies of queries in $\mathcal{Q}$ that involve F_{i_p} and F_{i_q} simultaneously.

Table 1. FUM of the Running Example

	F_1	F_2	F_3	F_4	F_5	F_6	F_7	F_8	Fr
Q_1	1	0	1	0	1	0	1	0	20
Q_2	1	1	1	1	0	0	0	0	35
Q_3	0	0	1	0	1	1	1	1	30
Q_4	1	1	1	1	1	1	1	1	15

Example 4. From the FUM shown in Table 1, the associated FAM is shown in Table 2.

Table 2. FAM of the Running Example

	F_1	F_2	F_3	F_4	F_5	F_6	F_7	F_8
F_1	-	50	70	50	65	15	35	15
F_2	50	-	50	50	15	15	15	15
F_3	70	50	-	50	65	45	65	45
F_4	50	50	50	-	15	15	1 5	15
F_5	65	15	65	15	-	45	65	45
F_6	15	15	45	15	45	-	45	45
F_7	35	15	65	15	65	45	-	45
F_8	15	15	45	15	45	45	45	-

4.5 $\mathcal{F}\&\mathcal{A}$ Algorithm

On the basis of the data partitioning phase and the data allocation phase described in Section 4.1 and Section 4.4, respectively, and the naive solution and improved solution to the PRDW design problem over database clusters provided in Section 4.2 and Section 4.3, respectively, for each fragmentation scheme candidate $\mathcal{FS}_C(D_j)$ of each dimensional table D_j in $\mathcal{D}$, the algorithm implementing our proposed methodology $\mathcal{F}\&\mathcal{A}$ performs the following steps:

1. Based on the FUM $\mathcal{M}_U$ and the FAM $\mathcal{M}_A$, generate groups of fragments G_z by means of the method presented in [15].
2. Compute the size of each fragment group G_z, as follows: $Size(G_z) = \sum_i Size(F_i)$, such that $Size(F_i)$ denotes the size of the fragment F_i.
3. Sort nodes in the target database cluster $\mathcal{DBC}$ by descendent ordering based on their storage capacities and processing powers.
4. Allocate "heavy" fragment groups on *powerful nodes* in $\mathcal{DBC}$, i.e. nodes with high storage capacity and high processing power, in a *round-robin* manner starting from the first powerful node. The allocation phase must *minimize* the total query processing cost due to evaluating queries in $\mathcal{Q}$ while *maximizing* the *productivity* of each node, based on the following theoretical formulation:

$$\sum_{l=0}^{L-1} f_l \times max_{0 \leq m \leq M-1} \left\{ \sum_{i=0}^{N_F-1} \frac{\mathcal{M}_U[l][i] \times \mathcal{M}_P[i][m] \times Size(F_i)}{P_m} \right\} \quad (4)$$

such that: (*i*) L denotes the number of queries against $\mathcal{DBC}$; (*ii*) M denotes the number of nodes of $\mathcal{DBC}$; (*iii*) N_F denotes the number of fragments belonging to the solution; (*iv*) $\mathcal{M}_U$ denotes the FUM; (*v*) $\mathcal{M}_P$ denotes the FPM; (*vi*) $Size(F_i)$ denotes the size of the fragment F_i; (*vii*) P_m denotes the processing power of the node N_m in $\mathcal{N}$. In formula (4), we implicitly suppose that the response time of any arbitrary query Q_l in $\mathcal{Q}$ is superiorly bounded by the time needed to evaluate Q_l against the *most-loaded node* in $\mathcal{DBC}$, thus we can consider it as a constant and omit it in formula (4).

5. Replicate on non-powerful nodes groups of fragments that require high computation time, in order to ensure a high performance.

5 Experimental Assessment and Results

In order to carefully evaluate the effectiveness and the efficiency of our proposed PRDW design methodology on database clusters, $\mathcal{F}\&\mathcal{A}$, we conducted an intensive experimental campaign. Our $\mathcal{F}\&\mathcal{A}$ algorithm (see Section 4.5) has been implemented by using *Java*, and experiments have been performed on an *Intel Pentium Core Duo* at 2.8 GHz equipped with 3 GB RAM.

As regards the setting of our experimental framework, we considered a simulated database cluster environment with 128 nodes. Storage capacity and processing power of each node have been generated according to a random distribution, thus obtaining a totally heterogenous database cluster environment.

As regards the data layer of our experimental framework, we considered the well-known benchmark *APB-1 release II* [4]. In detail, *APB-1* is characterized by one fact table *Sales* having 24,786,000 tuples, and the following four dimension tables, with respective number of tuples: *Product* (9,000 tuples), *Customer* (900 tuples), *Time* (24 tuples), and *Channel* (9 tuples).

As regards the query layer of our experimental framework, we considered a star query workload consisting of of 55 *single-block queries* (i.e., queries without nested sub-queries) characterized by 40 selection predicates defined on the following 9 distinct attributes: *Class*, *Group*, *Family*, *Line*, *Division*, *Year*, *Month*, *Retailer*, *All*. Domains of these attributes are split into the following number of sub-domains: 4, 2, 5, 2, 4, 2, 12, 4, 5, respectively. In our experimental assessment, we do not consider update queries, which are left for future work.

As regards the metrics of our experimental framework, we considered the *execution time* due to evaluating queries of the experimental query workload by gathering the total number of I/Os needed to this end divided by the *average* processing power of nodes. Here, we set the reference temporal unit determining the notion of processing power to seconds (see Section 3).

We performed several kinds of experiments, in order to obtain a "rich" and reliable experimental evaluation of the $\mathcal{F}\&\mathcal{A}$ algorithm. First, we compared our proposed methodology $\mathcal{F}\&\mathcal{A}$ against a classical iterative approach, where fragmentation and allocation are executed sequentially and without any iteration, still in a heterogeneous database cluster environment. The classical iterative approach is based on the hill climbing heuristic [8]. As regards the $\mathcal{F}\&\mathcal{A}$ algorithm, we set the fragmentation threshold W to 500 (see Section 3). We measured the query execution time versus the variation of the number of database cluster nodes M over the interval $[2 : 128]$. Figure 3 (a) shows the results obtained from the first experiment, and confirms to us that the combined approach outperforms the iterative one significantly. In the second experiment, we focused the attention on $\mathcal{F}\&\mathcal{A}$ solely, and we observed its performance in four different application scenarios which may arise in real-life database cluster environments: (i) heterogenous database cluster environments, according to the general guidelines of our experimental setting provided above; (ii) homogenous database cluster environments such that nodes have a "high" processing power (denoted by $P++$); (iii) homogenous database cluster environments such that nodes have a "low" processing power (denoted by $P--$); (iv) homogenous database cluster environments such that nodes have an "average" processing power (denoted by $P = AVG$). For all scenarios, we assumed a limited storage capacity, i.e. the following hypothesis holds: $\sum_{m=0}^{M-1} S_m > Size(DW)$, where S_m denotes the storage capacity of the node N_m in $\mathcal{N}$ and $Size(DW)$ denotes the size of the entire data warehouse, respectively. Figure 3 (b) shows the results obtained from the second experiment. As shown in Figure 3 (b), $\mathcal{F}\&\mathcal{A}$ performance reaches the best score in the case of scenario (ii), i.e. $P{+}{+}$, as expected. On the other hand, a collateral interesting phenomenon is represented by the fact that $\mathcal{F}\&\mathcal{A}$ performance over heterogenous database cluster environments outperforms $\mathcal{F}\&\mathcal{A}$ performance over the remaining two scenarios, i.e. $P--$ and $P = AVG$.

In the third experiment, we stressed the $\mathcal{F}\&\mathcal{A}$ performance under two different (heterogeneous database cluster) scenarios determined by the processing power of nodes, which is a fundamental factor in our research. According to the first scenario, the allocation phase of $\mathcal{F}\&\mathcal{A}$ has been performed by considering the processing power of nodes in the cost model (4), whereas in the second one the

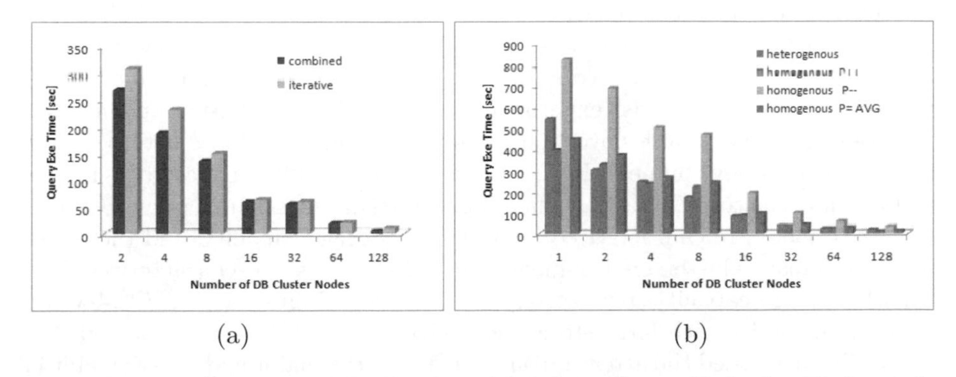

(a) (b)

Fig. 3. Query Performance vs the Number of Database Cluster Nodes for $\mathcal{F}\&\mathcal{A}$ and the Hill-Climbing-based Methodology in a Heterogeneous Environment (a) and for $\mathcal{F}\&\mathcal{A}$ over Four Different Database Cluster Environments Scenarios (b)

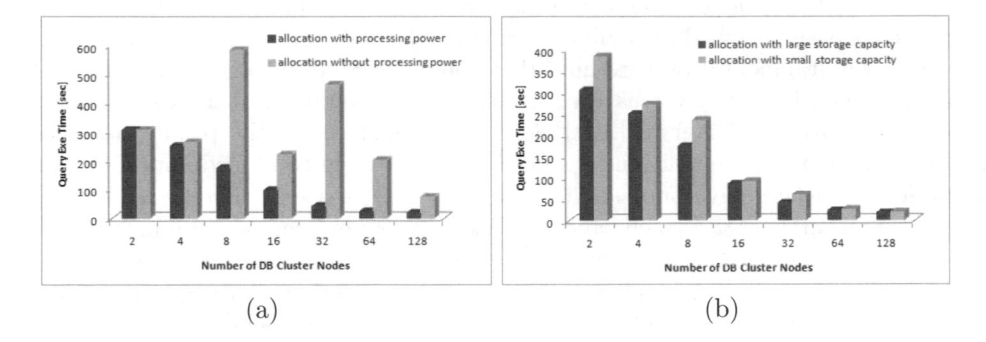

Fig. 4. $\mathcal{F}\&\mathcal{A}$ Query Performance vs the Number of Database Cluster Nodes in the dependence of Processing Power (a) and Storage Capacity (b) of Nodes

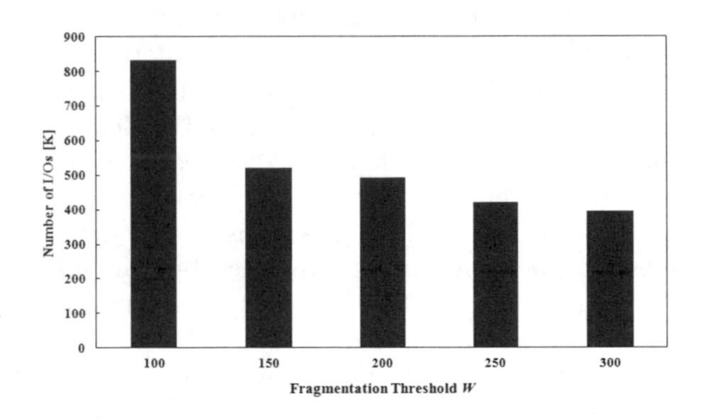

Fig. 5. Effect of the Fragmentation Threshold W on the Query Performance of $\mathcal{F}\&\mathcal{A}$

allocation has not considered the processing power of nodes. Figure 4 (a) shows the results obtained from the third experiment. Derived results show that, when the cost model (4) encompasses the processing power of nodes, $\mathcal{F}\&\mathcal{A}$ performance is higher as all the effective characteristics of nodes are taken into consideration. At the same, this confirms to us the effectiveness and the efficiency of $\mathcal{F}\&\mathcal{A}$. Finally, in the last experiment we focused the attention on the effect of storage capacity of nodes over the $\mathcal{F}\&\mathcal{A}$ performance, still in a heterogeneous database cluster environment. Here, we considered two different scenarios related to this critical factor of nodes, i.e. (heterogeneous) database cluster environments such that nodes are characterized by a "large" storage capacity, and (heterogeneous) database cluster environments such that nodes are characterized by a "small" storage capacity, respectively. As shown in Figure 4 (b), $\mathcal{F}\&\mathcal{A}$ works better when nodes with large storage capacity are considered, as expected.

Finally, we focused the attention on the effect of the maintenance constraint $\mathcal{W}$ (see Section 3) on the performance of $\mathcal{F}\&\mathcal{A}$ over heterogeneous cluster environments, still considering the main one developed in our experimental assessment. Here, we fixed the number of nodes to $M = 10$, and we ranged the fragmentation threshold W over the interval [100 : 300] in order to study how the $\mathcal{F}\&\mathcal{A}$ query performance varies accordingly. For each value of W, we run the $\mathcal{F}\&\mathcal{A}$ algorithm, and we estimated the total query processing cost due to evaluating queries of the target query workload in terms of number of I/Os. Figure 5 shows the obtained experimental results. From the analysis of Figure 5, it clearly follows that increasing the value of W improves the $\mathcal{F}\&\mathcal{A}$ query performance significantly, as this allows more (fragmentation) attributes to participate in the partitioning process. In addition to this, it should be noted that $\mathcal{F}\&\mathcal{A}$ query performance become stable starting from the cut-off value $W = 250$. This experimental result confirms to us the importance of carefully choosing the number of final fragments to be generated.

6 Conclusions and Future Work

In this paper, we have introduced and experimentally evaluated $\mathcal{F}\&\mathcal{A}$, an innovative PRDW design methodology on database clusters. The proposed methodology encompasses a number of advancements over state-of-the-art similar approaches, particularly (*i*) the fact it considers heterogeneous cluster nodes, i.e. nodes having heterogenous storage capacities and processing power, and (*ii*) the fact it performs the fragmentation and allocation phases simultaneously. As a secondary contribution of our research, we have provided a comprehensive experimental campaign where we demonstrated the effectiveness and the efficiency of our proposed approach. Future work is mainly oriented towards making our proposed design methodology able to deal with next-generation *Grid Data Warehouse Environments* [6].

References

1. Bellatreche, L., Benkrid, S.: A joint design approach of partitioning and allocation in parallel data warehouses. In: Pedersen, T.B., Mohania, M.K., Tjoa, A.M. (eds.) DAWAK 2009. LNCS, vol. 5691, pp. 99–110. Springer, Heidelberg (2009)
2. Bellatreche, L., Boukhalfa, K.: An evolutionary approach to schema partitioning selection in a data warehouse environment. In: Tjoa, A.M., Trujillo, J. (eds.) DaWaK 2005. LNCS, vol. 3589, pp. 115–125. Springer, Heidelberg (2005)
3. Bellatreche, L., Boukhalfa, K., Richard, P.: Data partitioning in data warehouses: Hardness study, heuristics and oracle validation. In: Song, I.-Y., Eder, J., Nguyen, T.M. (eds.) DaWaK 2008. LNCS, vol. 5182, pp. 87–96. Springer, Heidelberg (2008)
4. OLAP Council. Apb-1 olap benchmark, release ii (1998), `http://www.olapcouncil.org/research/bmarkly.htm`
5. Cuzzocrea, A., Darmont, J., Mahboubi, H.: Fragmenting very large XML data warehouses via k-means clustering algorithm. International Journal of Business Intelligence and Data Mining 4(3-4), 301–328 (2009)
6. Cuzzocrea, A., Kumar, A., Russo, V.: Experimenting the query performance of a grid-based sensor network data warehouse. In: Hameurlain, A. (ed.) Globe 2008. LNCS, vol. 5187, pp. 105–119. Springer, Heidelberg (2008)

7. Cuzzocrea, A., Serafino, P.: *LCS*-hist: taming massive high-dimensional data cube compression. In: 12th International Conference on Extending Database Technology, EDBT 2009 (2009)
8. Davis, L.D.: Bit-climbing, representational bias, and test suite design. In: Proceedings of the 4th International Conference on Genetic Algorithms (ICGE 1991), pp. 18–23 (March 1991)
9. DeWitt, D.J.D., Madden, S., Stonebraker, M.: How to build a high-performance data warehouse, `http://db.lcs.mit.edu/madden/high_perf.pdf`
10. Eadon, G., Chong, E.I., Shankar, S., Raghavan, A., Srinivasan, J., Das, S.: Supporting table partitioning by reference in oracle. In: SIGMOD 2008 (2008)
11. Furtado, P.: Experimental evidence on partitioning in parallel data warehouses. In: DOLAP, pp. 23–30 (2004)
12. Gupta, H.: Selection and maintenance of views in a data warehouse. Ph.d. thesis, Stanford University (September 1999)
13. Karlapalem, K., Pun, N.M.: Query driven data allocation algorithms for distributed database systems. In: Tjoa, A.M. (ed.) DEXA 1997. LNCS, vol. 1308, pp. 347–356. Springer, Heidelberg (1997)
14. Lima, A.B., Furtado, C., Valduriez, P., Mattoso, M.: Improving parallel olap query processing in database clusters with data replication. Distributed and Parallel Database Journal (2009) (to appear)
15. Navathe, S.B., Ra, M.: Vertical partitioning for database design: a graphical algorithm. In: ACM SIGMOD, pp. 440–450 (1989)
16. Özsu, M.T., Valduriez, P.: Principles of Distributed Database Systems, 2nd edn. Prentice Hall, Englewood Cliffs (1999)
17. Röhm, U., Böhm, K., Schek, H.-J.: Olap query routing and physical design in a database cluster. In: Zaniolo, C., Grust, T., Scholl, M.H., Lockemann, P.C. (eds.) EDBT 2000. LNCS, vol. 1777, pp. 254–268. Springer, Heidelberg (2000)
18. Röhm, U., Böhm, K., Schek, H.-J.: Cache-aware query routing in a cluster of databases. In: Proceedings of the International Conference on Data Engineering (ICDE), pp. 641–650 (2001)
19. Saccà, D., Wiederhold, G.: Database partitioning in a cluster of processors. ACM Transactions on Database Systems 10(1), 29–56 (1985)
20. Sarawagi, S.: Indexing olap data. IEEE Data Engineering Bulletin 20(1), 36–43 (1997)
21. Stöhr, T., Märtens, H., Rahm, E.: Multi-dimensional database allocation for parallel data warehouses. In: Proceedings of the International Conference on Very Large Databases, pp. 273–284 (2000)
22. Stöhr, T., Rahm, E.: Warlock: A data allocation tool for parallel warehouses. In: Proceedings of the International Conference on Very Large Databases, pp. 721–722 (2001)

Yet Another Algorithms for Selecting Bitmap Join Indexes

Ladjel Bellatreche[1] and Kamel Boukhalfa[2]

[1] LISI/ENSMA - Poitiers University, Futuroscope, France
bellatreche@ensma.fr
[2] USTHB University - Algiers- Algeria
boukhalk@gmail.com

Abstract. One of the fundamental tasks that data warehouse (DW) administrator needs to perform during the physical design is to select the right indexes to speed up her/his queries. Two categories of indexes are available and supported by the main DBMS vendors: (i) indexes defined on a single table and (ii) indexes defined on multiple tables such as join indexes, bitmap join indexes, etc. Selecting relevant indexes for a given workload is a NP-hard problem. A majority of studies on index selection problem was focused on single table indexes, where several types of algorithms were proposed: greedy search, genetic, linear programming, etc. Parallel to these research efforts, commercial DBMS gave the same attention to single table indexes, where automated tools and advisors generating recommended indexes for a particular workload and constraints are developed. Unfortunately, only few studies dealing with the problem of selecting bitmap join indexes are carried out. Due to the high complexity of this problem, these studies mainly focused on proposing pruning solutions of the search space by the means of data mining techniques. The lack of bitmap join index selection algorithms motivates our proposal. This paper proposes selection strategies for single and multiple attributes BJI. Intensive experiments are conducted comparing the proposed strategies using mathematical cost model and the obtained results are validated under Oracle using APB1 benchmark.

Keywords: Physical Design, Bitmap join index, Query performance.

1 Introduction

Queries defined on relational DW (called *star join queries*) are complex, since they involve several joins and selections. Indexes are a solid candidate to optimize such operations. Note that they are considered as the *pioneer* of the optimization techniques in database area. They represent an important part of any database system design as they can significantly impact workload performance by enabling quicker and more efficient access to data. In the DW context, when we talk about indexing, we refer to two different aspects: (i) *indexing techniques* and (i) *index selection problem*. A number of indexing strategies have been suggested for DWs that we propose to classify into two main categories: (1) *single table*

T.B. Pedersen, M.K. Mohania, and A M. Tjoa (Eds.): DaWaK 2010, LNCS 6263, pp. 105–116, 2010.

indexes and (2) *multiple table indexes.* A single table index is an index defined on one or several attributes of a single table, whereas a multiple table index involves several tables. A large spectrum of indexing techniques belonging to both categories has been proposed: *value-list index*, *projection index* [13], *bitmap index* [6], *data index* [11], *join index* [18], *star join index*, *bitmap join index* [13]. Note that single table indexes are not sufficient to optimize star join queries. A join index, considered as a multiple table index, is well adapted for such queries. It is the result of joining two tables on a join attribute and projecting the keys (or tuple identifiers) of the two tables. To join the two tables, we can use the join index to fetch the tuples from the tables followed by a join. In the relational DW, it is of interest to perform a multiple join (a star join) on the fact table and their dimension tables. Therefore, it will be helpful to build join indexes between the keys and the dimension tables and the corresponding foreign keys of the fact table. If the join indexes are represented in bitmap (called bitmap join indexes (BJI)), a multiple join could be replaced by a sequence of bitwise operations, followed by a relatively small number of fetch and join operations. An important characteristic of BJI is their ability to be compressed [19], where run-length compression is usually used to reduce the size of the bitmaps. Note that a BJI can be defined on only one attribute of a given dimension table (in this case it is called *single attribute BJI*) or on several attributes of the same or different dimension table (called *multiple attribute BJI*).

The index selection problem (ISP) has been studied since the early 70's and its importance during physical design is well recognized [9]. ISP consists in picking a set of indexes for given set of queries under some resources constraints (storage cost, maintenance overhead, etc.). It is a NP-hard problem. A large amount of studies dealing with this problem were proposed [7,10,16,8,14,12,20]. They are mainly focused on single table indexes. Two main types of algorithms were proposed to select them: (i) *heuristic algorithms*, such as greedy search [16,8], genetic, etc. and (ii) *integer linear programming approaches* to compute how close they get to the optimal solution [7,14]. Most academic selection approaches use mathematical cost models to guide the selection process and quantify the quality of the final indexes. Some industrial index tools use the cost models of their query optimizers to select indexes [20]. The single table index selection algorithms used by these tools are usually based on *greedy search* augmented with optimization techniques to reduce the number of index candidate they consider and the number of calls to the query optimizer [8]. Recently, DBMS vendors propose automated advisors generating recommended single table indexes for a particular workload and constraints [1,20].

So far, we realize that single table indexes received great attention from academic and industrial communities. This attention concerns both aspects of indexing: techniques and selection algorithms. Unfortunately, a little attention has been given to multiple table indexes and especially selection algorithms aspect. Most of studies related to BJI selection problem are mainly focused on pruning its search space [2,4,17]. In [2], a data mining algorithm *Close* [15] is used to perform the pruning. *Frequent itemsets* generated by Close are BJI candidate. Since,

BJI selection problem is constraint with a storage capacity, the authors propose a simple greedy algorithm to select a final configuration of BJI optimizing query processing cost and satisfying the storage constraint. The main drawback of this selection approach is that it considers only *frequencies of appearance of attributes* to generate frequent itemsets. In [4], we have shown using an example that the appearance frequencies cannot be the sole criteria to recommend BJI. Therefore, we proposed *DynaClose* algorithm which is an improvement of Close by adding other DW parameters such as frequencies of attributes, size of the dimension and fact tables, the system page size, etc to generate frequent itemsets. Once the pruning phase is done, a simple greedy algorithm is performed to select final BJI. In [4], horizontal partitioning technique (considered as an optimization structure), is used to prune the search space of BJI indexes. This work has been motivated by the existence of a strong similarity between horizontal partitioning and BJI - both optimize selections and joins and are concurrent to the same resource representing the selection attributes of dimension tables. The pruning process is done as follows: *if a restriction attribute is used to partition the DW, it will be automatically discarded from indexing process.* Similar work was developed in [17], but without considering BJI selection problem. It deals with parallel DW design, where algorithms for allocating fragments and BJIs are given. In [3], a tool (called *SimulPhd*) assisting, in iterative way, DW administrators (DWAs) in their physical design tasks are proposed. One of the functionalities of *SimulPhd* is the recommendation of BJIs based on *DynaClose* approach proposed in [4].

The lack of BJI selection algorithms motivates us to develop other strategies and to propose a strong evaluation comparing their efficiencies to optimize OLAP queries. Having a several BJI selection algorithms offers designers a large broad of choices during the physical design. Our proposed algorithms select both single and multiple attributes BJIs.

The paper is organized as follows. Section 2 presents background related to BJI and complexity of their selection problem. Section 3 presents BJI selection algorithms by describing in details their main steps. Section 4 presents intensive experiments using mathematical cost model and a validation under Oracle with data set of APB1 benchmark. Section 5 concludes the paper summarizing the main findings of our research, and proposing directions for future work.

2 Background

In this section, we present some BJI concepts, a formalization of their selection problem and its complexity.

BJI is used to pre-compute the joins between dimension table(s) and the fact table of relational DW modelled using a star schema [13]. Unlike standard bitmap index, where the indexed attributes belong to the table to be indexed, a BJI may be defined on one or more attributes belonging to various tables. More precisely, let A be an attribute of a given dimension table D with n distinct values $(v_1, v_2, \cdots, v_n)$ and m a number of instances of the fact table F. The construction of the BJI defined on F via the dimension attribute A is done as follows:

1. Create n vectors, where each one has m rows;
2. The ith bit of the vector corresponding to a value v_k is set to 1 if the ith tuple of the fact table is joined with a tuple of D having a value of its attribute A equal to v_k. It is set to 0 otherwise.

A BJI may be defined on one or several columns (attributes) of the same table or on more than one table. Besides disk saving (due to the binary representation and possible compression), BJIs speed up star join queries characterized by Boolean and COUNT operations. Note that BJI is defined on non key dimension attribute(s) with low cardinality [1] (called *indexable attributes*). An indexable attribute A_j of a given dimension table D_i is a column $D_i.A_j$ such that there is a condition of the form $D_i.A_j \ \ \theta \ \ Value$ in the WHERE clause. The operator θ must be among $\{=, <, >, \leq, \geq\}$ and $Value \in \ Domain(D_i.A_j)$.

The BJI selection is more difficult compare to single table indexes. This is due to the following points:

- in the context of the DW, the number of indexable attributes may be important, since star schemes used to model business intelligence applications are composed of thousand of dimension tables with various selection attributes (candidate for indexing),
- the fact that a BJI may be defined on a set of attributes belonging to one or several dimension tables increases the total number of BJIs. More formally, let $A = \{A_1, A_2, \cdots, A_K\}$ be the set of indexed attributes. Then, the possible number of BJIs that we should consider to select only one BJI grows exponentially: $\binom{K}{1}+\binom{K}{2}+...+\binom{K}{K} = 2^K-1$. To select more than one BJI, the number of possibilities is given by $\binom{2^K-1}{1} + \binom{2^K-1}{2} + ... + \binom{2^K-1}{2^K-1} = 2^{2^K-1}$,
- BJIs are not disjoint, since an indexable attribute may be found in two different BJIs.

Based on the above analysis, the problem of finding the set of BJIs that minimizes the total query processing cost while satisfying a storage constraint cannot be handled by first enumerating all possible BJIs and then computing the query cost for each candidate BJI. Due to this high complexity, we formalize it as an optimization problem with constraint as follows:

Given a DW with a set of dimension tables $\mathcal{D} = \{D_1, D_2, ..., D_d\}$ and a fact table F, a workload Q of queries $\mathcal{Q} = \{Q_1, Q_2, ..., Q_n\}$, where each query Q_i $(1 \leq i \leq n)$ has an access frequency, and a storage constraint $\mathcal{S}$, the aim of BJI selection problem is to find a set of BJIs among a pre-computed subset of all possible candidates which minimizes the cost of $\mathcal{Q}$ satisfies the storage requirements $\mathcal{S}$. We present in the next section our algorithms for selecting BJIs.

[1] The domain of this attribute should be an enumerated domain like *gender*.

3 Algorithms for Selecting BJIs

In this section, we present two algorithms one for selecting single attribute BJI and another for multiple attribute BJI. Note that single attribute BJIs are the first multiple table indexes proposed and supported by commercial DBMS [13]. The existing studies do not make this distinction. In this case, DWA shall wait the execution BJI selection algorithm to see whether the selected BJIs are defined on single or multiple attributes. In the real life, it suitable for DWA to have the choice to select her/him favourite selection strategy of BJIs.

Our algorithms use a cost model computing the number of inputs outputs required for executing a set of queries in the presence of BJIs [5].

3.1 Single Attribute BJI Selection

The algorithm for selecting a single-attribute BJI configuration is divided into three steps: (1) *identification of indexable attributes*, (2) *initialization of the configuration* and (3) *improving of the current configuration by adding new BJI.* In the first step, all queries are analyzed in order to extract the indexable attributes. These attributes are sorted based on their cardinality. The algorithm starts with an initial configuration consisting of a single attribute BJI with minimum cardinality, denoted by BJI_{min}. The initial configuration is iteratively improved by adding a BJI defined on other attributes not yet indexed and chosen from the ordered list of indexable attributes. The algorithm terminates when it arrives at a point, where it cannot see any more improvement of query processing cost and the storage space is consumed.

3.2 Multiple Attributes BJI Selection

By definition, single attribute BJI involves only one dimension table. Since OLAP queries cover several dimension tables involving selection predicates, the development of multiple attributes BJI selection algorithm becomes a necessity. For this purpose, we propose an *intuitive algorithm* for selecting such BJIs. It selects a BJI for each query having indexable attribute(s). Four steps characterize this algorithm: (1) *identification of indexable attributes*, (2) *construction of a configuration for each query*, (3) *construction of an initial configuration* and (4) *construction of a final configuration.*

Identification of indexable attributes. This step is done in the same way as in the previous algorithm.

Construction of a configuration of BJI by query. In this step, each query of the workload is associated to BJI involving its entire selection attributes candidate for indexation.

Example 1. Suppose the existence of five indexable attributes: *Time.Month*, *Time.Day*, *Product.Type*, *Customer.City* and *Customer.Gender* and ten queries

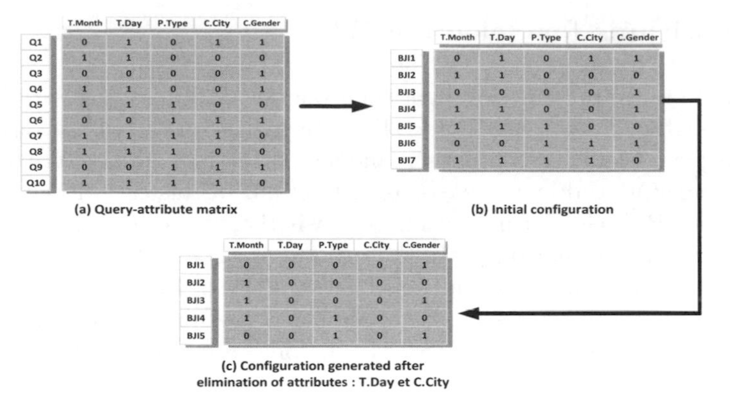

Fig. 1. Example of initial generated configuration and attribute elimination

$\{Q1, Q2, ..., Q10\}$. The indexable attributes used by each query are represented in the query-attribute matrix shown in Figure 1(a). In a query-attribute matrix the presence of an indexable attribute in a query is indicated by a 1 and absence by a 0. Applying the first step on this workload generates 10 IJB, each corresponding to a query (defined on all indexable attributes of that query).

Construction of an initial configuration. This configuration is computed as the union of all selected indexes. Note that the number of indexes of this configuration may be less than or equal to the number of queries of workload, because some queries share the same index and some do not have indexable attributes.

Example 2. The application of this step on the previous example generates an initial configuration consisting of 7 BJI, because queries (Q5, Q8), (Q6, Q9) and (Q7, Q10) respectively share the same indexes. The initial configuration is shown in Figure 1(b).

Construction of a final configuration. Recall that the initial configuration relaxes the storage constraint. If the storage cost required for storing all selected indexes does not exceed the storage capacity $\mathcal{S}$, our algorithm ends. Otherwise, some BJI should be reduced until the satisfaction of $\mathcal{S}$. To do so, we propose four elimination strategies:

1. *Elimination of attributes with high cardinality (HCA)*: The main cause of the explosion in the size of BJI is the cardinality of the indexed attributes. In this strategy, an attribute of high cardinality is eliminated from all indexes of the initial configuration. For this, attributes are sorted in descending order of their cardinality; they will be eliminated in that order until the size of the configuration meets the constraint of storage.
2. *Elimination of attributes belonging to small dimension tables (NLT)*: the principle of this strategy is keep BJI defined on largest dimension tables, since joins are costly, especially, when the size of involved tables are important.

3. *Elimination of the less used attributes (LUA)*: this strategy assumes that the most frequently used attributes should be indexed to satisfy most queries.
4. *Cost-based elimination (CBE)*: the main disadvantage of the above strategies is that they eliminate attributes without quantifying this elimination in terms of query processing reduction. To overcome this drawback, we propose CBE strategy that uses our cost model. An elimination of an attribute is *feasible* if it reduces significantly the query processing cost.

Example 3. Consider the 10 queries for which the *query-attribute matrix* is shown in Figure 1 (a) and the initial configuration is found in Figure 1 (b). Suppose that the cardinalities of attributes *Month*, *Day*, *Type*, *City* and *Gender* are respectively 12, 31, 5, 50 and 2. HCA strategy eliminates the following attributes in this order: *City*, *Day*, *Month*, *Type* and *Genre*. For instance, after the elimination of attributes *City* and *Day*, we obtain a configuration shown in Figure 1 (c). We see that these two selection attributes are not indexed (their column values are set to 0) and the number of BJI is decremented from 7 to 5, which reduces the storage cost of the resulting configuration.

4 Performance Study

Our BJI selection algorithms are implemented using Visual C++. All experiments were conducted on a Core 2 Duo machine with 2 GB of memory. We have developed a modular architecture to perform our experiments to facilitate their integration in our tool *SimulPhd* [3]. This architecture has five modules: (1) *meta-base querying module*, (2) *query management module*, (3) *BJI selection module*, (4) *indexation module* and (5) *query rewriting module*. The meta-base querying module contains information related to logical (list of tables, their attributes, length of each attribute, domain value of each attribute, etc.) and physical (attribute and table statistics, size of page of the disk, etc.) aspects of the DW. The queries management module allows a manual edition or an external importation of a workload. BJI selection module has as input a DW schema, a workload (Q) and a storage space (S) fixed by the administrator and it returns a configuration of BJI (CBJI) that minimizes the execution time of the workload and respecting the space constraint. Indexation module creates physically the selected BJI by using scripts directly on the DW. Finally, the query rewriting module forces query optimizer to use the selected indexes using Hint.

We use the star schema of the APB-1 benchmark[2]. It consists of one fact table (Actvars) and four dimension tables (ProdLevel, TimeLevel, CustLevel, and ChanLevel). We consider 12 candidates indexable attributes (C*lassLevel, GroupLevel, FamilyLevel, LineLevel, DivisionLevel, YearLevel, MonthLevel, QuarterLevel, RetailerLevel, CityLevel, GenderLevel* and *ALLLevel*) whose cardinalities are respectively: 605, 300, 75, 15, 4, 2, 12, 4, 99, 4, 2, 5. The used workload has 60 star join queries. We conduct our experiments as follows: (1) an evaluation using theoretical cost-model and (2) the obtained results are validated on Oracle 10g using the data set of our benchmark.

[2] http://www.olapcouncil.org/research/bmarkly.htm)

4.1 Theoretical Evaluation

We have performed several experiments using a theoretical cost-model that estimates the number of inputs outputs (in terms of pages) required to execute the 60 queries. We implemented three selection algorithms: (1) single attribute BJI selection algorithm (MI), (2) multi-attributes BJI selection algorithm with four pruning strategies HCA, NLT, LUA and CBE and (3) data mining algorithm (DM) presented in [2]. To enable DM to select a better configuration, you must set the value of *minsup* that represents the minimum support of frequent itemsets. For this, we performed experiments for different values of *minsup*. For each value we execute DM algorithm and the size of obtained BJI are estimated. In the same time, the cost of executing queries is computed in presence of BJI selected for each value of *minsup*. Figures 2 and 3 show the results. It is clear that when minsup is small, many indexes are created and thus occupy more space. When *minsup* is high, few indexes are generated, and thus less space is occupied. The difficulty to use DM algorithm is that DWA should identify the value of *minsup* that gives a good compromise between performance and storage space.

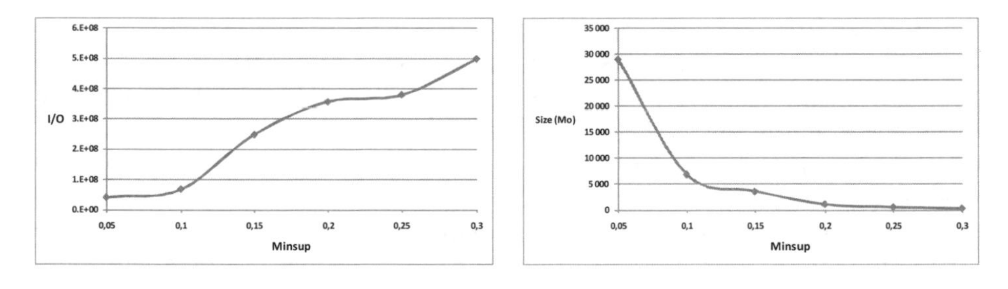

Fig. 2. Minsup vs Performance

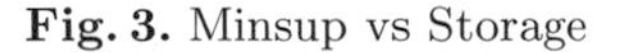
Fig. 3. Minsup vs Storage

Note that multi-attribute BJI selection algorithm generates an initial configuration that needs a storage space of 26.4 GB. This motivates the development of our pruning strategies to reduce the storage cost.

In the first experiment, we vary storage space (from 0 to 4 GB) and we estimate the query processing cost for each strategy in order to measure the impact of space on query performance. For DM algorithm, minimum support is set to 0.25. Figure 4 summarizes obtained results. The best performances are obtained by the MI and CBE algorithms. This is due to the fact that they are based on a cost-model which considers several parameters: selectivity factor of selection predicate, attribute cardinalities, size of dimensions tables, etc. Other strategies use only a unique parameter to perform the pruning process. For example, the LUA algorithm considers only the frequency of attributes which is not sufficient to obtain a good performance. Another interesting result is that when the storage space is reduced, DM gives best result. This is because the other algorithms are penalized by the storage space, therefore, only few BJIs

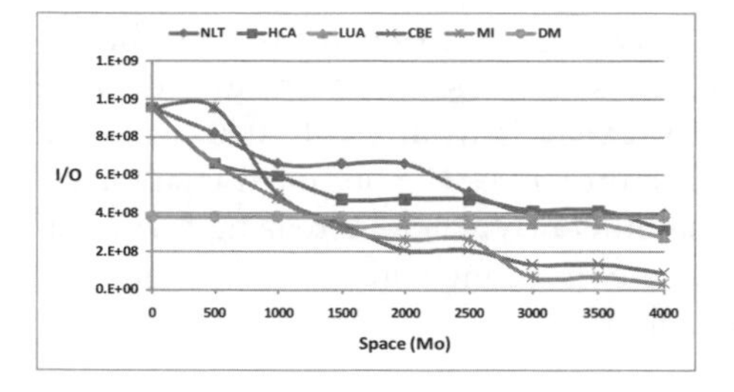

Fig. 4. Query Performance vs. Storage Space

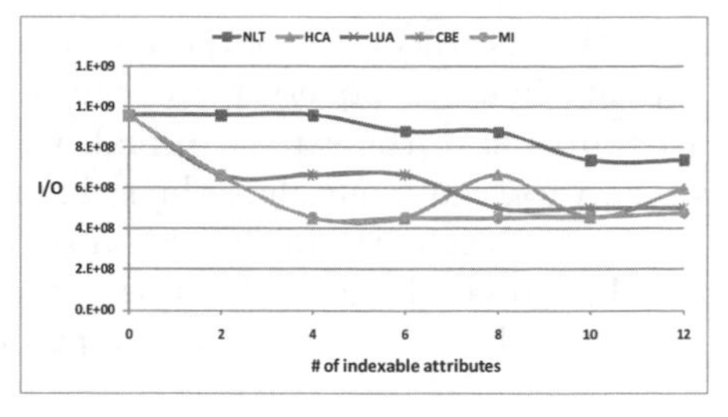

Fig. 5. Performance vs Number of attributes

are generated. Based on these results, we can conclude that the selection of optimized BJI shall be done using a cost model incorporating logical and physical parameters of the DW.

In a second experiment, we compare the performance of the proposed strategies by varying the number of selection attributes (from 1 to 12). The results of this experiment are presented in Figure 5. We realize that the query performance increases proportionally to the number of these attributes. The better performance is obtained when all query attributes are indexed. In this case, several joins between the fact and dimension tables are saved. In our experiments, when the number of candidate attributes is greater than 8, no significative improvement is observed. This is due to the storage constraint.

We have study the effect of varying the number of dimension tables in the BJI selection process. We run the different algorithms with 1, 2, 3 and 4 dimensions tables (Figure 6). We notice that the performance increases proportionally to the increase of the number of dimension tables. Indeed, join between tables are precalculated using the created BJI. However, we note that the NLT strategy does not follow this rule. When only one table is used it gives better result. This is because when the number of tables increases the NLT strategy removes smallest tables which are sometimes benefit for indexing. The overall performance remains stable from a certain number of dimension tables, this is because the storage cost is consumed by all selected indexes.

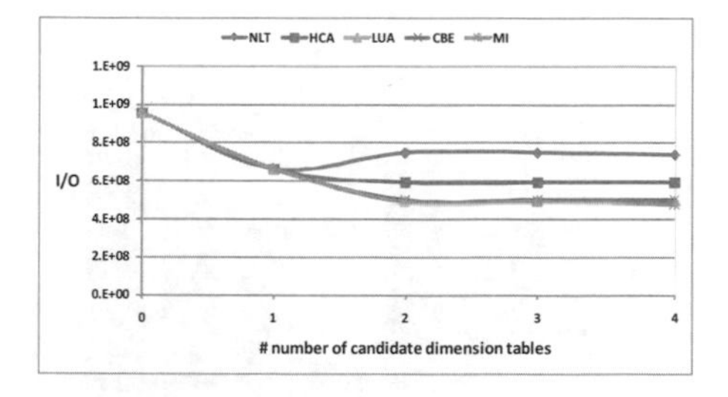

Fig. 6. Performance vs. Number of dimension tables

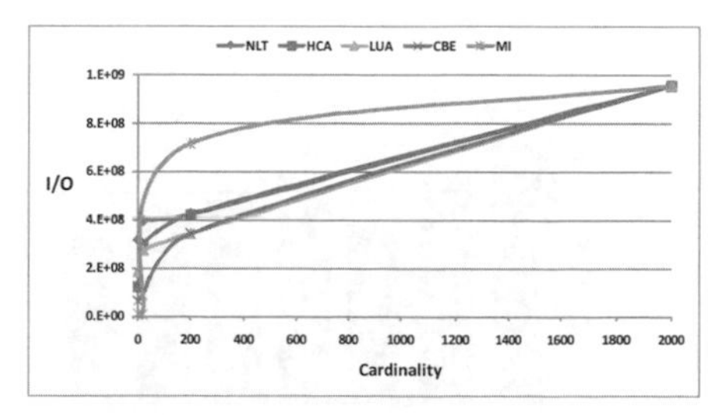

Fig. 7. Performance vs. Cardinality

In the last theoretical experiment, we study the effect of cardinality of indexable attributes on query performance. To do so, we assume that all indexable attributes have the same cardinality, and we vary it from 2 till 2000. For each value, we execute our algorithms. Figure 7 summarizes the obtained results. We notice that the performance deteriorates significantly when increasing cardinalities. Indexes on high cardinalities are storage-consuming and to perform queries, large BJI should be loaded in the main memory.

4.2 Validation on Oracle 10g

To validate our approach, we create a DW schema of APB1 benchmark and we populate its tables using generator program provided by that benchmark. We consider 60 star join queries using 12 indexable attributes. These queries have different shapes: COUNT(*) queries with and without aggregation, queries using aggregation function as Sum, Min, Max, etc. and queries having dimension attributes in the SELECT clause. We run our selection algorithms with a storage space of 3 GB using our mathematical cost model. For the DM algorithm, the minimum support is fixed to 0.25. The generated BJI by each algorithm are used to execute our queries using hint. To make sure that query optimizer considers the selected BJI, we use Explain Plan Tool provided by Oracle that shows in details the execution plan of each query.

Figures 8 and 9 show respectively the execution time of the workload and the cost reduction using the BJI configuration generated by each strategy. The obtained results confirm the utility of BJI for the COUNT(*) queries. On the other hand, the queries that get less benefit from the created BJI are those using dimensions attributes in the SELECT clause and those having any selection attribute in the selected BJI. Consequently, these queries need additional joins between the dimension tables and the fact table. The best gain in response time is obtained when multiple attributes BJI covering all indexable attributes of the queries are used. MI and DM approaches give better results. This is because they generate BJI defined on single attribute covering respectively 3 and 4 dimension tables. For the multi-attribute BJI, the best result is obtained when CBE strategy is used, since it uses a cost-model which incorporates several parameters. This result shows the quality of our mathematical cost model. Another

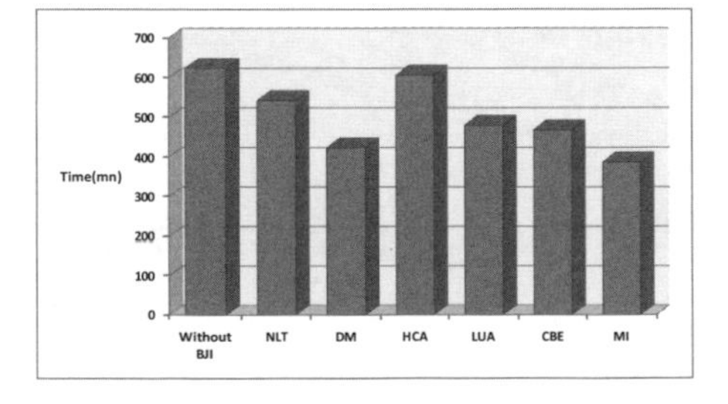

Fig. 8. Workload execution time

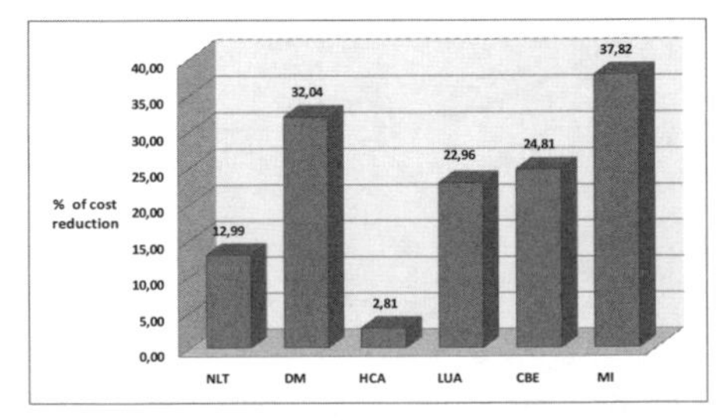

Fig. 9. Percentage of cost reduction

interesting result concerns NLT and HCA strategies that are outperformed by other strategies. Note that HCA strategy eliminates attributes with high cardinality, but sometimes creating a BJI on attribute with low cardinality can benefit for some queries but not all. For instance, DM algorithm proposes to create two BJI on attributes *RetailerLevel* and *MonthLevel* having important cardinalities compare to other attributes (99 and 12 respectively) to achieve good results since they are used by several queries of the workload. All results obtained on ORACLE DBMS confirm the theoretical ones.

5 Conclusion

Indexes are one of the pioneer optimization techniques. They represent an important part of any database system design as they can significantly impact workload performance by enabling quicker and more efficient access to data. The importance of indexes was amplified as query optimizers became sophisticated to cope with complex OLAP queries. Several indexing techniques were proposed. Selecting right indexes is a crucial issue for query optimization. In DW context, studies on selection indexes were mainly concentrated on single table indexes. In this paper, we have motivated the need to develop indexing algorithms for selecting bitmap join indexes to optimize star join queries. We propose two main types of algorithms: (i) one for selecting indexes defined on only one attribute of a dimension table and (ii) another for selecting indexes defined on several attributes of same or different dimension tables. We conducted several experiments using theoretical cost model and we propose a comparison between existing and our proposed algorithms. The obtained indexing schemes generated by our algorithms are validated on Oracle using the data set of APB1 benchmark.

References

1. Agrawal, S., Chaudhuri, S., Kollar, L., Marathe, A., Narasayya, V., Syamala, M.: Database tuning advisor for microsoft sql server 2005. In: Proceedings of the International Conference on Very Large Databases, pp. 1110–1121 (2004)
2. Aouiche, K., Boussaid, O., Bentayeb, F.: Automatic Selection of Bitmap Join Indexes in Data Warehouses, pp. 64–73 (August 2005)
3. Bellatreche, L., Boukhalfa, K., Alimazighi, Z.: Simulph.d.: A physical design simulator tool. In: Bhowmick, S.S., Küng, J., Wagner, R. (eds.) DEXA 2009. LNCS, vol. 5690, pp. 263–270. Springer, Heidelberg (2009)
4. Bellatreche, L., Missaoui, R., Necir, H., Drias, H.: A data mining approach for selecting bitmap join indices. Journal of Computing Science and Engineering 2(1), 206–223 (2008)
5. Boukhalfa, K.: De la conception physique aux outils d'administration et de tuning des entrepts de donnes. Phd. thesis, Poitiers University, France (2009)
6. Chan, C.Y., Ioannidis, Y.E.: Bitmap index design and evaluation. In: Proceedings of the ACM SIGMOD International Conference on Management of Data, pp. 355–366 (June 1998)

7. Chaudhuri, S.: Index selection for databases: A hardness study and a principled heuristic solution. IEEE Transactions on Knowledge and Data Engineering 16(11), 1313–1323 (2004)
8. Chaudhuri, S., Narasayya, V.: An efficient cost-driven index selection tool for microsoft sql server. In: Proceedings of the International Conference on Very Large Databases, pp. 146–155 (August 1997)
9. Chaudhuri, S., Narasayya, V.: Self-tuning database systems: A decade of progress. In: Proceedings of the International Conference on Very Large Databases, pp. 3–14 (September 2007)
10. Choenni, S., Blanken, H.M., Chang, T.: On the selection of secondary indices in relational databases. Data Knowledge Engineering 11(3), 207–238 (1993)
11. Datta, A., Ramamritham, K., Thomas, H.: Curio: A novel solution for efficient storage and indexing in data warehouses. In: Proceedings of the International Conference on Very Large Databases, pp. 730–733 (September 1999)
12. Gupta, H., Harinarayan, V., Rajaraman, A., Ullman, J.: Index selection for olap. In: Proceedings of the International Conference on Data Engineering (ICDE), pp. 208–219 (April 1997)
13. O'Neil, P., Quass, D.: Improved query performance with variant indexes. In: Proceedings of the ACM SIGMOD International Conference on Management of Data, pp. 38–49 (May 1997)
14. Papadomanolakis, S., Ailamaki, A.: An integer linear programming approach to database design. In: ICDE Workshops, pp. 442–449 (2007)
15. Pasquier, N., Bastide, Y., Taouil, R., Lakhal, L.: Discovering frequent closed itemsets. In: Beeri, C., Bruneman, P. (eds.) ICDT 1999. LNCS, vol. 1540, pp. 398–416. Springer, Heidelberg (1998)
16. Sanjay, A., Surajit, C., Narasayya, V.R.: Automated selection of materialized views and indexes in microsoft sql server. In: Proceedings of the International Conference on Very Large Databases, pp. 496–505 (September 2000)
17. Stöhr, T., Märtens, H., Rahm, E.: Multi-dimensional database allocation for parallel data warehouses. In: Proceedings of the International Conference on Very Large Databases, pp. 273–284 (2000)
18. Valduriez, P.: Join indices. ACM Transactions on Database Systems 12(2), 218–246 (1987)
19. Wu, K., Shoshani, A., Stockinger, K.: Analyses of multi-level and multi-component compressed bitmap indexes. ACM Transactions on Database Systems 35(1) (2010)
20. Zilio, D.C., Rao, J., Lightstone, S., Lohman, G.M., Storm, A., Garcia-Arellano, C., Fadden, S.: Db2 design advisor: Integrated automatic physical database design. In: Proceedings of the International Conference on Very Large Databases, pp. 1087–1097 (August 2004)

Speeding Up Queries in Column Stores

A Case for Compression

Christian Lemke[1,2], Kai-Uwe Sattler[2], Franz Faerber[1], and Alexander Zeier[3]

[1] SAP AG, Walldorf, Germany
c.lemke@sap.com, franz.faerber@sap.com
[2] Ilmenau Univ. of Technology, Ilmenau, Germany
kus@tu-ilmenau.de
[3] Hasso-Plattner-Institute, Potsdam, Germany
alexander.zeier@hpi.uni-potsdam.de

Abstract. BI accelerator solutions like the SAP NetWeaver database engine TREX achieve high performance when processing complex analytic queries in large data warehouses. They do so with a combination of column-oriented data organization, memory-based processing, and a scalable multiserver architecture. The use of data compression techniques further reduces both memory consumption and processing time. In this paper we study query operators like scan and aggregation on compressed data structures implemented in TREX.

1 Introduction

Recent years have seen growing demands on data warehousing and OLAP technologies being able to handle terabytes of data and complex analytic queries from several hundred users simultaneously. Furthermore, more and more customers need ad-hoc and realtime evaluation of queries that make materialization of results and precalculation of reports more or less useless.

In order to meet these requirements, the limiting factor of disk I/O in database systems has to be addressed. Basically, in the area of data warehousing there are currently three main approaches: *(1)* Reduce the amount of data to be read from disk and read it as fast as possible. Besides index structures, column-oriented data organization shows great potential. *(2)* Avoid disk access completely by keeping and processing data in memory. *(3)* Exploit the computing capacity of a large number of inexpensive servers by using parallel processing.

In the field of Business Intelligence (BI) technologies some or all of these approaches are currently combined in the concept of BI accelerator solutions or analytical engines. An example is the SAP NetWeaver BWA based on TREX. TREX runs in a scalable multiserver architecture on blade servers. Processing is performed completely in main memory, fact and dimension tables are organized column-wise (vertically partitioned) and the columns are partitioned horizontally among the server nodes. The combination of parallelism and main memory processing allows interactive execution of analytical queries without pre-aggregation.

T.B. Pedersen, M.K. Mohania, and A M. Tjoa (Eds.): DaWaK 2010, LNCS 6263, pp. 117–129, 2010.

However, since RAM is still expensive compared to hard disk drives and cannot be enlarged indefinitely, very large data warehouse installations require a large number of server nodes. Furthermore, scanning the entire memory of a node does not allow to exploit the benefit of L2 caches. Therefore, data compression techniques are used to reduce the data volume and hence improve the cache utilization. But, compressing data only is not the silver bullet: If data processing as part of query evaluation requires an expensive decompression or additional memory space, the benefit of compression is mitigated or even lost.

Thus, query operators should process compressed data directly without computing-intensive decompression. In our work, we investigate such strategies for implementing query operators. Based on the TREX infrastructure and the discussion of several compression schemes for in-memory column stores we present strategies for filters, scans and aggregations which exploit these compressed data structures. The results from our experimental evaluation show that depending on the data distribution significant performance improvements can be achieved.

2 Related Work

The first work that suggests working directly on compressed data as long as possible was [4]. Its focus is on joins but also exact match comparisons for selections and duplicate elimination (grouping) is described. They use domain coding as lightweight compression in a row store implementation and specify the needed properties of compression techniques for an efficient query processing. The decompression of values cannot be avoided for most aggregation functions and when the data has to be displayed to the user. To reduce the repeated decompression of fields [10] introduce an extended iterator model.

Raman and Swart showed in [7] that fast processing is also possible on heavyly compressed data in a row store. They present a novel Huffman coding scheme to evaluate equality and range predicates on compressed data without full dictionary access. Before working on compressed data can take place, the compressed records and fields have to be extracted. Afterwards index scan, hash join, merge join, grouping and count (distinct) aggregation can work directly on the Huffman codes. For the min and max computation the codes of each code length need to be decompressed for comparison. This is because of the coding scheme where only the codes with the same length are ordered.

By contrast in [2,13] the data is always decompressed for query execution.

One of the first papers about working on compressed data in column stores was [3]. Here the scan and join operators in queries are executed in a main memory database on domain coded data. An additional speedup is gained by using multidimensional hash tables as indexes.

Based on the column-oriented C-Store [9] Abadi et al. [1] introduced an architecture for a query executor that works directly on compressed data. In contrast to this work, we use data structures that represent the data of a whole column and not only parts of it.

There is also research on avoiding decompression while querying for other database architectures. In [5] linearized multidimensional arrays are used to efficiently aggregate data whereas [6] concentrate on joins in a binary relational database where triples are stored.

Another approach to speed up the query execution on compressed data structures is the use of modern hardware. [12] as an early paper uses SIMD instructions in a column store to exploit parallelism and eliminate branch mispredictions. Zhou and Ross consider sequential scan, aggregation, index operations and joins. For a single compressed tuple instead of several values of one field [8] evaluates a conjunction of equality and range predicates using SIMD.

3 Data Structures for Column Compression

In the following we briefly describe some of the data structures used in TREX for storing compressed column values. These data structures are pure main memory structures and the main goals are *(1)* an effective compression scheme to reduce the memory consumption and *(2)* allow to process queries without decompressing the data. Particularly, the latter goal requires efficient access both to individual values and also to blocks of values.

The basis for all the following techniques is *domain coding* [10,1,13]. For domain coding, all values from a column are stored in lexicographical order in a dictionary. The original column is replaced by an index vector that stores only bit-compressed pointers to the dictionary. To minimize the bit lengths, a total of u distinct values appearing in n rows are coded using $n\lceil\log_2 u\rceil$ bits. Figure 1(a) shows an example of domain coding for the sample data *Aachen, Aachen, Aachen, Karlsruhe, Aachen, Aachen, Leipzig, Münster*, where each value is represented in the index vector by two bits. Note, that the column `pos` is given in the figures only for illustrative purposes and not physically stored. The use of integers instead of the original values has the advantages that it reduces the data volume to be processed and allows to exploit hardware optimization for integer processing. Furthermore, multiple values can be read at once into the CPU cache and processed in parallel with special SIMD processor commands.

Based on this scheme *prefix coding* can be applied as a simple compression technique. Here, repetitions of the same value at the start (prefix p) of a column are deleted and replaced by one value and its frequency. For a column with n elements and u_{col} distinct values, $(n-p)\lceil\log_2 u_{col}\rceil + 64$ bits are required. Fig. 1(b) shows an example of the original uncompressed index vector and its compressed version that is constructed using prefix coding.

If the most frequent value appears not only in the prefix but also scattered among the other values, then *sparse coding* can be applied to achieve a good compression (Fig. 1(c)). Here the positions of all appearances f of the most frequent value are recorded in an additional bit vector and the original values or deleted from the index vector. It is possible to use prefix coding for the bit vector, which for a large prefix p can further improve the compression ratio. With this technique only $(n-f)\lceil\log_2 u_{col}\rceil + (n-p) + 64$ bits are needed.

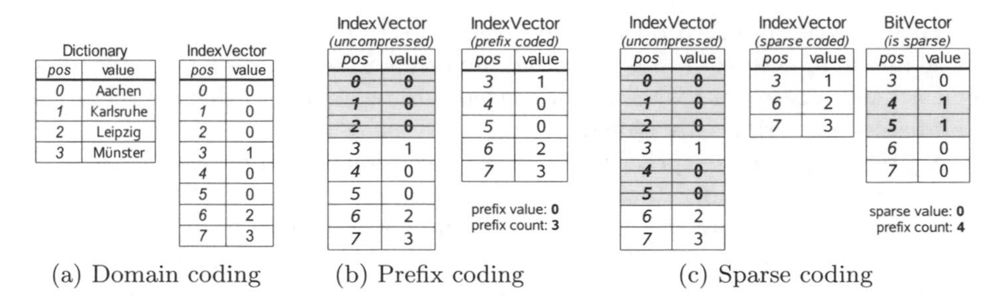

(a) Domain coding (b) Prefix coding (c) Sparse coding

Fig. 1. Examples of domain, prefix and sparse coding

All the techniques described in the following will use prefix coding and work on blocks of data containing minimal numbers of distinct values in order to achieve a good compression ratio. In *cluster coding* only blocks with a single distinct value are compressed by storing only the single value. Additionally, a bit vector is needed to indicate which blocks are compressed, in order to be able to reconstruct the original column. In this paper, we do not further consider how to determine the optimal block size and its impact on the compression rate and query runtime. In any case the number of elements should be an integral power of two, so that instead of multiplication and modulo operation we can exploit fast bit operations. Figure 2(a) shows an example in which the block size is two and the compressed values are shown in gray boxes.

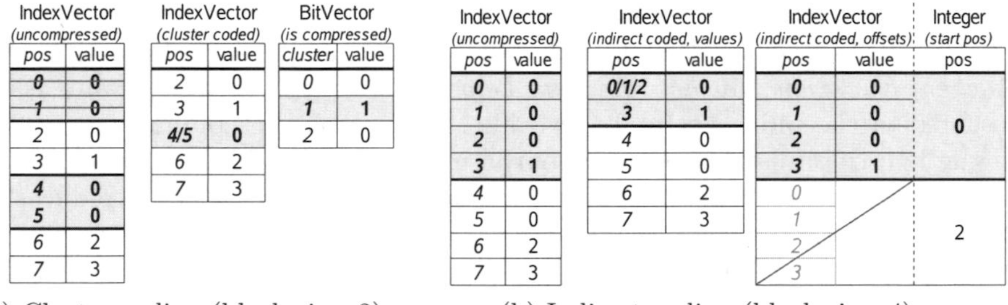

(a) Cluster coding (block size: 2) (b) Indirect coding (block size: 4)

Fig. 2. Examples of cluster and indirect coding

If the data blocks contain more than one but only few distinct values, *indirect coding* can be used. Here domain coding is applied to appropriate blocks, which adds the indirection of a separate mini-dictionary for each block. To reduce the number of dictionaries and hence the memory consumption, one dictionary can be used for a continuous sequence of blocks as long as any new entries in the dictionary do not increase the number of bits required to code the entries. For a column with u_{col} distinct values, a block with k values and u_{block} distinct values benefits from indirect coding if and only if the dictionary and the references take less space than the original (domain coded) data: $u_{block}\lceil\log_2 u_{col}\rceil + k\lceil\log_2 u_{block}\rceil < k\lceil\log_2 u_{col}\rceil$.

Figure 2(b) illustrates the data structures used in the implementation. Here one block contains four values and the compressed elements are shown again in gray boxes. The dictionaries and the uncompressed data are stored in the middle index vector, *values*, and each block is addressed with a start position. In the data structure on the right, compressed blocks have their own index vector, *offsets*, which points to the individual values.

Finally, a slightly modified variant of *run length coding* can be used, which compresses sequences of repeated values to a single value for each run together with the number of repetitions. In order to calculate the start position of each value in the run, we add up the frequencies of the previous values. However, this can result in a high overhead, so instead we decided to reduce the compression slightly by storing the start position and not the number of repeats.

To speed up single-value accesses we introduce two inverted index structures: a blocked and a signature index. The *blocked index* stores for each value a list of blocks in which this value occurs and is therefore only applicable if using cluster, indirect or run length coding. For the *signature index* the data is split into a fixed number of parts and for each value the occurrences in that parts are stored in a bit vector which significantly reduces the amount of storage.

4 Query Operators for Compressed Columns

In this section we will describe in detail how the compressed data structures are used in the standard query operators. We designed the compression techniques for an efficient direct access to the data and gaining significant performance improvements over uncompressed data structures especially for the scan operator. In the evaluation we will show that the choice of the best compression technique depends on the data and can be done automatically.

4.1 Basic Operators

The basic operator in a data warehouse environment is the *scan* operator with optional filter predicates. In scenarios with mass data and non selective predicates, the performance of the scan operator is very critical. The *get* operator provides random access to a column by retrieving the value (or a reference to the dictionary) of a given row id. This operation is needed for projections and selective filters. The performance of the operators depends on the result materialization data structures (like bit vector or integer vector), the used compression technique and many more factors. In the following we will present the scan and get operators for the most complex compression techniques: sparse and indirect coding.

Sparse Coding. As mentioned above, the sparse coding maintains a bit vector B_{nf} for the most frequent value v_{f} of a column. A bit is set if the value of the corresponding row is not v_{f}, otherwise the bit is unset. For all compression techniques a prefix offset p specifies the first row id, which is different from the prefix value v_{p} that is omitted. The values that are different from v_{f} are stored in an index vector I_{nf}.

Scan Operator. The scan starts by testing the predicate for v_p when there is a prefix p. The same is done for v_f whose corresponding row ids can easily be extracted from B_nf. If necessary, I_nf is also scanned and the corresponding row ids are calculated by counting the unset bits in B_nf:

1. estimate the current row id by assuming that no values are removed (row id = relative row id)
2. determine the number n_nf of non frequent values (unset bits)
3. repeat the following steps until n_nf is higher than the relative row id
   (a) increase the current row id
   (b) if the current value is not v_f then increase n_nf
4. calculate the absolute row id with current row id + p

In order to accelerate the calculation of unset bits a data structure is introduced, which stores the number of set bits for every block of rows (i.e. all 128 rows) as a sum of all previous blocks.

Get Operator. The get operator first checks if the requested value is in the prefix. If the row id is higher than p then B_nf is tested. If the corresponding bit is unset then the result is v_f, otherwise the position of the requested value in I_nf is calculated and the requested value is extracted from there.

The sparse coding shows very good performance characteristics if the sparse value is dominant in a way that it covers more than 90% of the rows. Otherwise the costs of the indirect access over the bit vector is too high as shown in the evaluation.

Indirect Coding. By neglecting the prefix, the indirect coding is a mixture of variable and fixed size coding. Per block of rows the references to a local block dictionary are of fixed size, but each block (e.g. 1024 rows) has its own dictionary which references to the global dictionary of the column. For blocks where local dictionaries are not suitable (e.g. with as many distinct values as rows) no local dictionary is used, but the references to the global dictionary are stored without further indirection. Because the references in the local dictionaries and the references in blocks without local dictionaries are of fixed size, they are stored in an index vector I_v. A second vector V_b stores information for each block, like the start position in I_v and the references to the local dictionary (if it exists).

Scan Operator. The scan operator scans I_v by testing the predicate for every value and storing the hit positions in a temporary structure. For each entry in this structure the following is done:

1. determine the corresponding block in V_b by checking the start position
2. if the block is uncompressed (no local dictionary) then calculate the absolute row id with (block number * block size) + match pos - start pos
3. if the block is compressed (local dictionary)
   (a) calculate the local dictionary reference with match pos - start pos
   (b) scan the local references for that reference
   (c) calculate the absolute row id from the local reference hit positions

Get Operator. Retrieving the value of a given row id starts by calculating the block number in V_b. If the block is uncompressed, the position in I_v is row id - (block number * block size) + start position. If the block is compressed, the position in the local references is row id - (block number * block size). The final position in I_v is the local reference plus the start position.

Hardware Optimization. Another approach to speed up data processing is the implementation of the scan and decompression operations using SIMD instructions. The SIMD implementation used in our work is extensively evaluated in [11] and we will only investigate the influence on our compressed data structures.

4.2 Aggregate Operators

Though, in this work we focus on the efficient implementation of scan operators because they are most critical regarding performance, we discuss in the following aggregation operators, too.

We start with a description of single column aggregations and grouping. These aggregation operators are implemented as part of the scans such that filter predicates can be calculated without additional efforts. Furthermore, some aggregate functions such as `min` and `max` can be evaluated using the dictionary only.

Because columns are partitioned horizontally among all servers, aggregate operators are evaluated in two steps: a scan phase that is performed in parallel on all partitions followed by a merge phase. During the scan phase, partial aggregates are computed for each partition which are then merged into the final aggregates. For sparse coding the scan phase is performed in the following way. We assume a single column grouping with `count` as aggregate function. In order to collect the counts per group an array G is used that is indexed by the values of the dictionary (i.e. the positions of the actual values in the dictionary). Let I denote the index vector storing the column values in sparse coding and B the corresponding bit vector.

1. Read the first value v_0 from the dictionary and set $G[v_0] = p$
2. Determine the number n of bits set in B and let $G[v_0] = G[v_0] + n$
3. Scan the index vector for all values v_i and let $G[v_i] = G[v_i] + 1$

After the aggregate array G_i of all partitions are calculated, they are merged in a straightforward way. This step is simplified by the same ordering of all grouping arrays because on all partitions the same dictionary is used.

For other compression schemes this approach has to be slightly modified. For example, for cluster coding the bit vector is scanned in parallel to the index vector. If a bit i is set, the corresponding entry i of the index vector is skipped and $G[v_0] += \mathit{blocksize}$. Indirect coding requires an additional step to process the index vectors of the individual blocks.

This single-column aggregation scheme can be extended to the multi-column case by maintaining a single group array for all grouping columns $G[c_1, \ldots, c_m]$ and perform the scans on these columns in parallel. Then, for each tuple $v_1, \ldots, v_m$ the corresponding entry in G is updated as described above.

5 Experimental Evaluation

This section presents the results of evaluating the scan operator on different synthetic datasets without a query optimizer. We use a micro benchmark to analyze special properties of the compression techniques, which can be found also in real datasets. The data consists of one column with 10 million data items and 4472 unique integer values. One exception is the **single** dataset which contains only one unique value. In the **linear skew** distribution the value i occurs $i+1$ times contiguously and in the **uniform** distribution all values occur equally often. For the **sparse** dataset the items are consecutively numbered values and a very frequent (sparse) value is added. The position of the sparse data items can be grouped at the top, at the bottom or evenly scattered. In the **blocked** data, blocks with one or more unique values are generated. We have 2236 single-value blocks and 2236 multi-value blocks with 447 unique values where the block size is for both cases 2237 values.

The results of our experiments are shown relative to the domain coded data (index vector) to exclude effects of different data types. Furthermore the cluster and indirect coding use a block size of 1024 values in our implementation. If not stated otherwise the used scan operator gets a value range as input and writes the results in a vector. Furthermore we only evaluated the blocked index of the presented inverted index structures.

All experiments were performed on an Intel® Xeon® processor X5650 with 2.67 GHz. The scalability of the index vector if using more cores has already been shown in [11]. So because the memory bandwidth required by the compressed data is lower, we concentrated on single-core measurements.

5.1 Experiments without SSE (Streaming SIMD Extensions)

First we want to show that with an optimal compression, not only the memory size is reduced but also queries can be accelerated. The dictionary coded **single** dataset needs with 1 bit per value around 1221 KiB. Because the presented compression techniques omit the prefix they need less than a kibibyte to store the values. Even in sparse coding the bit vector will not be stored.

Figure 3 overviews the memory consumption of the distributions we focused on in our evaluation. The **linear skew** distribution in Fig. 3(b) shows that if the most frequent value is below a certain threshold the additional costs of the bit vector in the sparse coding leads to an increased memory size. Because a value occurs in less 1024-blocks, the additional memory used for the inverted indexes is low. The run length coding can adapt best to the different-sized single-value blocks which leads to significant reduction. Using the **uniform** distribution the cluster coding performs better because of the bigger single-value blocks. For the other techniques there are only small differences in the memory size.

If the most frequent value in the **sparse** dataset is evenly scattered (Fig. 3(d)) the sparse coding reduces memory size most and only cluster and indirect coding without inverted indexes can exploit the block of the most frequent value at the end. This single-value block arises from the scattered generation and a sparse

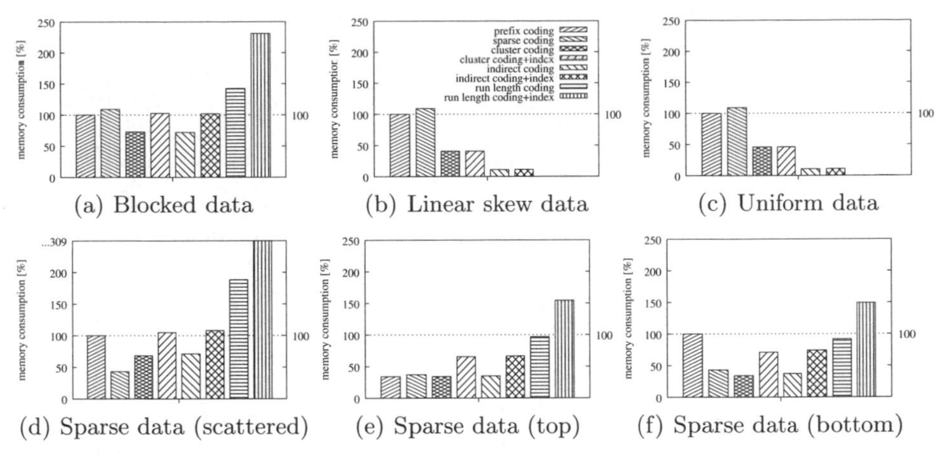

(a) Blocked data (b) Linear skew data (c) Uniform data

(d) Sparse data (scattered) (e) Sparse data (top) (f) Sparse data (bottom)

Fig. 3. Memory consumption

occurrence of more than 50% (in our case 66%). Because of the many value changes the run length coding performs worst. For better readability we omit a part of the bar graph and specify the maximum value on the y-axis.

If the most frequent value is at the top (Fig. 3(e)) all compression techniques apply prefix coding which leads to a significant memory reduction. Sparse coding needs slightly more memory because of the additional bit vector. Because of many value changes and less big single-value blocks in the **blocked** dataset (Fig. 3(a)) only cluster and indirect coding can reduce the memory consumption.

To show the possible savings in execution time we do a scan for value 0 and value 1 in Fig. 4. Value 0 is the most frequent value in the sparse dataset. In Fig. 4(a) with the blocked dataset the speed of the query correlates to the size of the data structures except when using inverted indexes. Here the total size of the index structure is big but only the blocks specified in the inverted index need to be scanned which results in a very fast execution. For the linear skew dataset Fig. 4(b) shows that of all blocked compression techniques only cluster coding cannot adapt well to the small single-value blocks and that is why it performs worse. If the most frequent value is scattered in the sparse dataset (Fig. 4(c)) it is the worst case distribution. Here just the sparse coding can compete with a query on the only dictionary coded data because of the reduced amount of data to scan. In Fig. 4(d) one can see clearly that less memory consumption does not always imply a faster execution. Also the row id reconstruction overhead has to be taken into account like in the case of applying sparse coding.

If the result size is small like in the scans showed in Fig. 4(f) and 4(e) the row id reconstruction and result vector resize costs are negligible compared to the savings. Because for value 1 there are no single-value blocks, the very frequent single-value access to two index vectors is too expensive. That is also the case if the amount of data to scan is less (Fig. 4(f)).

All measurements show that when using inverted indexes scan times for single values always decrease. Run length coding is only worth if there are few but big single-value blocks and the fasted technique if using an inverted index. Cluster

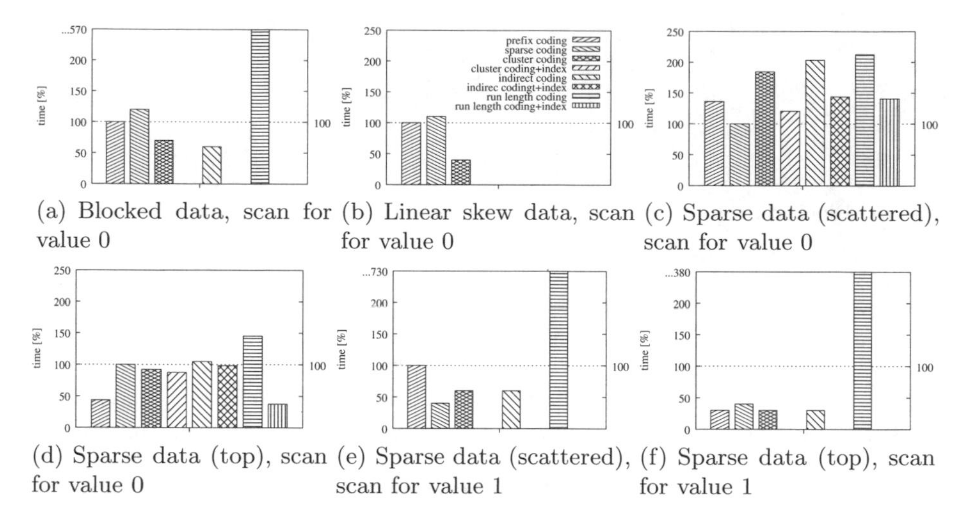

(a) Blocked data, scan for value 0 (b) Linear skew data, scan for value 0 (c) Sparse data (scattered), scan for value 0

(d) Sparse data (top), scan for value 0 (e) Sparse data (scattered), scan for value 1 (f) Sparse data (top), scan for value 1

Fig. 4. Single-value scans

coding on the other hand has the best memory-speed tradeoff for all evaluated data distributions.

We further expect that using inverted indexes is only faster than the baseline if the requested values are located in few blocks. Figure 5(a) shows the times of the experiments where we varied in how many blocks value 0 occurs by using a step size of 100 blocks. The data distribution is based on the sparse dataset and value 0 is occurring 9766 times which corresponds to the number of blocks with a size of 1024 values. Because of the many value changes and the small amount of occurrences of the value the run length coding is constantly around ten times slower than the baseline and for this reason we cut the graph for better readability. If using the inverted indexes for the cluster and indirect coding the scans become slower the more blocks have to be considered.

To determine if it is worth to spend more memory for the inverted index we calculate a cost-benefit ratio by dividing the time gained by the additional memory needed. Figure 5(b) shows that for run length coding the higher memory consumption always results in an increased speed. For the two other techniques it depends strongly on the distribution of the value and at some point the additional memory used is counter-productive. To deal with this problem in our implementation we do a full table scan if an indexed value occurs in too many blocks.

Next we want to show that the more values are queried the slower the query becomes. In this experiments we use the uniform data distribution and do a scan for 1, 2, ..., all values. The results are shown for the slowest, an average and the fastest compression technique with and without using an inverted index in Fig. 6. One can see that the scans using run length coding are always faster and sparse coding always slower than the baseline. The outliers in the graph result from copying memory when resizing the result vector using a doubling strategy.

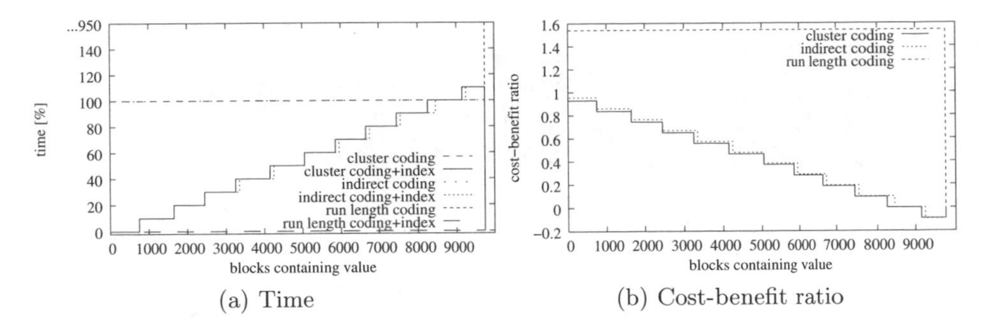

(a) Time (b) Cost-benefit ratio

Fig. 5. Increasing number of blocks containing the requested value

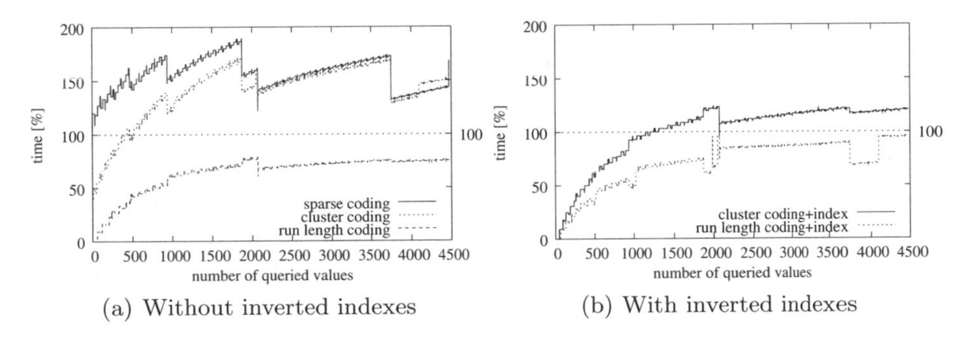

(a) Without inverted indexes (b) With inverted indexes

Fig. 6. Multi-value scans

Furthermore the dictionary coded algorithms use another initial vector size because of the missing prefix handling. The cluster coding is only worthwhile if the number of requested values is small because of the increasing row id reconstruction costs. Another reason for the increasing time needed if using a inverted index are the increased number of blocks to consider for the scan.

5.2 Experiments with SSE

Finally, we show in Fig. 7 the performance improvements possible when using SSE as SIMD implementation on the compressed data structures. The baseline is the scan times measured without active SSE. The results show that the biggest saving in time is achieved for the domain and prefix coded data because here always full table scans take place and SSE is optimized for mass data processing. If the result size is increasing as in Fig. 7(b) the performance gain is decreasing because of the reconstruction and resizing costs. There is no or just a little benefit by using SSE in combination with the inverted indexes because here the amount of data to be scanned is already significantly reduced. Also the sparse coding technique cannot exploit the advantages of SSE. Reasons are the reduced data size if the most frequent value occurs often (Fig. 7(b) and (c)) and that most of the time is spent for row id reconstruction (Fig. 7(a)). The times for cluster

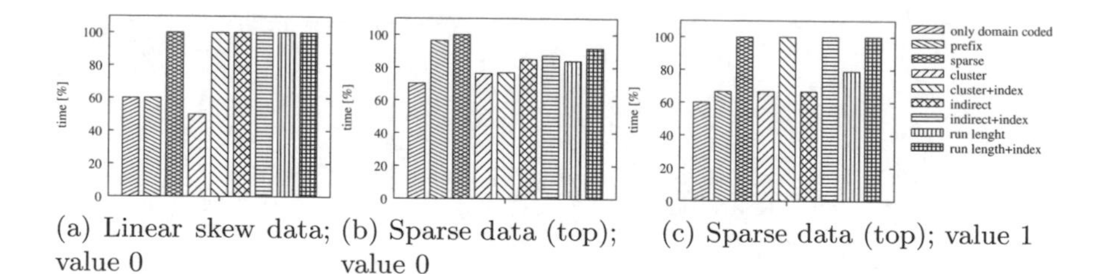

(a) Linear skew data; value 0
(b) Sparse data (top); value 0
(c) Sparse data (top); value 1

Fig. 7. Single-value scans using SSE

coding when using the linear skew (and uniform) dataset are reduced as there are few compressed clusters which leads to more data that has to be scanned.

6 Conclusion

Data compression is an important technique to reduce memory consumption and query processing time in data warehouse systems. However, the real benefit of compression can be only leveraged if the decompression effort can be minimized. To tackle this problem, we have presented in this paper several dictionary-based compression schemes as well as query operator implementations for scans and aggregates which work directly on the compressed data. Our experimental results show that depending on the data characteristics and appropriate compression techniques significant improvements in query processing time can be achieved, but require a careful choice of the compression scheme. We partly implemented an optimizer that determines an optimal row order, the appropriate compression techniques and inverted index structures depending on the data distribution.

References

1. Abadi, D.J., Madden, S.R., Ferreira, M.C.: Integrating compression and execution in column-oriented database systems. In: Proc. SIGMOD, pp. 671–682 (2006)
2. Chen, Z., Gehrke, J., Korn, F.: Query optimization in compressed database systems. In: Proc. SIGMOD, pp. 271–282 (2001)
3. Cockshott, W.P., McGregor, D., Wilson, J.: High-performance operations using a compressed database architecture. The Computer Journal 41(5), 283–296 (1998)
4. Graefe, G., Shapiro, L.D.: Data compression and database performance. In: Proc. ACM/IEEE-CS Symp. on Applied Computing, pp. 22–27 (1991)
5. Li, J., Srivastava, J.: Efficient aggregation algorithms for compressed data warehouses. IEEE TKDE 14(3), 515–529 (2002)
6. O'Connell, S.J., Winterbottom, N.: Performing joins without decompression in a compressed database system. SIGMOD Rec. 32(1), 6–11 (2003)
7. Raman, V., Swart, G.: How to wring a table dry: Entropy compression of relations and querying of compressed relations. In: Proc. 32nd VLDB, pp. 858–869 (2006)

8. Raman, V., Swart, G., Qiao, L., Reiss, F., Dialani, V., Kossmann, D., Narang, I., Sidle, R.: Constant-time query processing. In: Proc. 24th ICDE, pp. 60–69 (2008)
9. Stonebraker, M., et al.: C-store: A column-oriented dbms. In: Proc. 31st VLDB, pp. 553–564 (2005)
10. Westmann, T., Kossmann, D., Helmer, S., Moerkotte, G.: The implementation and performance of compressed databases. SIGMOD Rec. 29(3), 55–67 (2000)
11. Willhalm, T., Popovici, N., Boshmaf, Y., Plattner, H., Zeier, A., Schaffner, J.: Simd-scan: Ultra fast in-memory table scan using on-chip vector processing units. PVLDB 2(1), 385–394 (2009)
12. Zhou, J., Ross, K.A.: Implementing database operations using simd instructions. In: Proc. SIGMOD, pp. 145–156 (2002)
13. Zukowski, M., Héman, S., Nes, N., Boncz, P.: Super-scalar ram-cpu cache compression. In: Proc. 22nd ICDE, p. 59 (2006)

Mining Non-redundant Information-Theoretic Dependencies between Itemsets*

Michael Mampaey**

Dept. of Mathematics and Computer Science, University of Antwerp
michael.mampaey@ua.ac.be

Abstract. We present an information-theoretic framework for mining dependencies between itemsets in binary data. The problem of closure-based redundancy in this context is theoretically investigated, and we present both lossless and lossy pruning techniques. An efficient and scalable algorithm is proposed, which exploits the inclusion-exclusion principle for fast entropy computation. This algorithm is empirically evaluated through experiments on synthetic and real-world data.

1 Introduction

The discovery of rules from data is a popular task in data mining. Mining association rules in transactional datasets has received a lot of attention especially [2,3,4]. The objective of association rule mining is to find highly confident rules between sets of items frequently occurring together. This has been generalized to, among others, relational tables with categorical or numerical attributes [5]. In this context, much attention has gone to the discovery of (approximate) functional dependencies in relations [6,7]. A functional dependency $A \Rightarrow B$ between two sets of attributes is said to hold, if any two tuples agreeing on the attributes of A also agree on the attributes of B. Often it is desirable to also find rules that 'almost' hold. Typically, an error is associated with a functional dependency, which describes how well the relation satisfies that dependency, commonly this is the minimum relative number of tuples that need to be removed from the relation for the dependency to hold (known as g_3 [6]). These tuples can be thought of as being the exceptions to the rule. However, the fact that $A \Rightarrow B$ has a low error, does not necessarily imply that B strongly depends on A, in fact, A and B might even be independent.

Therefore, in this paper we take an information-theoretic approach to mining dependencies in binary data. We will describe the dependence of a rule based on the mutual information between consequent and antecedent. Furthermore, we use the entropy of a rule or itemset to describe its complexity. We present an algorithm called μ-Miner, which efficiently mines rules with a high dependence and a low complexity.

On top of this, we investigate what kinds of redundancy can occur in the collection of all low entropy, high dependence rules. For traditional association rules, several types of

* An extended version of this paper is available as a technical report [1].

** Michael Mampaey is supported by the Institute for the Promotion of Innovation through Science and Technology in Flanders (IWT-Vlaanderen).

T.B. Pedersen, M.K. Mohania, and A M. Tjoa (Eds.): DaWaK 2010, LNCS 6263, pp. 130–141, 2010.

redundancy have been presented [8,9]. We look at lossless closure-based redundancy in the context of this paper, as well as lossy pruning methods based on some information theoretical properties. These techniques are then integrated into our algorithm.

2 Related Work

The discovery of exact and approximate functional dependencies from relations has received a lot of attention in the literature. The TANE algorithm proposed by Huhtala et al. [7] finds exact and approximate functional dependencies in a relation, which have a low g_3 error. TANE is a breadth-first algorithm that works with tuple partitions induced by attribute sets, which can be constructed in linear time with respect to the size of the relation. If the partition induced by XY does not refine the partition induced by X, then $X \Rightarrow Y$ is a functional dependency. The error of an approximate dependency can also be computed using these partitions. The main difference with our work is the way that the strength of a dependency is measured, but also that TANE only mines minimal rules, i.e. rules of the form $X \Rightarrow Y$ for which $|Y| = 1$ and there is no $X' \subsetneq X$ such that $X' \Rightarrow Y$ is an (approximate) functional dependency. On top of this, we also consider the complexity of dependencies.

Dalkilic and Robertson [10] use conditional entropy to determine the strength of dependencies in relational data. Their work examines several of their properties and information inequalities from a theoretical viewpoint, without focussing on rule discovery.

Heikinheimo et al. mine all low entropy sets from binary data, as well as trees based on these sets [11]. A breadth-first mining algorithm is proposed that exploits the monotonicity of entropy, after which a Bayesian tree structure is imposed on the itemsets. Its nodes correspond to the items, and the directed edges express the conditional entropy between them. The paper also discusses high entropy sets, argued to be potentially interesting due to the lack of correlation among their items.

Jaroszewicz and Simovici use information theoretic measures to assess the importance of itemsets or association rules on top of the traditional support/confidence based mining framework [12]. They use Kullback-Leibler divergence to determine the redundancy of confident association rules. Given the supports of some subsets of an association rule, its most likely confidence is computed (using a maximum entropy model); if the actual confidence is close to the estimate, the rule is considered to be redundant.

3 Preliminaries and Notation

Below we establish some notation and introduce some concepts that are used later on. We are given a set of items $\mathcal{I}$. A dataset $\mathcal{D}$ is a bag of transactions t, which are subsets of $\mathcal{I}$. The collection of all transactions is denoted $\mathcal{T}$. We write single items as x, y, z and itemsets as X, Y, Z. For the sake of brevity, we use the shorthands xyz for a set $\{x, y, z\}$, XY for the union of sets $X \cup Y$, and X-Y for set difference $X \setminus Y$. A rule between two itemsets X and Y is written as $X \Rightarrow Y$, where both sets are assumed to be either disjoint ($X \cap Y = \emptyset$) or inclusive ($X \subseteq Y$), depending on the context. The support *supp* and frequency *fr* of the pair $(X = v)$, with $v \in \{0,1\}^{|X|}$, are the absolute and relative number of transactions $t \in \mathcal{D}$ for which $t_X = v$, respectively.

The notion of entropy as a measure of information was introduced by Shannon [13]. Given a discrete random variable X with values v in a domain $dom(X)$, and a probability distribution p, the entropy of X is defined as $H(X) = -\sum_{v\in dom(X)} p(v)\log_2 p(v)$. For brevity, if $dom(X) = \{0,1\}$ we also write $H(f)$ where $f = fr(X = 1) \in [0,1]$. The entropy expresses the amount of information that is contained in a random variable, expressed in bits. Alternatively, it can be seen as a measure of its complexity or of its uncertainty. The mutual information between two discrete random variables X and Y is measured as the divergence of the joint distribution $p(v,u)$ from the product distribution $p(v)p(u)$; $I(X,Y) = \sum_{v\in dom(X),u\in dom(Y)} p(v,u)\log_2 \frac{p(v,u)}{p(v)p(u)}$. If X and Y are independent then the mutual information is zero and vice versa. Mutual information can conveniently be expressed in terms of entropy, as $I(X,Y) = H(Y) - H(Y \mid X) = H(X) + H(Y) - H(X,Y)$, where $H(Y \mid X)$ is the conditional entropy of Y given X, and $H(X,Y)$ is the joint entropy of X and Y.

4 Strong Dependence Rules

In this section we define our interestingness measures for itemsets and rules, and explore some of their properties.

4.1 Definitions

In the following, an itemset X is seen as a discrete random variable with $dom(X) = \{0,1\}^{|X|}$. For the probability distribution p, the frequency distribution fr is taken.

Definition 1 (Rule Entropy). *Let X and Y be two disjoint itemsets. The entropy h of the rule $X \Rightarrow Y$ is defined as the joint entropy of X and Y: $h(X \Rightarrow Y) = H(X \cup Y)$.*

It is easy to see that for any set X it holds that $0 \le h(X) \le \log_2 |dom(X)| = |X|$.

Definition 2 (Rule Dependence). *Let X and Y be two disjoint itemsets. We define the dependence μ of the rule $X \Rightarrow Y$ as*

$$\mu(X \Rightarrow Y) = \frac{I(X,Y)}{H(Y)}.$$

By dividing by $H(Y)$ we obtain a normalized, asymmetric variant of mutual information ranging between 0 and 1. When X an Y are independent then $\mu(X \Rightarrow Y) = 0$, this means that X tells us nothing about Y. On the other hand, $\mu(X \Rightarrow Y) = 1$ if and only if X fully determines Y; in this case the rule is called exact.

4.2 Properties

We describe some useful properties of h and μ which we exploit to construct an efficient set and rule mining algorithm later on. Due to space restrictions proofs are omitted but can be found in the technical report [1].

Theorem 1 (Monotonicity of Entropy). *Let X and X' be two itemsets. If $X \subseteq X'$, then $h(X) \leq h(X')$.*

Using the monotonicity of h, it is possible to efficiently traverse the search space of all itemsets in a typical Apriori-like breadth-first algorithm, or a memory-efficient depth-first algorithm as μ-Miner does.

Theorem 2 (Antecedent Monotonicity). *Let X, X' and Y be itemsets with $X \subseteq X'$, then $\mu(X \Rightarrow Y) \leq \mu(X' \Rightarrow Y)$.*

Theorem 2 implies that rules containing all of the items in $\mathcal{I}$ have the highest dependence. However, the entropy of such rules is also very high, and hence they will be pruned. Furthermore, some of the items might be independent of the other ones, and it is quite likely that such large rules display some sort of redundancy as described in Section 5.

Theorem 3 (Partial Monotonicity). *Let $X \Rightarrow Y$ be a rule, where X and Y are disjoint. There exists an item $y \in Y$ such that $\mu(X \Rightarrow Y) \leq \mu(Xy \Rightarrow Y\text{-}y)$.*

This last theorem allows us to systematically and efficiently construct all rules with a high dependence from a given low entropy set, in a levelwise fashion. This can be achieved without having to consider the exponential number of possible rules that can possibly be constructed from that itemset.

4.3 Closedness

Due to the explosion of patterns commonly encountered in classic frequent itemset mining, one often turns to mining a condensed representation of a collection of frequent itemsets. Such pattern collections are typically much smaller in magnitude, can be discovered faster, and it is possible to infer other patterns from them. One example are maximal frequent itemsets [14,15]. Two other popular condensed representations are closed and non-derivable frequent itemsets, which we extend to our framework.

The concept of closedness is well-studied for support based itemset mining [16]. An itemset is closed with respect to support if all of its proper supersets have a strictly smaller support. A closure operator can be defined that maps an itemset to its (unique) smallest closed superset, i.e. its closure. Similarly, we can consider itemsets that are closed with respect to entropy. We formally do this as follows. The set inclusion relation ($\subseteq$) defines a partial order on the powerset $\mathcal{P}(\mathcal{I})$ of all itemsets. Furthermore, a partial order, i.e. refinement ($\sqsubseteq$), can be defined on the set $\mathcal{Q}(\mathcal{T})$ of all transaction partitions. A given itemset $X \in \mathcal{P}(\mathcal{I})$ partitions $\mathcal{T}$ into equivalence classes according to the value of X in all transactions, and conversely a partition in $\mathcal{Q}(\mathcal{T})$ corresponds to an itemset in $\mathcal{P}(\mathcal{I})$. (The entropy of an itemset is computed using the sizes of the equivalence classes in its corresponding partition.) Let us call these two mapping functions i_1 and i_2. It can be shown that i_1 and i_2 form a Galois connection between $(\mathcal{P}(\mathcal{I}), \subseteq)$ and $(\mathcal{Q}(\mathcal{T}), \sqsubseteq)$. The composition $cl := i_2 \circ i_1$ defines a closure operator on $\mathcal{P}(\mathcal{I})$, which satisfies the following properties for all itemsets X.

$$\begin{cases} X \subseteq cl(X) & \text{(extension)} \\ cl(X) = cl(cl(X)) & \text{(idempotency)} \\ X \subseteq X' \Rightarrow cl(X) \subseteq cl(X') & \text{(monotonicity)} \end{cases}$$

Definition 3. *We call an itemset $X \subseteq \mathcal{I}$ closed if $X = cl(X)$. Conversely, the set X is called a generator if for all $X' \subsetneq X$ it holds that $cl(X') \neq X$.*

It holds that all proper supersets of a closed itemset have a strictly higher entropy and $h(X) = h(cl(X))$. All proper subsets of a generator have strictly lower entropy. Note that the rule $X \Rightarrow Y$ is exact if and only if $X \subseteq XY \subseteq cl(X)$. Furthermore, if an exact rule $X \Rightarrow Y$ is minimal, then X is a generator.

4.4 Derivability

The idea of itemset derivability was introduced by Calders and Goethals [17]. Given the supports of all proper subsets of an itemset $(X = 1)$, it is possible, using the inclusion-exclusion principle, to derive tight lower and upper bounds on its support. If these bounds coincide we know $supp(X = 1)$ exactly, and $(X = 1)$ is called derivable (with respect to support). The set of all frequent itemsets can thus be derived from the set of all non-derivable frequent itemsets. Similarly, we can define the derivability of the entropy of an itemset.

Definition 4. *We call X h-derivable if its entropy can be determined exactly from the entropies of its proper subsets.*

The set of all non-derivable itemsets is downward closed. Interestingly, an itemset X is h-derivable if and only if it is derivable with respect to support.

5 Rule Redundancy

Mining all low entropy, high dependence rules can yield a very large set of patterns, which is not desirable for a user who wants to analyze them. Typically, this collection contains a lot of redundant rules. In this section we investigate how we can characterize and prune such rules. We define two types of redundancy, one that is based on the closure of itemsets, and one that is based on the superfluous augmentation of the antecedent or consequent of a rule.

5.1 Closure-Based Redundancy

As mentioned in the previous section, rules of the form $X \Rightarrow cl(X)$ are always exact. It should be clear that combining an arbitrary rule with an appropriate exact rule yields a new rule with identical entropy and dependence. For instance, if $A \Rightarrow B$ is exact, then $\mu(A \Rightarrow C) = \mu(AB \Rightarrow C)$.

Theorem 4. *Let $X \Rightarrow Y$ and $X' \Rightarrow Y'$ be two rules, where $X \subseteq X' \subseteq cl(X)$ and $Y \subseteq Y' \subseteq cl(Y)$. Then $h(X \Rightarrow Y) = h(X' \Rightarrow Y')$ and $\mu(X \Rightarrow Y) = \mu(X' \Rightarrow Y')$.*

Since the entropy and dependence of such larger rules can be inferred using the smaller rule and the closure operator, we call them redundant. These minimal rules are constructed using generators.

Definition 5 (Closure-based Redundancy). *A rule $X \Rightarrow Y$ is redundant with respect closure if*

$$\begin{cases} X \text{ is not a generator or } |Y| > 1 & \text{if } \mu(X \Rightarrow Y) = 1, \\ XY \text{ is not a generator} & \text{if } \mu(X \Rightarrow Y) < 1. \end{cases}$$

Note that if XY is a generator, then X and Y are also generators, since the set of all generators is downward closed.

5.2 Augmentation Redundancy

Here we define a stricter kind of redundancy that prunes rules which have items unnecessarily added to their antecedents or consequents.

Antecedent Redundancy. Suppose we have two rules with a common consequent, $X \Rightarrow Y$ and $X' \Rightarrow Y$, with $X' = X \cup \{x\}$. Theorem 2 tells us that $\mu(X \Rightarrow Y) \leq \mu(X' \Rightarrow Y)$. Even though the latter rule has a higher dependence, it might be redundant if x does not make a real contribution to the rule. For instance, if X and x are independent, then $\mu(X' \Rightarrow Y)$ is simply the sum of $\mu(X \Rightarrow Y)$ and $\mu(x \Rightarrow Y)$. To characterize this type of redundancy we use the chain rule of mutual information, $I(Xx, Y) = I(X, Y) + I(x, Y \mid X)$, where the last term is the conditional mutual information (which can be written as $H(x \mid X) - H(x \mid XY)$). It is known that I does not behave monotonically with conditioning. In the case where X and x are independent, we have $I(x, Y \mid X) = I(x, Y)$. If X already explains part of the dependency between x and Y, then $I(x, Y \mid X) < I(x, Y)$, meaning there is an "overlap" between X and x. Otherwise, if $I(x, Y \mid X) > I(x, Y)$, this means that under knowing X, the mutual information between x and Y increases. Intuitively, it means that Xx tells us more about Y than the sum of X and x separately. This motivates the following definition.

Definition 6 (Antecedent Redundancy). *A rule $X \Rightarrow Y$ is redundant with respect to antecedent augmentation, if there exists an item $x \in X$ such that*

$$\begin{cases} \mu(X \Rightarrow Y) \leq \mu(X\text{-}x \Rightarrow Y) + \mu(x \Rightarrow Y), \text{or} \\ X\text{-}x \Rightarrow Y \text{ is redundant.} \end{cases}$$

It follows that $\mu(X \Rightarrow Y) > \sum_{x \in X} \mu(x \Rightarrow Y)$ if the rule $X \Rightarrow Y$ is non-redundant.

Consequent Redundancy. Consider the rule $X \Rightarrow Y$ and an item $y \notin XY$. Unlike in the previous section, μ is not monotonic with respect to augmentation of the consequent, so in general the dependence of $X \Rightarrow Yy$ can either be higher or lower that that of $X \Rightarrow Y$. An increase in μ means that adding y to Y increases the mutual information $I(X, Y)$ more than it increases the entropy $H(Y)$. Put differently, the relative increase in uncertainty from $H(Y)$ to $H(Yy)$, is surpassed by the increase of the amount of information X gives about Y and Yy. X gives relatively less information about Yy than it does about Y.

Definition 7 (Consequent Redundancy). *A rule $X \Rightarrow Y$ is redundant with respect to consequent augmentation, if there exists an item $y \in Y$ such that*

$$\begin{cases} \mu(X \Rightarrow Y) \leq \mu(X \Rightarrow Y\text{-}y), \text{or} \\ X \Rightarrow Y\text{-}y \text{ is redundant.} \end{cases}$$

Algorithm 1. μ-Miner

Input: Binary dataset $\mathcal{D}$, thresholds $h_{\max}$ and $\mu_{\min}$
Output: Non-redundant, low entropy, high dependence rules
1. $\mathcal{P} \leftarrow \{\{x\} \subset \mathcal{I}; h(x) \leq h_{\max}\}$
2. SetMine($\mathcal{P}$, $h_{\max}$, $\mu_{\min}$)

It follows that if $X \Rightarrow Y$ is non-redundant, then $\forall Y' \subset Y : \mu(X \Rightarrow Y') < \mu(X \Rightarrow Y)$.

Relation to Closure-based Redundancy. It turns out that augmentation redundancy is strictly stronger than closure-based redundancy, as stated in the theorem below.

Theorem 5. *If a rule $X \Rightarrow Y$ is redundant with respect to closure, then it is also redundant with respect to antecedent augmentation or consequent augmentation.*

6 The μ-Miner Algorithm

In this section we present μ-Miner (see Algorithm 1). As input it expects a dataset $\mathcal{D}$, a maximum entropy threshold $h_{\max}$, and a minimum dependence threshold $\mu_{\min}$. The algorithm efficiently mines low entropy itemsets, and from these sets strong dependence rules are constructed. Further, μ-Miner prunes rules that are closure redundant or augmentation redundant. Computation of entropy and dependence, and the checking of redundancy is performed by doing some simple arithmetic operations and lookups, and only one scan of the database is required.

6.1 Mining Itemsets

In the SetMine function (Algorithm 2), itemsets with a low entropy are mined. This recursive function takes a collection of itemsets with a common prefix as input, initially this is the set of all low entropy singletons. The search space is traversed in a depth-first, right-most fashion. This is less memory-intensive than a breadth-first approach, and the right-most order ensures that when an itemset is considered, all of its subsets will already have been visited in the past (lines 1&3), a fact we exploit later. This implies that we need to impose an order on the itemsets, e.g. a simple lexicographical ordering. In our implementation of μ-Miner we use a heuristic ordering based on the entropy of the items, such that large subtrees of the search space are rooted by sets which are expected to be have a high entropy, allowing us to prune larger parts of the subspace.

6.2 Efficiently Computing Entropy

A straightforward method to compute $h(X)$ is to perform a scan over the database to obtain the frequencies of $(X = v)$ for all values $v \in \{0,1\}^{|X|}$. In total there are 2^k such frequencies for $k = |X|$, however, at most $|\mathcal{D}|$ of them are nonzero and hence this method requires $O(|\mathcal{D}|)$ time. If the database fits in memory this counting method is

Algorithm 2. SetMine

Input: Itemset collection $\mathcal{P}$, thresholds $h_{\max}$ and $\mu_{\min}$
1. **for** X_1 in $\mathcal{P}$ in descending order **do**
2. $\quad\mathcal{P}' \leftarrow \emptyset$
3. $\quad$**for** $X_2 < X_1$ in $\mathcal{P}$ **do**
4. $\quad\quad X \leftarrow X_1 \cup X_2$
5. $\quad\quad$Compute and store $fr(X = 1)$
6. $\quad\quad h(X) \leftarrow$ EntropyQIE(X)
7. $\quad\quad$**if** X is not a generator **then**
8. $\quad\quad\quad$Report corresponding exact rule(s)
9. $\quad\quad$**else if** $h(X) \leq h_{\max}$ **then**
10. $\quad\quad\quad\mathcal{P}' \leftarrow \mathcal{P}' \cup \{X\}$
11. $\quad\quad\quad$RuleMine(X, $\mu_{\min}$)
12. $\quad$SetMine($\mathcal{P}'$)

perfectly feasible, otherwise it becomes too expensive, since database access is required for each candidate itemset. Another option is to use the partitioning technique used by TANE [7]. For each itemset a partition of the transactions is explicitly computed in $O(|\mathcal{D}|)$ time, and then the sizes of the sets in the partition can be used to compute $h(X)$.

μ-Miner uses a different entropy computation method that does not require direct database access, and has a lower complexity, which is beneficial especially for large datasets. We start from a simple right-most depth-first itemset support mining algorithm similar to Eclat [4], and store the supports in a trie (line 5). When we have mined the support of $(X = 1)$, the frequencies of all $(X = v)$ are computed with the stored supports of all subsets, by using quick inclusion-exclusion (Algorithm 3), which takes $O(k \cdot 2^{k-1})$ time [18]. However, we can again use the fact that at most $|\mathcal{D}|$ frequencies are nonzero, hence this counting method is $O(\min(k \cdot 2^{k-1}, |\mathcal{D}|))$. The advantage of our method is that it is fast and it does not require database access. The disadvantage is that the supports of all mined itemsets must be stored, which may be a problem if memory is scarce and $h_{\max}$ is set rather high. Note that if we were to restrict ourselves to mining only non-derivable itemsets, we know that $k \leq \lceil \log_2 |\mathcal{D}| \rceil$ [17]. In this case the total number of frequencies we need to store is $O(|\mathcal{I}|^{\log_2 |\mathcal{D}|})$ in the worst case, which is polynomial in $|\mathcal{I}|$ for a fixed database size, and polynomial in $|\mathcal{D}|$ for a fixed number of items.

6.3 Mining Non-redundant Dependence Rules

For each low entropy itemset, RuleMine (Algorithm 4) is called to generate high dependence rules. It starts with rules whose consequent is a singleton, and then moves items from the antecedent to the consequent. By using the partial monotonicity property from Theorem 3, not all 2^k possible rules need to be considered. Since we have the entropies of all subsets available to us, we can compute the dependence by performing just a few lookups. If a rule is found to have high dependence, it is checked whether the rule is

Algorithm 3. EntropyQIE

Input: Candidate itemset $X \subset \mathcal{I}$
Output: $h(X)$, the entropy of X
1. **for all** $X' \subseteq X$ **do**
2. $\quad p(X') \leftarrow fr(X' = 1)$
3. **for all** $x \in X$ **do**
4. $\quad$ **for all** $X' \subseteq X$ with $x \in X'$ **do**
5. $\quad\quad p(X'\text{-}x) \leftarrow p(X'\text{-}x) - p(X')$
6. **return** $h(X) = -\sum_{X' \subseteq X} p(X') \cdot \log_2 p(X')$

Algorithm 4. RuleMine

Input: Low entropy itemset X; dependence threshold $\mu_{\min}$
Output: Non-redundant strong dependence rules based on X
1. $\mathcal{L} \leftarrow \{X\text{-}x \Rightarrow x; x \in X\}$
2. **while** $\mathcal{L} \neq \emptyset$ **do**
3. $\quad$ **for** $A \Rightarrow B$ in $\mathcal{L}$ **do**
4. $\quad\quad$ Compute $\mu(A \Rightarrow B)$
5. $\quad\quad$ **if** $\mu(A \Rightarrow B) \geq \mu_{\min}$ and $A \Rightarrow B$ is non-redundant **then**
6. $\quad\quad\quad$ Report $A \Rightarrow B$
7. $\quad \mathcal{L} \leftarrow \{A\text{-}a \Rightarrow Ba; A \Rightarrow B \in \mathcal{L}$, using Theorem 3$\}$

redundant (line 6). Again, since we have the entropies of all subsets available, these redundancy checks can be performed quite efficiently.

7 Experimental Evaluation

We perform experiments on several datasets to evaluate the efficiency of μ-Miner. We also investigate the effect of our pruning techniques. The algorithm is implemented in C++[1], and the experiments were executed on a machine with a 2.2GHz CPU and 2GB of memory, running Linux. More experiments can be found in the technical report [1].

7.1 Datasets

First, we have some benchmark datasets taken from the FIMI Repository [19]: CHESS, MUSHROOM and PUMSB, containing 3196, 8124 and 49046 transactions respectively. The original PUMSB dataset contains 2112 items, in our experiments we used the 100 most high entropy items. MUSHROOM originally has 119 items, for our experiments we removed items with frequencies higher than 0.9 or lower than 0.1. These datasets are used to test the efficiency of μ-Miner.

Second, we generated a SYNTHETIC dataset which contains an embedded pattern. We use it to evaluate the scalability of μ-Miner with respect to the size of the database. The dataset consists of 1 000 000 transactions and has 16 items. The 15 first items are independent and have random frequencies between 0.1 and 0.9. The last item equals the

[1] The source code of μ-Miner is publicly available at `http://www.adrem.ua.ac.be`

sum of the other two modulo 2, i.e. the rule $\{1, \ldots, 15\} \Rightarrow \{16\}$ is an exact one. Note that this dependency is also minimal.

7.2 Experiments

First, we perform some experiments on the benchmark datasets. To begin with, we set the value of $h_{\max}$ to a fixed value (1.5 for CHESS, 2 for PUMSB, and 3.5 for

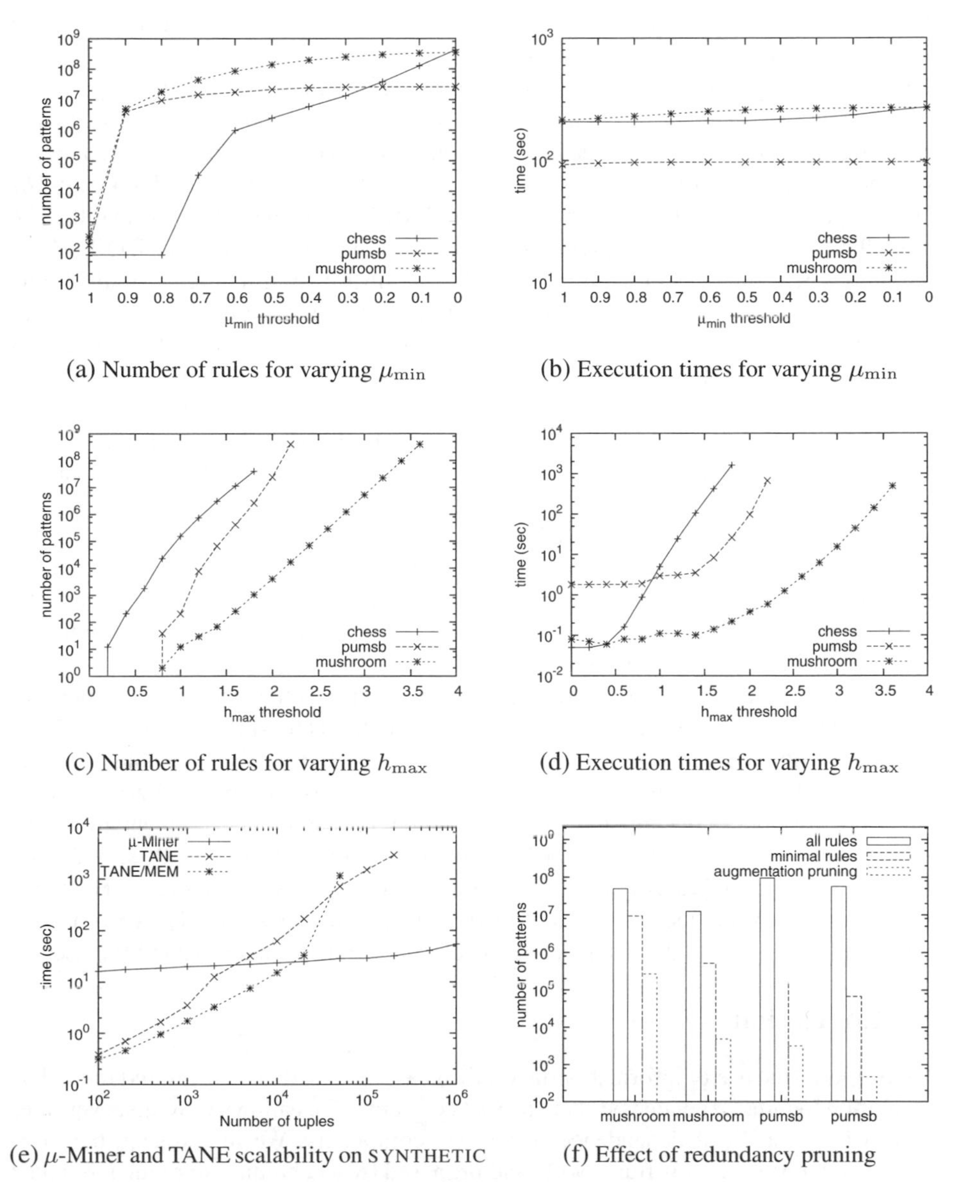

(a) Number of rules for varying $\mu_{\min}$

(b) Execution times for varying $\mu_{\min}$

(c) Number of rules for varying $h_{\max}$

(d) Execution times for varying $h_{\max}$

(e) μ-Miner and TANE scalability on SYNTHETIC

(f) Effect of redundancy pruning

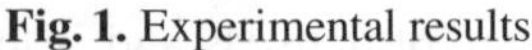

Fig. 1. Experimental results

MUSHROOM), and we vary the minimum dependence threshold $\mu_{\min}$ between 0 and 1. As can be seen in Figure 1a, the number of rules increases roughly exponentially when $\mu_{\min}$ is decreased. Noticeably, the execution times stay roughly constant as $\mu_{\min}$ decreases, as shown in Figure 1b. This is not surprising, since most computations are performed in the itemset mining phase, and the computation of μ involves just a few lookups. Next, we set the value of $\mu_{\min}$ to a fixed value (in this case 0.4 for all datasets), and gradually increase the maximum entropy threshold $h_{\max}$ from zero upward. In Figure 1c we see that for very low values of $h_{\max}$ no rules are found. Then, the number of rules increases exponentially with $h_{\max}$, which is to be expected. In Figure 1d we see that this trend also translates to the execution times. For lower thresholds ($h_{\max} \leq 1$) the runtimes stay roughly constant, because they are dominated by I/O time.

Secondly, we evaluate the scalability of μ-Miner with respect to the size of the database using the SYNTHETIC dataset. The aim is to discover the embedded functional dependency, in order to do this we set $\mu_{\min}$ to 1, and $h_{\max}$ sufficiently high (say, 16). We compare μ-Miner with the TANE and TANE/MEM implementations from [20]. The main TANE algorithm stores partitions to disk level per level, while the TANE/MEM variant entirely operates in main memory. The number of transactions is gradually increased from 10^2 to 10^6 and the runtimes are reported in Figure 1e. We see that all algorithms scale linearly with $|\mathcal{D}|$, although the slope is much steeper for TANE and TANE/MEM, while the execution time of μ-Miner increases very slowly. At around ± 3000 transactions, μ-Miner overtakes TANE in speed, and at 10^5 transactions our algorithm is already two orders of magnitude faster. The TANE/MEM algorithm is faster up to ± 10000 transactions, but cannot handle any datasets much larger than that due to heavy memory consumption. This observed difference in speed can be explained entirely by the counting method. TANE explicitly constructs a partition of size $O(|\mathcal{D}|)$ for each itemset (and stores these to disk or in memory level per level), while our algorithm computes the required sizes of the partitions without actually constructing them. The increase in the execution time of μ-Miner can be accounted for almost entirely by the increase in time it takes to read the data file.

Next, let us investigate how redundancy pruning affects the size of the output. We experimented on the MUSHROOM and the PUMSB datasets for different values of $h_{\max}$ (3 for MUSHROOM and 1.5 for PUMSB) and $\mu_{\min}$ (0.2 and 0.8 for both datasets). The results are shown in Figure 1f. For the MUSHROOM dataset pruning all non-minimal rules already reduces the output by roughly an order of magnitude. Augmentation pruning reduces the output by an additional two orders of magnitude. For the PUMSB dataset pruning non-minimal rules reduces the output by three orders of magnitude. Here augmentation pruning reduces the output in size even further, by roughly two orders of magnitude. This makes the result collection of rules far more manageable for a user.

8 Conclusions

We proposed the use of information-theoretic measures based on entropy and mutual information to mine dependencies between sets of items. This allows us to discover rules with a high statistical dependence, and a low complexity. We investigated the problem of redundancy in this framework, and proposed two techniques to prune redundant rules. One is based on the closure of itemsets and is lossless, while the other, shown to be stronger, is lossy and penalizes the augmentation of rules with unrelated items. We presented our algorithm μ-Miner, which mines such dependencies and applies the presented pruning techniques. Several experiments showed that μ-Miner is efficient and scalable: it can easily handle datasets with millions of transactions and does not require a large amount of memory. Furthermore, our pruning techniques were shown to be very effective in reducing the size of the output by several orders of magnitude.

References

1. Mampaey, M.: Mining non-redundant information-theoretic dependencies between itemsets. Technical Report, University of Antwerp (2010)
2. Agrawal, R., Imielinski, T., Swami, A.: Mining association rules between sets of items in large databases. ACM SIGMOD Record 22(2), 207–216 (1993)
3. Han, J., Pei, J., Yin, Y.: Mining frequent patterns without candidate generation. ACM SIGMOD Record 29(2), 1–12 (2000)
4. Zaki, M., Parthasarathy, S., Ogihara, M., Li, W., et al.: New algorithms for fast discovery of association rules. In: Proceedings of KDD (1997)
5. Srikant, R., Agrawal, R.: Mining quantitative association rules in large relational tables. ACM SIGMOD Record 25(2), 1–12 (1996)
6. Kivinen, J., Mannila, H.: Approximate inference of functional dependencies from relations. Theoretical Computer Science 149(1), 129–149 (1995)
7. Huhtala, Y., Karkkainen, J., Porkka, P., Toivonen, H.: TANE: An efficient algorithm for discovering functional and approximate dependencies. The Computer Journal 42(2), 100–111 (1999)
8. Zaki, M.J.: Generating non-redundant association rules. In: Proceedings of KDD, pp. 34–43 (2000)
9. Balcázar, J.L.: Minimum-size bases of association rules. In: Proceedings of ECML PKDD, pp. 86–101 (2008)
10. Dalkilic, M.M., Robertson, E.L.: Information dependencies. In: Proceedings of ACM PODS, pp. 245–253 (2000)
11. Heikinheimo, H., Hinkkanen, E., Mannila, H., Mielikäinen, T., Seppänen, J.K.: Finding low-entropy sets and trees from binary data. In: Proceedings of KDD, pp. 350–359 (2007)
12. Jaroszewicz, S., Simovici, D.A.: Pruning redundant association rules using maximum entropy principle. In: Chen, M.-S., Yu, P.S., Liu, B. (eds.) PAKDD 2002. LNCS (LNAI), vol. 2336, pp. 135–147. Springer, Heidelberg (2002)
13. Shannon, C.E.: A mathematical theory of communication. Bell System Technical Journal 27, 379–423 (1948)
14. Bayardo Jr., R.: Efficiently mining long patterns from databases. In: Proceedings of ACM SIGMOD, pp. 85–93 (1998)
15. Gouda, K., Zaki, M.: Efficiently mining maximal frequent itemsets. In: Proceedings of IEEE ICDM, pp. 163–170 (2001)
16. Pasquier, N., Bastide, Y., Taouil, R., Lakhal, L.: Discovering frequent closed itemsets for association rules. In: Proceedings of ICDT, pp. 398–416 (1999)
17. Calders, T., Goethals, B.: Non-derivable itemset mining. Data Mining and Knowledge Discovery 14(1), 171–206 (2007)
18. Calders, T., Goethals, B.: Quick inclusion-exclusion. In: Bonchi, F., Boulicaut, J.-F. (eds.) KDID 2005. LNCS, vol. 3933, pp. 86–103. Springer, Heidelberg (2006)
19. Goethals, B.: Frequent itemset mining implementations repository, `http://fimi.cs.helsinki.fi/data`
20. Huhtala, Y., Karkkainen, J., Porkka, P., Toivonen, H.: TANE homepage, `http://www.cs.helsinki.fi/research/fdk/datamining/tane`

Discovery and Application of Functional Dependencies in Conjunctive Query Mining

Bart Goethals[1], Dominique Laurent[2], Wim Le Page[1]

[1] University of Antwerp, Dept of Mathematics and Computer Science B-2020 Antwerp
[2] ETIS-CNRS-ENSEA-Université de Cergy-Pontoise F-95000 Cergy-Pontoise

Abstract. We present an algorithm for mining frequent queries in arbitrary relational databases, over which functional dependencies are assumed. Building upon previous results, we restrict to the simple, but appealing subclass of simple conjunctive queries. The proposed algorithm makes use of the functional dependencies of the database to optimise the generation of queries and prune redundant queries. Furthermore, our algorithm is capable of detecting previously unknown functional dependencies that hold on the database relations as well as on joins of relations. These detected dependencies are subsequently used to prune redundant queries. We propose an efficient database-oriented implementation of our algorithm using SQL, and provide several promising experimental results.

1 Introduction

The discovery of recurring patterns in databases is one of the main topics in data mining and many efficient solutions have been developed for different classes of patterns and data collections. Almost all techniques, however, work on so called transaction databases [1]. Not only for itemsets, but also in the case of trees [20] and graphs [12,15,19], the database consists of a collection of transactions, and a frequent pattern is discovered if it occurs in enough such transactions. Even in the multi-relational case, as considered in the WARMR system [4], the database can be seen as a collection of transactions in which each transaction consists of a small relational database. A query is then called frequent if it gives a non-empty answer in enough of such databases.

Obviously, many relational databases are not suited to be converted into such a transactional format and even if this would be possible, a lot of information implicitly encoded in the relational model would be lost after conversion. Recently, we have considered association rule mining on arbitrary relational databases by combining pairs of queries which could reveal interesting properties in the database [8,13]. Intuitively, we pose two queries on the database such that one query is more specific than the other (w.r.t. query containment). Then, if the number of tuples in the output of both queries is almost the same, a potentially interesting discovery is revealed.

To illustrate, consider the well known Internet Movie Database [11] containing almost all possible information about movies, actors and everything related to that, and consider the following queries: first, we ask for all actors that have starred in a movie of the genre 'drama'; then, we ask for all actors that have starred in a movie of the genre 'drama', but that also starred in a (possibly different) movie of the genre 'comedy'.

T.B. Pedersen, M.K. Mohania, and A M. Tjoa (Eds.): DaWaK 2010, LNCS 6263, pp. 142–156, 2010.

Now suppose the answer to the first query consists of 1000 actors, and the answer to the second query consists of 900 actors. Obviously, these answers do not necessarily reveal any significant insights on themselves, but when combined, it reveals the potentially interesting pattern that actors starring in 'drama' movies typically (with a probability of 90%) also star in a 'comedy' movie. Of course, this pattern could also have been found by first preprocessing the database, and creating a transaction for each actor containing the set of all genres of movies he or she appeared in. Similarly, a pattern like: 77% of the movies starring Ben Affleck, also star Matt Damon, could be found by posing the query asking for all movies starring Ben Affleck, and the query asking for all movies starring both Ben Affleck and Matt Damon. Again, this could also be found using frequent set mining methods, but this time, the database should have been differently preprocessed in order to find this pattern. Furthermore, it is even impossible to preprocess the database only once in such a way that the above two patterns would be found by frequent set mining, as they are counting different types of transactions: actors in the first example and movies in the second example.

Also truly relational patterns can be found which can not be found using typical set mining techniques, such as 80% of all movie directors that have ever been an actor in some movie, also star in at least one of the movies they directed. This is expressed by two queries of which one asks for all movie directors that have ever acted, and the second one asks for all movie directors that have ever acted in one of their own movies.

The Conqueror algorithm recently developed by Goethals et al. [8] has shown to discover interesting association rules over a simple, but appealing subclass of conjunctive queries, called *simple conjunctive queries*. Furthermore, the algorithm had an efficient database-oriented implementation in SQL. One challenge that remained to be solved in this approach, was the huge number of generated patterns. Part of the volume is inherently due to the relational setting, but a substantial part, however, is due to redundancies induced by dependencies embedded in the data.

Jen et al. [13], studied the problem of mining all frequent queries from a single relational table. They considered projection-selection queries, and assumed that the table to be mined satisfies a set of functional dependencies. A pre-ordering over queries was defined, and shown to be anti-monotonic towards the support measure. Moreover, this pre-ordering induces an equivalence relation and two equivalent queries are shown to have the same support. Therefore, one computation per equivalence class allows to know the support of all queries in that class. In [14], this work has been generalised to several tables in the case where the database operates over a star schema. The challenge however remains to generalise the theory to arbitrary relational databases.

Clearly, the combination of the approaches in [13] and [8] would resolve the issues posed, *i.e.,* mining non redundant simple conjunctive queries (thus including arbitrary joins), given a collection of functional dependencies over the relations of an arbitrary relational database. This is one major contribution of this paper.

Moreover, combining these techniques also results in new opportunities. That is, next to the given functional dependencies, we introduce a novel technique to discover previously unknown functional dependencies, and immediately exploit them for reducing the number of frequent queries in the output. Furthermore, we do so not only for the relations of the database, but also for *any join* of relations. This is the second contribution of this paper, and several experiments clearly show the benefits of this approach, thus making the discovery of simple conjunctive queries a feasible and attractive method towards the exploration of arbitrary relational databases.

The paper is organised as follows: In Section 2, we recall the basic concepts and definitions used in this work and we briefly review from [13] how functional dependencies are used to compare queries. We present our algorithm Conqueror$^+$ in Section 3, combining the two approaches [8,13], and in Section 4, we report experiments, showing that Conqueror$^+$ clearly outperforms Conqueror. We conclude in Section 6.

2 Formal Model

2.1 Background

We consider a fixed attribute set U and a relational database schema $\mathcal{D} = \{R_1, \ldots, R_n\}$ over U in which, for $i = 1, \ldots, n$, R_i is a relation name associated with a subset of U, called the *schema* of R_i and denoted by $sch(R_i)$. Without loss of generality, we assume that, for all distinct i and j in $\{1, \ldots, n\}$, $sch(R_i) \cap sch(R_j) = \emptyset$. In order to make this assumption explicit, for all i in $\{1, \ldots, n\}$, every A in $sch(R_i)$ is referred to as $R_i.A$.

We also assume that we are given functional dependencies over $\mathcal{D}$. More precisely, each R_i is associated with a set of functional dependencies over $sch(R_i)$, denoted by $\mathcal{FD}_i$, and the set of all functional dependencies defined in $\mathcal{D}$ is denoted by $\mathcal{FD}$.

As in [8], the queries of interest in our approach, are conjunctive projection-selection-join queries whose joins are expressed using a conjunction of selection conditions of the form $R_i.A = R_j.A'$. We note that by doing so, all possible equi-joins can be considered, which would not the case using the universal relation associated to the given database. Moreover, we recall from [8] that such a conjunctive condition F induces a partition $blocks(F)$ of U, where every block β of $blocks(F)$ is a maximal set of attributes such that for all $R_i.A$ and $R_j.A'$ in β, $R_i.A = R_j.A'$ is a consequence of F. In such a case, we say that R_i and R_j are *connected through* F.

Definition 1. *Denoting by R the cartesian product $R_1 \times \ldots \times R_n$, let $Q = \pi_X \sigma_F R$ where $F = \bowtie(Q) \wedge \sigma(Q)$, such that $\bowtie(Q)$ and $\sigma(Q)$ are respectively conjunctions of selection conditions of the form $R_i.A = R_j.A'$ and $R_k.A = a$, where i, j and k are in $\{1, \ldots, n\}$ and a is in $dom(A)$. $Q = \pi_X \sigma_F R$ is said to be a* simple conjunctive query *if all relation names occurring in X or in $\sigma(Q)$ are connected through $\bowtie(Q)$.*

Given a simple conjunctive query $Q = \pi_X \sigma_F R$, the set X is denoted by $\pi(Q)$ and the tuple defined by the conjunctive selection condition $\sigma(Q)$ is denoted by Q^σ.

We call Q a join query *if $\sigma(Q)$ is the empty condition and if $\pi(Q)$ is the set of all attributes of all relation names occurring in $\bowtie(Q)$. Given a simple conjunctive query Q, we denote by $J(Q)$ the join query such that $\bowtie(J(Q)) = \bowtie(Q)$.*

To simplify notation, given a simple conjunctive query Q, the corresponding partition of U, $blocks(\bowtie(Q))$ is simply denoted by $blocks(Q)$. We emphasise that, according to Definition 1, considering simple conjunctive queries avoids computing cartesian products. We illustrate this definition below.

Example 1. Let us consider a database schema $\mathcal{D}$ consisting of two relation names R_1 and R_2 with the following schemas: $sch(R_1) = \{A, B\}$ and $sch(R_2) = \{C, D, E\}$.

According to Definition 1, R denotes the cartesian product $R_1 \times R_2$. Since $sch(R_1) \cap sch(R_2)$ is clearly empty, in this example and in the forthcoming examples dealing with $\mathcal{D}$, we do not prefix attributes with relation names. For example, $R_1.A$ is denoted by A.

The query $Q = \pi_{AD}\sigma_{(A=B)\wedge(E=e)}R$ is *not* a simple conjunctive query because R_1 and R_2 are not connected through the condition $A = B$. Computing the answer to this query requires to consider explicitly the cartesian product $R_1 \times R_2$.

On the other hand, $Q_1 = \pi_{AD}\sigma_{(A=C)\wedge(E=e)}R$ is a simple conjunctive query such that $\bowtie(Q_1) = (A = C)$, $\pi(Q_1) = AD$, $\sigma(Q_1) = (E = e)$ and $Q^\sigma = e$. Moreover, $J(Q_1) = \pi_{ABCDE}\sigma_{(A=C)}R$, and $blocks(Q_1)$ contains four blocks, namely: $\{A, C\}$, $\{B\}$, $\{D\}$ and $\{E\}$. In this case, computing the answer to Q_1 does not require to consider the cartesian product $R_1 \times R_2$, since R_1 and R_2 are joined through $A = C$. □

We now define as in [8] the *support* of a query, and when a query is said to be *frequent*.

Definition 2. *Given an instance $\mathcal{I}$ of $\mathcal{D}$ and a simple conjunctive query Q, the answer to Q in $\mathcal{I}$ is denoted by $Q(\mathcal{I})$ and is seen as a set in which no duplicates are allowed.*

The support *of Q in $\mathcal{I}$, denoted $support_{\mathcal{I}}(Q)$ or simply $support(Q)$, is the cardinality of the answer to Q in $\mathcal{I}$. Given a minimum support threshold minsup, Q is said to be* frequent *if $support(Q) > minsup$.*

To end the preliminaries, we mention the strong relationship between support and functional dependency, as stated by the following proposition whose easy proof is omitted.

Proposition 1. *Let T be a relational table over the attribute set $sch(T)$ and let X and X' be subsets of $sch(T)$. T satisfies $X \rightarrow X'$ if and only if $support(\pi_{XX'}T) = support(\pi_X T)$.*

In the context of Example 1, for $Q = \pi_{AD}\sigma_{(A=C)}R$ and $Q' = \pi_A\sigma_{(A=C)}R$, considering an instance $\mathcal{I}$ of $\mathcal{D}$ for which $support(Q) = support(Q')$ indicates that $\sigma_{(A=C)}R(\mathcal{I})$ satisfies the functional dependency $A \rightarrow D$. Consequently, for every conjunctive selection condition S, the queries $Q_S = \pi_{AD}\sigma_{(A=C)\wedge S}R$ and $Q'_S = \pi_A\sigma_{(A=C)\wedge S}R$ also have the same support. Thus, computing the support of Q'_S is redundant, assuming that the support of Q_S is known.

We recall that one of the main contributions of this paper is to discover functional dependencies in order to avoid computing unnecessary supports.

2.2 Query Comparison

Inspired by [13], we compare queries based on functional dependencies.

Definition 3. *Let $Q_1 = \pi_{X_1}\sigma_{F_1}R$ and $Q_2 = \pi_{X_2}\sigma_{F_2}R$ be two simple conjunctive queries. Denoting by Y_i the schema of Q_i^σ, for $i = 1, 2$, $Q_1 \preceq Q_2$ holds if*

1. *$\bowtie(Q_1) \subseteq \bowtie(Q_2)$,*
2. *$J(Q_2)(\mathcal{I})$ satisfies $X_1Y_2 \rightarrow X_2$ and $Y_2 \rightarrow Y_1$, and*
3. *the tuple $Q_1^\sigma Q_2^\sigma$ is in $\pi_{Y_1Y_2}J(Q_2)(\mathcal{I})$.*

Example 2. In the context of Example 1, assume that $\mathcal{FD}_1 = \emptyset$ and $\mathcal{FD}_2 = \{C \to D, E \to D\}$, and let $Q_1 = \pi_{AD}\sigma_{(A=C)\wedge(E=e)}R$ and $Q_2 = \pi_C\sigma_{(A=C)\wedge(D=d)}R$.

We have $\bowtie(Q_1) = \bowtie(Q_2)$ and $J(Q_1) = J(Q_2) = \pi_{ABCDE}\sigma_{(A=C)}R$. Then, if $\mathcal{I}$ is an instance of $\mathcal{D}$, $J(Q_2)(\mathcal{I})$ satisfies $\mathcal{FD}$. Moreover, due to the equality defining $\bowtie(Q_2)$, $J(Q_2)(\mathcal{I})$ also satisfies $A \to C$ and $C \to A$. Therefore, $J(Q_2)(\mathcal{I})$ satisfies $CE \to AD$ and $E \to D$, and so, if $de \in \pi_{DE}J(Q_2)(\mathcal{I})$, by Definition 3, $Q_2 \preceq Q_1$. □

It can be seen from [13] that $\preceq$ is a pre-ordering and that the support of queries is anti-monotonic with respect to $\preceq$. In other words, for all Q_1 and Q_2 such that $Q_1 \preceq Q_2$, we have $support(Q_2) \leq support(Q_1)$. Anti-monotonicity is used in our algorithms to prune infrequent queries, in much the same way as in Apriori [1].

Moreover, the pre-ordering $\preceq$ induces an equivalence relation, denoted by $\sim$, defined as follows: given two simple conjunctive queries Q_1 and Q_2, $Q_1 \sim Q_2$ holds if $Q_1 \preceq Q_2$ and $Q_2 \preceq Q_1$. As a consequence of anti-monotonicity, if $Q_1 \sim Q_2$ holds then $support(Q_1) = support(Q_2)$. Thus, only *one* computation per equivalence class modulo $\sim$ allows to know the support of *all* queries in that class.

In order to characterize equivalence classes modulo $\sim$, we denote by X^+ the closure of a relation schema X with respect to a given set of functional dependencies FD. Then, based on [13], it can be seen that for $Q_1 = \pi_{X_1}\sigma_{F_1}R$ and $Q_2 = \pi_{X_2}\sigma_{F_2}R$, $Q_1 \sim Q_2$ holds if and only if $\bowtie(Q_1) = \bowtie(Q_2)$, $(X_1Y_1)^+ = (X_2Y_2)^+$, $Y_1^+ = Y_2^+$ and $Q_1^\sigma Q_2^\sigma \in \pi_{Y_1Y_2}J(Q_1)(\mathcal{I})$.

Now, given a query Q, the representative of the equivalence class of Q considered in this paper is the query Q^+, such that $\pi(Q^+) = \pi(Q)^+$, $\bowtie(Q^+) = \bowtie(Q)$ and $\sigma(Q^+)$ is the selection condition corresponding to the super tuple of Q^σ, denoted by $(Q^\sigma)^+$, defined over $sch(Q^\sigma)^+$, and that belongs to $\pi_{sch(Q^\sigma)^+}J(Q)(\mathcal{I})$.

Moreover, if $\pi(Q) \subseteq sch(Q^\sigma)$ then the support of Q is 1, which is meant to be less than the minimum support threshold. Therefore, the queries Q of interest are such that

$$\pi(Q) = \pi(Q)^+, \; sch(Q^\sigma) = sch(Q^\sigma)^+, \text{ and } sch(Q^\sigma) \subset \pi(Q).$$

In what follows, such queries are said to be *closed queries* and the closed query equivalent to a given query Q is denoted by Q^+.

It is important to notice that, considering only such queries in our algorithms, reduces the size of the output set of frequent queries.

Example 3. Referring back to the queries Q_1 and Q_2 of Example 2, it is easy to see that they do *not* satisfy the restrictions above. For instance, as $sch(Q_1^\sigma) = E$ and $\pi(Q_1) = AD$, the inclusion $sch(Q_1^\sigma) \subset \pi(Q_1)$ is not satisfied. It can be seen that none of these queries are closed, and thus, none of them is considered in our algorithms. But as $J(Q_1)(\mathcal{I})$ satisfies $C \to D$, $E \to D$, $A \to C$ and $C \to A$, the closed queries Q_1^+ and Q_2^+ defined below are processed instead.

$$Q_1^+ = \pi_{ACDE}\sigma_{(A=C)\wedge(E=e)}R \text{ and } Q_2^+ = \pi_{ACDE}\sigma_{(A=C)\wedge(E=e)\wedge(D=d)}R.$$

We also note that Q_1 and Q_2 would not be considered either in [8], as in there, $\pi(Q_i)$ $(i = 1, 2)$ is required to contain all attributes from the same block of $blocks(Q_i)$ but no attributes from $\sigma(Q_i)$. Thus, in [8], $Q_1' = \pi_{ACD}\sigma_{(A=C)\wedge(E=e)}R$ and $Q_2' = \pi_{AC}\sigma_{(A=C)\wedge(E=e)\wedge(D=d)}R$ are processed instead. As $Q_i \sim Q_i' \sim Q_i^+$ for $i = 1, 2$, these queries have the same support. □

3 Mining Queries under Functional Dependencies

3.1 Algorithm Conqueror$^+$

In this section, we present our algorithm called Conqueror$^+$ (given as Algorithm 1) for mining frequent queries. We mention in this respect that frequent simple conjunctive queries $\pi_X \sigma_F R$ are mined in much the same way as the Conqueror algorithm [8], that is, according to the following steps:

- **Join loop:** Generate all instantiations of F, without constants, in a breadth-first manner, using restricted growth to represent partitions [8]. Every partition gives rise to a join query JQ and functional dependencies of its ancestors are inherited.
- **Projection loop:** For each generated partition, all projections of the corresponding join query JQ are generated in a breadth-first manner, and their frequency is tested against the given instance $\mathcal{I}$. During this loop, functional dependencies are discovered and used to prune the search space.
- **Selection loop:** For each frequent projection-join query, constant assignments are added to F in a breadth-first manner, as in Conqueror. Moreover, here again, functional dependencies are used to prune the search space.

As in the Conqueror algorithm, attributes are ordered, so as candidate queries are generated at most once in the different loops: This ordering is implicit lines 1 and 12 in Algorithm 1 (the k-th element in the string refers to the k-th attribute according to the ordering), and is explicitly used line 17 in Algorithm 2 and line 10 in Algorithm 3.

As an important difference with the Conqueror algorithm, a (possibly empty) set of functional dependencies $\mathcal{FD}$ can be specified as input. This set is first used for the relations of the database instance (line 3 of Algorithm 1) and then augmented during the projection loop (line 15 of Algorithm 2).

3.2 Join Loop

The generation of joins is done in much the same way as in Conqueror ([8]), by generation of restricted growth strings [18]. Such a restricted growth string represents a partition of the attributes, and such a partition maps to a join.

For example, referring back to Example 1, the set U of all attributes occurring in $\mathcal{D}$ is $\{A, B, C, D, E\}$. Then, the restricted growth string 12231 represents the condition $(A = E) \wedge (B = C)$, which corresponds to the partition $\{\{A, E\}, \{B, C\}, \{D\}\}$.

As in the Conqueror algorithm, we include a check against the user defined most specific join, which allows a user to specify the sensible joins in the database (see line 11, Algorithm 1). By default, however, every possible join of every attribute pair is considered. A new addition to the join loop is the inheritance of functional dependencies shown on lines 13-14, and discussed in detail in Section 3.5.

3.3 Projection Loop

Compared to the Conqueror algorithm, one major change in the projection loop is the fact that the generation of selections is now performed after all projections are generated (line 22, Algorithm 2) so as to be able to immediately use the discovered functional

Algorithm 1. Conqueror$^+$

Input: Database $\mathcal{D}$, Set of functional dependencies $\mathcal{FD}$, Minimum support threshold *minsup*
Output: Frequent Queries FQ

```
1: ⋈(Q) := "1" // initial restricted growth string
2: for all R_i in D do
3:    FD_Q := FD_i
4:    push(Queue, R_i)
   // Join Loop
5: while not Queue is empty do
6:    JQ := pop(Queue)
7:    if ⋈(JQ) does not represent a cartesian product then
8:       FQ := FQ ∪ ProjectionLoop(JQ)
9:    children := RestrictedGrowth(⋈(JQ), m)
10:   for all rgs in children do
11:      if join defined by rgs is not more specific than the user most specific join then
12:         ⋈(JQC) := rgs
13:         for all PJQ such that ⋈(JQC) = ⋈(PJQ) ∧ (R_i.A = R_j.A') do
14:            FD_JQC := FD_JQC ∪ FD_PJQ
15:            if ⋈(PJQ) = "1" then
16:               FD_JQC := FD_JQC ∪ {R_i.A → R_j.A', R_j.A' → R_i.A}
17:         blocks(JQC) := blocks(JQ) where the blocks containing R_i.A and R_j.A' are merged
18:         push(Queue, JQC)
19: return FQ
```

dependencies to prune redundant queries. The functional dependency discovery is performed lines 13-16 of Algorithm 2 and is discussed in Section 3.5.

We point out that, according to lines 17-20 of Algorithm 2, candidate projection queries are generated by removing blocks in $blocks(JQ)$, because attributes in a given block are mutually dependent. However, it might be the case that removing such a block does not result in a closed projection schema. This is why, line 9 of Algorithm 2, we check whether π(PQ) is closed; if not, the projection query is simply queued without any further processing. This however induces complications in the monotonicity check line 10 of Algorithm 2, because projections over non closed schemas are not processed. To cope with this difficulty, if PQ is such that π(PQ) is closed, for every predecessor PPQ of PQ, the closure of π(PPQ) under $\mathcal{FD}_{\mathrm{Q}}$ is computed. The check is passed if all corresponding projection queries are in FPQ.

Also notice that the function blocks(Q) returns the set of connected blocks of a restricted growth string, *i.e.*, the connected part of the partition $blocks(Q)$. We require such blocks to form a single connected component, so as to avoid considering cartesian products, as stated in Definition 1. Clearly, line 7 in Algorithm 1 prunes these queries.

3.4 Selection Loop

In the selection loop of our new algorithm, marked queries are not considered, since they are redundant (line 23, Algorithm 2). When adding blocks to the selection condition, the closure is taken, ensuring no redundant queries are generated (line 13, Algorithm 3).

Algorithm 2. ProjectionLoop

Input: Conjunctive Query Q
1: **if** $\bowtie$(Q) = "1" **then**
2: $\quad\pi$(Q) := $sch(R_i)$ // Q is the query R_i
3: **else**
4: $\quad\pi(Q)$:= **union of** blocks(Q)
5: push(*Queue*, Q)
6: FPQ := $\emptyset$
7: **while not** *Queue* is empty **do**
8: $\quad$PQ := pop(*Queue*)
9: $\quad$**if** π(PQ) is closed **then**
10: $\quad\quad$**if** monotonicty(PQ) **then**
11: $\quad\quad\quad$**if** support(PQ) > *minsup* **then**
12: $\quad\quad\quad\quad$FPQ := FPQ $\cup$ {PQ}
13: $\quad\quad\quad\quad$**for all** PPQ in FPQ such that ($\nexists$ PPQ$'$ $\in$ FPQ : π(PQ) $\subset$ π(PPQ$'$) $\subset$ π(PPQ)) **do**
14: $\quad\quad\quad\quad\quad$**if** support(PQ) = support(PPQ) **then**
15: $\quad\quad\quad\quad\quad\quad\mathcal{FD}_{\mathrm{Q}} := \mathcal{FD}_{\mathrm{Q}} \cup \{\pi(\mathrm{PQ}) \rightarrow \pi(\mathrm{PPQ}) \setminus \pi(\mathrm{PQ})\}$
16: $\quad\quad\quad\quad\quad\quad$mark PQ
17: $\quad$**for all** $\beta > lastremoved$(PQ) **do**
18: $\quad\quad\pi$(PQC) := π(PQ) with block β removed
19: $\quad\quad lastremoved$(PQC) = β
20: $\quad\quad$push(*Queue*, PQC)
21: FQ := FQ $\cup$ FPQ
22: **for all** PQ $\in$ FPQ **do**
23: $\quad$**if** PQ is not marked **then**
24: $\quad\quad$FQ := FQ $\cup$ SelectionLoop(PQ)
25: **return** FQ

However, closing of these sets of blocks requires to reorder the queue of candidates in order to use the Apriori-trick. The following example illustrates this point.

Example 4. Considering the attributes A, B and C, along with the functional dependency $A \rightarrow B$, the generation of sets for the selection results in the generation-tree (a) shown below. Indeed, the addition of A entails that B must also be added so as to consider closed schemas only.

However, because of the monotonicity property, we need to consider B before AB (since the selection according to B is less restrictive than that according to AB). We accomplish this by reordering the candidate queue, to ensure B is considered before AB and BC is considered before ABC, as shown in the generation-tree (b) below. □

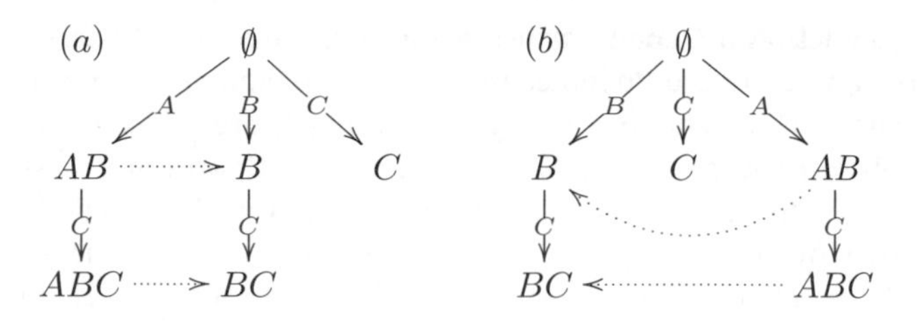

Moreover, as stated previously, line 14 of Algorithm 3 ensures that $\sigma(Q)$ is a strict subset of $\pi(Q)$. However, not all strict subsets of $\pi(PQ)$ are considered, since we only have to consider assignments over closed schemas under $\mathcal{FD}_{JQ}$ (see line 13, Algorithm 3). Furthermore, in line 14 of Algorithm 3, we make sure that the corresponding closure has not been processed previously, which can happen since a closed set can be generated from several non-closed sets.

Then, in lines 7-8 of Algorithm 3, the obtained queries are processed against $\mathcal{I}$ using the same strategy as in [8]. The instantiation of constant values in Algorithm 3 is performed analogously to Conqueror by performing SQL queries in the database. For further details, we therefore refer the reader to [8].

Algorithm 3. SelectionLoop

Input: Conjunctive Query Q
1: push(*OrderedQueue*,Q)
2: **while** not *OrderedQueue* is empty **do**
3: CQ := pop(*OrderedQueue*)
4: **if** σ(CQ) $= \emptyset$ **then**
5: *toadd* := all blocks of π(Q)
6: **else if** monotonicty(CQ) **then**
7: **if** exist frequent constant values for σ(CQ) in $\mathcal{I}$ **then**
8: FQ := FQ $\cup$ instances of CQ
9: *uneq* := all blocks of π(Q) $\notin \sigma$(CQ)
10: *toadd* := all blocks B in *uneq* $>$ last of σ(CQ)
11: **for all** $B_i \in$ *toadd* **do**
12: σ(CQC) := σ(CQ) with B_i added
13: σ(CQC) := closure of σ(CQC) under $\mathcal{FD}_Q$
14: **if** σ(CQC) has not been generated before **and** σ(CQC) is different than π(Q) **then**
15: push(*OrderedQueue*, CQC)
16: **return** FQ

3.5 Handling and Discovering Functional Dependencies

In this section, we show that, according to our algorithms:

1. A given join query is associated with the set of all functional dependencies satisfied by its predecessor join queries.
2. Only join and projection queries over *closed* relation schemas are processed.
3. Considering given functional dependencies along with *discovered* functional dependencies preserves the above property.

Handling Functional Dependencies. A given join query JQ is associated with a set of functional dependencies, denoted by $\mathcal{FD}_{JQ}$, and built up in Algorithm 1 as follows.

First, when $\bowtie$(Q) is the restricted growth string 1, every instantiated relation $R_i(\mathcal{I})$ in the database is pushed in *Queue* (lines 2 and 5, Algorithm 2), associated with the set $\mathcal{FD}_i$ (see line 3, Algorithm 1). Then, the restricted growth strings represent a join condition of the form $(R_i.A = R_j.A')$. Denoting by JQ the corresponding join query, if $R_i = R_j$ then $JQ(\mathcal{I})$ satisfies $\mathcal{FD}_i$ (since JQ is a selection of R_i) along with $R_i.A \rightarrow R_i.A'$ and $R_i.A' \rightarrow R_i.A$. Thus, $\mathcal{FD}_{JQ}$ is set to $\mathcal{FD}_i \cup \{R_i.A \rightarrow R_i.A', R_i.A' \rightarrow R_i.A\}$. Similarly, if $R_i \neq R_j$, then JQ is a join of R_i and R_j, and so, $JQ(\mathcal{I})$ satisfies $\mathcal{FD}_i \cup \mathcal{FD}_j$, as well as $R_i.A \rightarrow R_j.A'$ and $R_j.A' \rightarrow R_i.A$. Thus, we set $\mathcal{FD}_{JQ} = \mathcal{FD}_i \cup \mathcal{FD}_j \cup \{R_i.A \rightarrow R_j.A', R_j.A' \rightarrow R_i.A\}$ (see lines 13-16 of Algorithm 1). At this stage, $\pi(JQ)$ is either $sch(R_i)$ (if $R_i = R_j$) or $sch(R_i) \cup sch(R_j)$ (if $R_i \neq R_j$), and so, $\pi(JQ)$ is closed under $\mathcal{FD}_{JQ}$.

In the general case, at a given level, the join query JQ is generated from join queries PJQ in the previous level by setting $\bowtie(JQ)$ to $\bowtie(PJQ) \wedge (R_i.A = R_j.A')$, and by augmenting $\pi(PJQ)$ accordingly. Therefore, $JQ(\mathcal{I})$ satisfies the dependencies of $\mathcal{FD}_{PJQ}$, and thus, $\mathcal{FD}_{JQ}$ *is set to be the union of all* $\mathcal{FD}_{PJQ}$ *where* PJQ *allows to generate* JQ (see lines 13-14 of Algorithm 1). Consequently, assuming that $\pi(PJQ)$ is closed under $\mathcal{FD}_{PJQ}$ clearly entails that $\pi(JQ)$ is closed under $\mathcal{FD}_{JQ}$.

Thus, for every join query JQ, $\pi(JQ)$ is closed under those functional dependencies of $\mathcal{FD}_{JQ}$ that belong to $\mathcal{FD}$ or that are obtained through the connected blocks of $blocks(JQ)$. Moreover, the discovered functional dependencies in the projection loop of JQ preserve this property, because these new dependencies are defined with attributes in $\pi(JQ)$ only. Thus, *for every join query* JQ, $\pi(JQ)$ *is closed under* $\mathcal{FD}_{JQ}$.

Then, the check performed line 9 of Algorithm 2 ensures that only those projection-join queries PQ such that $\pi(PQ)$ is closed under $\mathcal{FD}_{JQ}$ are considered. We note that for performing this check, it is enough to make sure that there is no dependency $X \rightarrow Y$ in $\mathcal{FD}_{JQ}$ such that $X \subseteq \pi(PQ)$ and $Y \not\subseteq \pi(PQ)$.

Discovering Functional Dependencies. Functional dependencies, other than those in $\mathcal{FD}$, are *discovered* in the projection loop (see lines 13-16 of Algorithm 2) as follows. At a given level, a projection-join query PQ is generated from the projection-join queries PPQ of the previous level by removing blocks from $\pi(PPQ)$. Thus, by Proposition 1, if $support(PQ) = support(PPQ)$ (see line 14 of Algorithm 2), $JQ(\mathcal{I})$ satisfies $\pi(PQ) \rightarrow \pi(PPQ) \setminus \pi(PQ)$. The dependency is thus added to $\mathcal{FD}_{JQ}$ and PQ is marked, since $\pi(JQ)$ is no longer closed (see lines 15 and 16 of Algorithm 2).

Notice that, as projection-join queries are generated in a breadth-first manner, the 'best' functional dependencies (*i.e.,* those with minimal left-hand side) are discovered last, during the projection loop. However, by doing so, we mark all queries that do not have to be processed in the selection loop. The following example illustrates this point.

Example 5. In the context of Example 1, let us consider the projection loop associated to the join query $JQ = \pi_{ABCDE}\sigma_{(A=C)}R$. In this case, $blocks(JQ) = \{\{A, C\}, \{B\}, \{D\}, \{E\}\}$. Assuming that all projections are frequent and that $JQ(\mathcal{I})$ satisfies $A \rightarrow D$, the following dependencies are found: $ACBE \rightarrow D$, $ACE \rightarrow D$, $ACB \rightarrow D$ and $AC \rightarrow D$. Consequently, the queries $\pi_{ACBE}(JQ)$, $\pi_{ACE}(JQ)$, $\pi_{ACB}(JQ)$ and $\pi_{AC}(JQ)$ are marked, and so, are not processed by the selection loop.

We note that, $A \rightarrow D$ is actually not found, because $\mathcal{FD}_{JQ}$ contains $A \rightarrow C$ and $C \rightarrow A$, which enforces A and C to appear together in the projections. Of course, $A \rightarrow D$ is a consequence of $AC \rightarrow D$ and $A \rightarrow C$ that now belong to $\mathcal{FD}_{JQ}$. □

The output of the projection loop is processed in the selection loop of Algorithm 3 as follows: for every *non marked* frequent projection-join query PQ, selections over closed schemas are generated breadth-first by assigning constant values to some of the attributes in $\pi(PQ)$.

4 Experimental Results

The Conqueror$^+$ algorithm was written in Java using JDBC to communicate with a sqlite relational database. Experiments were run on a standard computer with 2GB RAM and a 2.16 GHz processor. We also note that this implementation not only computes frequent queries, but also association rules. The issue of association rules is not addressed in this paper, due to lack of space. We performed experiments using Conqueror$^+$ and compared it to Conqueror [8]. We used the backend database of an online quiz website [2] and a snapshot of the Internet Movie Database (IMDB) [11]. The characteristics of these databases are shown in Table 1.

Table 1. Number of tuples per attribute in the QuizDB and IMDB databases

(a) Quiz database

Attribute	Tuples
SCORES.*	868755
SCORES.SCORE	14
SCORES.NAME	31934
SCORES.QID	5144
SCORES.DATE	862769
SCORES.RESULTS	248331
SCORES.MONTH	12
SCORES.YEAR	6
QUIZZES.*	4884
QUIZZES.QID	4884
QUIZZES.TITLE	4674
QUIZZES.AUTHOR	328
QUIZZES.CATEGORY	18
QUIZZES.LANGUAGE	2
QUIZZES.NUMBER	539
QUIZZES.AVERAGE	4796

(b) IMDB

Attribute	Tuples
ACTORS.*	45342
ACTORS.AID	45342
ACTORS.NAME	45342
GENRES.*	21
GENRES.GID	21
GENRES.NAME	21
MOVIES.*	71912
MOVIES.MID	71912
MOVIES.NAME	71906
ACTORMOVIES.*	158441
ACTORMOVIES.AID	45342
ACTORMOVIES.MID	54587
GENREMOVIES.*	127115
GENREMOVIES.GID	21
GENREMOVIES.MID	71912

4.1 Impact of Dependency Discovery

We performed four types of experiments with functional dependencies. As a first type, we executed the regular Conqueror. The second type, denoted 'disc' in Figure 1, is Conqueror$^+$ where discovery of dependencies is enabled, but the set of initial provided dependencies is empty. The third type, denoted 'given', is Conqueror$^+$ provided with a set of initial dependencies, but without any discovery of functional dependencies. For QuizDB we provided the key dependencies of the QUIZZES and SCORES relations, and for IMDB we provided the key dependencies for ACTORS, GENRES and MOVIES. The fourth type, denoted as 'given+disc', is Conqueror$^+$ provided with these dependencies as well as discovery of new functional dependencies.

As can be seen in Figure 1a, Conqueror$^+$ with discovery greatly outperforms Conqueror in runtime. This is due to the large reduction in number of queries generated which is clear from the figure. Adding an initial set of (key) functional dependencies results in a small gain in runtime, due to a small reduction in number of queries generated. Similarly, providing a set of dependencies whilst also discovering new ones, results in a small relative gain. We also observe that the exponential behavior of query generation is still present, but only for low support values. Furthermore, for Conqueror$^+$ with discovery, runtime remains almost linear for a large portion of the support values, while for Conqueror, it is increasing rapidly.

The experiments on the IMDB shown in Figure 1b show similar results, but in this case, the impact of discovery is smaller. The small amount of attributes in the database reduces the impact of the use of functional dependencies. It is however clear that also in this case, the discovery of functional dependencies reduces the exponentiality of query generation and has an almost linear runtime. Likewise the small impact of providing key dependencies as input to the algorithm, is comparable to QuizDB.

We also performed some time analysis to determine the cost of functional dependency discovery. The results for an experiment using QuizDB are shown in Figure 2. It is clear that the time needed for the discovery of functional dependencies (shown as 'fdisc' in Figure 2a) is negligible in comparison to the time gained in the selection loop (shown as 'sel'). Adding discovery also requires extra time in the join loop (shown as 'join'), but again, the gain in the selection loop outweighs this. Looking at the partitioning of time

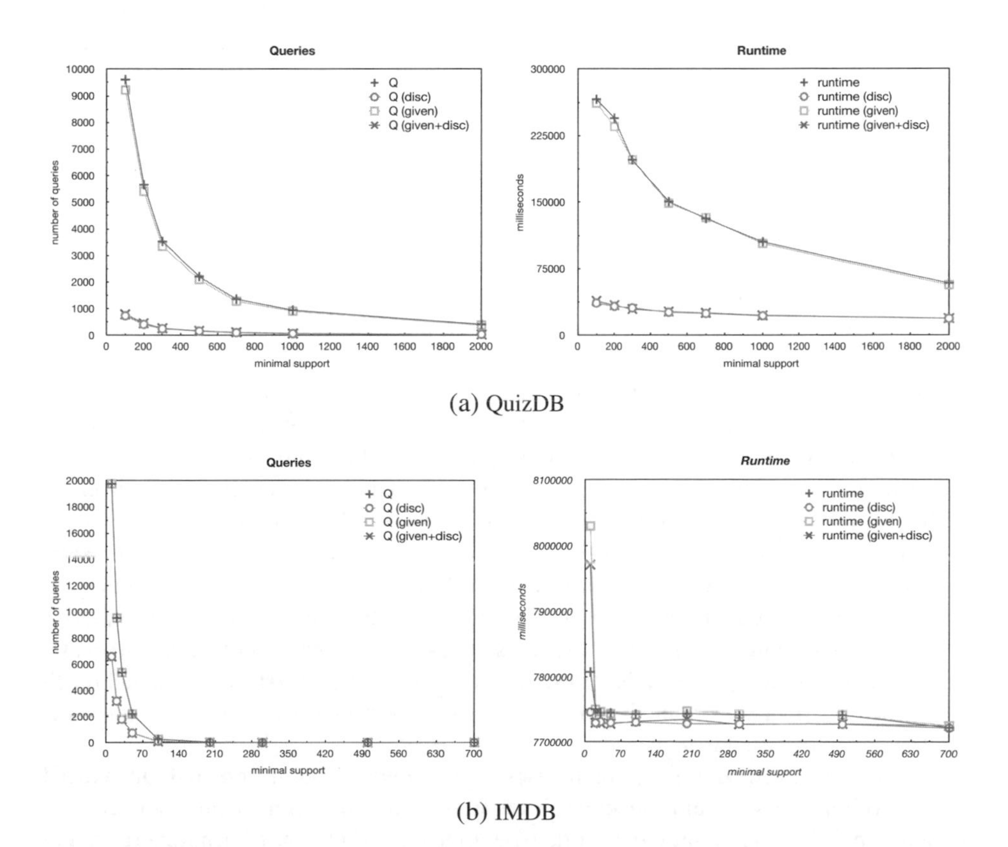

Fig. 1. Results for Conqueror, Conqueror$^+$ with a set of Functional Dependencies given but no detection (given), Conqueror$^+$ with only detection of Functional Dependencies (disc), and Conqueror$^+$ both with given Functional Dependencies and detection.

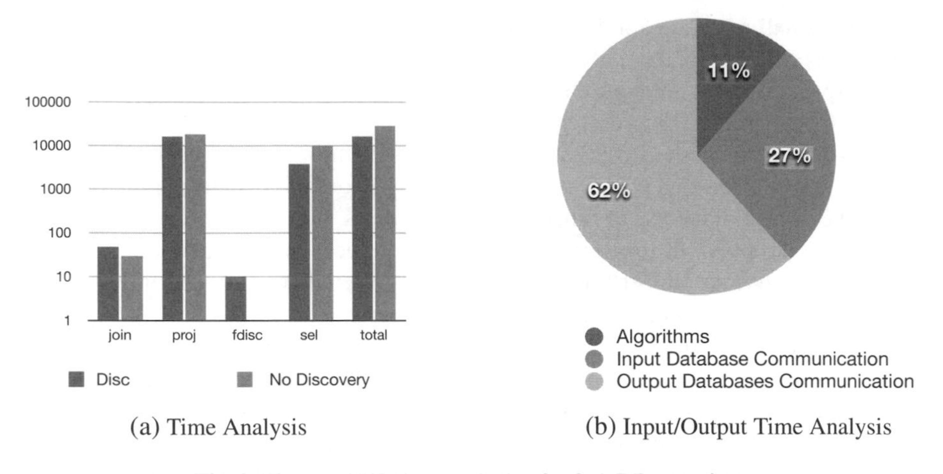

(a) Time Analysis

(b) Input/Output Time Analysis

Fig. 2. Time and I/O time analysis of a QuizDB experiment

in Figure 2b, we clearly see that most time is spent in output and input. Since functional dependency discovery in Conqueror$^+$ greatly reduces output and input, we get a large reduction in runtime as was observed in Figure 1.

5 Related Work

Mining frequently occurring patterns in arbitrary relational databases has been the topic of several research efforts. Dehaspe and Toivonen developed the WARMR algorithm [4], that discovers association rules in over a limited type of Datalog queries in an Inductive Logic Programming setting. The input to their algorithm consists of a collection of databases, and then, queries are generated in a level-wise manner, and each candidate query is evaluated against all of these databases. The frequency of a query is the number of databases for which it gives a nonempty answer. Therefore, the interpretation of frequent queries is incomparable to the conjunctive queries considered in this paper.

In [6], we studied a strict generalization of WARMR. The notion of *diagonal containment* provided an excellent tool to compare queries with different sets of projected attributes. Unfortunately, the search space is infinite and there exist no most general and no most specific patterns. However, the subclass of tree-shaped conjunctive queries defined over a single binary relation representing a graph was studied, showing that these tree queries are powerful patterns, useful for mining graph-structured data [7,10]. In [8], we considered conjunctive queries over several relations, allowing a more efficient algorithm, called Conqueror.

Considering projection-selection queries over a single relation, Jen et al. introduced a new notion of query equivalence [13], taking functional dependencies into account, which is not the case in previous work. The approach of [13] has been generalised in [14] to databases defined over multiple relations, organised according to a star schema.

All approaches other than those discussed just above and dealing with mining frequent queries ([5,9,16,17]) are far more restrictive than ours. Indeed, whereas our approach considers several tables and all possible ways to count supports as distinct values over all possible attribute sets, all these approaches consider a fixed relation to be mined, along with a fixed characterisation of how to count supports. For instance, in [9,16] tuples are counted, Turmeaux et al. [17] characterize counting by tuple values over a given attributes, whereas Diop et al. [5] characterize counting by a query, called the reference. Notice that all these approaches (except for [17]) are also restricted to conjunctive queries, as is the case in this paper.

6 Concluding Remarks

The contribution of this paper is threefold. First, we combined the results of different prior work resulting in a new algorithm for mining association rules over simple conjunctive queries in arbitrary relational databases, over which functional dependencies are assumed. The algorithm makes use of the functional dependencies of the database to optimise the generation of frequent queries and prune redundant queries. Second, our new algorithm is capable of detecting new functional dependencies that were not given but that hold on the database relations or on any join of these relations. Third, these newly detected dependencies are used to prune even more redundant queries. We implemented our algorithm, and we showed that it greatly outperforms our previous methods and efficiently reduces the amount of queries generated.

Several new opportunities for future work exist. First, the additional use of key and foreign key contraints is an issue that we are currently investigating. Other appealing related constraints are *conditional functional dependency*, introduced by Fan et al. [3]. As these constraints generalise standard functional dependencies using selections, it seems interesting to investigate how they could be used and discovered in our framework.

References

1. Agrawal, R., Mannila, H., Srikant, R., Toivonen, H., Verkamo, A.I.: Fast discovery of association rules. In: Advances in Knowledge Discovery and Data Mining, pp. 309–328. AAAI-MIT Press (1996)
2. Bocklandt, R.: `http://www.persecondewijzer.net`
3. Bohannon, P., Fan, W., Geerts, F., Jia, X., Kementsietsidis, A.: Conditional functional dependencies for data cleaning. In: ICDE, pp. 746–755 (2007)
4. Dehaspe, L., De Raedt, L.: Mining association rules in multiple relations. In: Džeroski, S., Lavrač, N. (eds.) ILP 1997. LNCS, vol. 1297, pp. 125–132. Springer, Heidelberg (1997)
5. Diop, C.T., Giacometti, A., Laurent, D., Spyratos, N.: Composition of mining contexts for efficient extraction of association rules. In: Jensen, C.S., Jeffery, K., Pokorný, J., Šaltenis, S., Bertino, E., Böhm, K., Jarke, M. (eds.) EDBT 2002. LNCS, vol. 2287, pp. 106–123. Springer, Heidelberg (2002)
6. Goethals, B., Van den Bussche, J.: Relational association rules: getting warmer. In: Hand, D.J., Adams, N.M., Bolton, R.J. (eds.) Pattern Detection and Discovery. LNCS (LNAI), vol. 2447, pp. 125–139. Springer, Heidelberg (2002)
7. Goethals, B., Hoekx, E., Van den Bussche, J.: Mining tree queries in a graph. In: ACM KDD, pp. 61–69 (2005)
8. Goethals, B., Le Page, W., Mannila, H.: Mining association rules of simple conjunctive queries. In: SIAM-SDM, pp. 96–107 (2008)

9. Han, J., Fu, Y., Wang, W., Koperski, K., Zaiane, O.: Dmql: A data mining query language for relational databases. In: SIGMOD-DMKD 1996, pp. 27–34 (1996)
10. Hoekx, E., Van den Bussche, J.: Mining for tree-query associations in a graph. In: IEEE ICDM, pp. 254–264 (2006)
11. IMDB (2008), `http://imdb.com`
12. Inokuchi, A., Washio, T., Motoda, H.: An Apriori-based algorithm for mining frequent substructures from graph data. In: Zighed, D.A., Komorowski, J., Żytkow, J.M. (eds.) PKDD 2000. LNCS (LNAI), vol. 1910, pp. 13–23. Springer, Heidelberg (2000)
13. Jen, T.Y., Laurent, D., Spyratos, N.: Mining all frequent selection-projection queries from a relational table. In: EDBT 2008, pp. 368–379. ACM Press, New York (2008)
14. Jen, T.Y., Laurent, D., Spyratos, N.: Mining frequent conjunctive queries in star schemas. In: International Database Engineering and Applications Symposium (IDEAS), pp. 97–108. ACM Press, New York (2009)
15. Kuramochi, M., Karypis, G.: Frequent subgraph discovery. In: IEEE ICDM, pp. 313–320 (2001)
16. Meo, R., Psaila, G., Ceri, S.: An extension to sql for mining association rules. Data Mining and Knowledge Discovery 9, 275–300 (1997)
17. Turmeaux, T., Salleb, A., Vrain, C., Cassard, D.: Learning caracteristic rules relying on quantified paths. In: Lavrač, N., Gamberger, D., Todorovski, L., Blockeel, H. (eds.) PKDD 2003. LNCS (LNAI), vol. 2838, pp. 471–482. Springer, Heidelberg (2003)
18. Weisstein, E.W.: Restricted growth string. In: From MathWorld – A Wolfram Web Resource (2009), `http://mathworld.wolfram.com/RestrictedGrowthString.html`
19. Yan, X., Han, J.: gspan: Graph-based substructure pattern mining. In: IEEE ICDM, p. 721 (2002)
20. Zaki, M.J.: Efficiently mining frequent trees in a forest. In: ACM KDD, pp. 71–80 (2002)

Using Transitivity to Increase the Accuracy of Sample-Based Pearson Correlation Coefficients

Taylor Phillips, Chris GauthierDickey*, and Ramki Thurimella*

University of Denver
Department of Computer Science
taylorphillips@du.edu, {chrisg,ramki}@cs.du.edu

Abstract. Pearson product-moment correlation coefficients are a well-practiced quantification of linear dependence seen across many fields. When calculating a sample-based correlation coefficient, the accuracy of the estimation is dependent on the quality and quantity of the sample. Like all statistical models, these correlation coefficients can suffer from overfitting, which results in the representation of random error instead of an underlying trend.

In this paper, we discuss how Pearson product-moment correlation coefficients can utilize information outside of the two items for which the correlation is being computed. By introducing a transitive relationship with one or more additional items that meet specified criterion, our Transitive Pearson product-moment correlation coefficient can significantly reduce the error, up to over 50%, of sparse, sample-based estimations. Finally, we demonstrate that if the data is too dense or too sparse, transitivity is detrimental in reducing the correlation estimation errors.

1 Introduction

Statistical models are used in the day-to-day lives of modern humans. Alleviating traffic congestion, predicting weather patterns, or investing in the stock market are all common examples of such models. When insufficient quantities of data are used by these models, they exhibit a phenomenon known as *overfitting*. This overfitting causes the models to display random error instead of an underlying trend, which in turn makes it difficult to utilize the results in a sensible fashion.

We propose an algorithm that reduces the effects of overfitting by using information in the data set other than that which the statistical model was built to utilize. For the purpose of discussion, we focus on a particular, ubiquitous example of a statistical model that is susceptible to overfitting known as the *Pearson product-moment correlation coefficient* (PMCC). This PMCC measures the correlation, or linear dependence, between two vectors, i and j, and relies

* This research was funded in part by the National Science Foundation under Grant No. DUE–0911991. Any opinions, findings and conclusions or recommendations expressed in this material are those of the author(s) and do not necessarily reflect those of the National Science Foundation.

T.B. Pedersen, M.K. Mohania, and A M. Tjoa (Eds.): DaWaK 2010, LNCS 6263, pp. 157–171, 2010.

solely on the intersection of those two vectors. Our proposed algorithm works by finding transitive neighbors, ks, such that the ks are the vectors in the data set most similar to i. These ks are then used to form estimates for i's relationship with j, allowing our algorithm to incorporate auxiliary information that is normally disregarded. Note that statistically PMCC does not exhibit transitivity: i.e., if X and Y are correlated and Y and Z are correlated, then X and Z are *not necessarily* correlated. Our goal, however, is to try to exploit those cases where the transitive relationship exists.

The existing approaches to alleviating the effects of overfitting do not address this issue directly. Instead, techniques *specific* to particular applications have been developed. Our work, which uses the idea of transitivity, could in theory be applied to improve estimations of many statistical models that use sparse data.

To quantify the performance of the algorithms presented in this paper, the Netflix Prize data set was used. This readily available data consists of approximately 100 million user-movie pairs. The results demonstrate that our Transitive Pearson product-moment correlation coefficient algorithm can reduce the error by up to 50% of the PMCC approximations in sparse data sets.

The notion of utilizing transitivity in statistical models to reduce the effects of sparse data is both abstract and powerful. The algorithm proposed in this paper is important because it is the first to demonstrate a significant reduction of error in sparse, sample-based PMCC estimations. PMCCs find uses in education, psychology, physics, mathematics, economics, and finance, all of which can suffer from overfitting and can subsequently benefit from the ideas presented in this paper. Further, this notion of neighbor transitivity used by our algorithm could be extended to reduce the error of other statistical models operating on sparse data.

2 Background

Collaborative filtering (CF) is the process by which users rate material in a collection or database of materials so that other users may use those ratings to help them select materials they are interested in [8]. The first opportunities for collaborative filtering came through the Internet where users were able to provide real time feedback on a product or service. Tivo [2] and Amazon were some of the first commercial entities to take advantage of it and an analysis of collaborative filtering in Amazon demonstrated a 20% increase in sales attributed to personalization through CF [11]. Adomavicius and Tuzhilin provide an extensive survey of collaborative filtering techniques [1].

Recently, the Netflix Prize offered $1,000,000 to anyone who could improve the accuracy of Netflix's proprietary movie recommendation algorithm by 10% [13,16,12,7,14]. Their algorithm was designed to recommend movies to customers based on how a specific customer rated his previously viewed movies. This same task of recommending movies could also be looked as the task of predicting a rating for an unseen movie, and then recommend movies with the highest predicted rating. Thus, the goal of the Netflix Prize was to improve the accuracy of the predictions of unseen movies.

Bell and Koren's work towards the Netflix Prize focused on nearest-neighbor models, latent-factor models, and combining those models together [3,5,10]. A synopsis of the research by Bell et al. on the Netflix Prize can be found in [6,4]. While being related to our own work, we are more interested in estimating the measures of similarity employed by these various algorithms.

The most widely used correlation coefficient in the Netflix Prize was the Pearson product-moment correlation coefficient (PMCC) [3,5,10,9,15], where it was used directly as a measure of similarity or to recommend an alternative. PMCCs are used as a measure of similarity in numerous applications of many different models, including matrix factorization and the nearest neighbor models.

Like any statistical model, the usefulness of the correlation coefficients is contingent on having sufficient data. While the Netflix dataset is large, the ratings are sparse and thus the challenge is to find relationships between users, movies and their ratings. Using the dataset, the PMCC is used to measure the correlation between two user's ratings. Thus, to compute the correlation coefficient for two users, an overlap in movies seen is needed to draw any conclusion about the relation between those two users. Our algorithm does not have this requirement and is able to make estimations of correlation coefficients when there is absolutely no data of this kind. This in turn allows models that make use of Pearson correlation coefficients to improve estimations of sample-based data and even make predictions that simply were not possible without this technique.

3 Motivation

Quantifying a relationship between users or items is an important component of collaborative filtering, with *similarity* being a measure of this relationship [3,5,10,9,15]. Similarity is then used to weight different opinions proportionally to the similarity of opinion between items. Determining the distribution of the weight from similarity is a specific focus in [5,10], which demonstrate how crucial similarity and weighting are when forming approximations from sparse data. There are multiple interpretations of similarity, but one commonly accepted method is correlation (linear or otherwise) [3,9,15].

3.1 Pearson Product-Moment Correlation Coefficient

The *Pearson product-moment correlation coefficient* (PMCC) is a measure of linear dependence between two vectors, i and j, in the range of [-1,1]. A PMCC of 1 indicates an exact positive correlation, -1 indicates an exact negative correlation, while 0 indicates there is no linear relationship. The formula for the PMCC of i and j can take many forms, one of which is shown in Eq. 1. The PMCC of i and j, dubbed r_{ij}, is based only on $i \cap j$ which are the points of data common between both variables. We will refer to $|i \cap j|$ or $|r_{ij}|$ as the *direct sample size* of i and j.

$$r_{ij} = \frac{\sum_x (i_x - \bar{i})(j_x - \bar{j})}{\sqrt{\sum_x (i_x - \bar{i})^2}\sqrt{\sum_x (j_x - \bar{j})^2}} \tag{1}$$

PMCCs are a standard for measuring linear dependence and thus, play a role in many fields ranging from math and statistics to social sciences and psychology. These correlations are, however, limited by the classic phrase that correlation does not not equal causation. This means that, for example, although temperature and humidity are negatively correlated, it does not imply that the increase in temperature caused the reduction in humidity. Correlations can still provide insight because they demonstrate that historical data indicates that there simply is a negative correlation, regardless of cause.

Other measures of similarity used in collaborative filtering include Euclidean distance and the Cosine similarity. Euclidean distance, defined as $d(i,j) = \sqrt{\sum_{x=1}^{n} i_x^2 - j_x^2}$, finds a natural usage when dealing with spatial proximities. The Cosine similarity finds the angle between vectors just as PMCCs find the slope between vectors and takes the form $cos(\theta) = \frac{i \cdot j}{||i|| ||j||}$.

3.2 Overfitting

PMCC can be limited in practice due to its susceptibility to *overfitting.* Overfitting, also referred to as inductive bias, is a symptom exhibited by statistical models that causes them to display random error instead of an underlying relationship. This means that a statistical model can indicate a relationship that is not true as a result of insufficient data. PMCCs are often used in sample-based scenarios which can have a small set of data that is not guaranteed to be representative of the theoretical, complete set of data. For example, if a statistical model relies on a single point of data, that point could be an outlier causing the model to predict incorrectly.

An alternative way of understanding overfitting is rooted in the law of large numbers. The law of large numbers states that the more data points that exist for a random variable, the more likely that data is to be representative of the expected value of that random variable. Rolling a single dice has six possible outcomes or values, all of which are equally likely. Because each outcome is equally likely, the expected value can be computed as the average of all outcomes, which for this example is $(1+2+3+4+5+6)/6 = 3.5$. If the dice is rolled only once yielding a one, there will be significant error if that single roll is assumed to be indicative of all possible rolls. Furthermore, statisticians are able to provide a confidence interval using the law of large numbers. That is, a sample size and interval may be specified such that the sample mean will fall within the specified interval the desired percentage of the time.

This idea may be applied analogously to the Netflix Prize where a given user's rating is the random variable in question. If we have only one rating, that rating is not necessarily representative of the long term opinions of that user. The law of large numbers states that the more ratings we have, the more likely that data will be representative of the long term. Thus, when dealing with an incomplete set of data, as in collaborative filtering, it is important to understand and account for this overfitting.

Various heuristics exist to curb overfitting and arrive at a more conservative estimate. This may be beneficial, or even necessary in some situations, but the ability of these techniques to significantly improve the estimations is limited. Such heuristics can be as simple as skewing the original, overfitted value towards the mean of all values. In this paper, we propose an algorithm that minimizes the effects of overfitting of sample-based PMCCs by using information other than $i \cap j$. The myriad of applications of PMCCs can also suffer from overfitting and can subsequently benefit from the ideas presented in this paper.

4 Algorithms

In this section we describe two heuristics and our Transitive PMCC algorithm. For the following sections we assume there is a universe of vectors, for which a PMCC could be computed between any two vectors using Eq. 1. The algorithms will operate on some original PMCC, r_{ij}, and yield a new PMCC, r'_{ij}, that is intended to replace the r_{ij} for all subsequent applications.

4.1 Heuristics

The two heuristics presented in this section dampen the effects of overfitting by reducing the reliance on the data specific to r_{ij}. This is done by using a linear combination of the original r_{ij} and some given constant. The first heuristic algorithm, HeuristicA, takes two constants α and C where α is the linear combination weight given to r_{ij} and $1 - \alpha$ is given to C. For our purposes we choose $C = 0$, indicating that the more weight C gets, the more it would transform r_{ij} to zero which, for PMCCs, means that there is no linearly dependent relationship between i and j. This choice of C curbs overfitting by skewing the actual r_{ij} towards this conservative value. HeuristicA is descibed formally in Eq. 2.

$$HeuristicA_{ij} = \alpha r_{ij} + (1 - \alpha)C = \alpha r_{ij} \tag{2}$$

Note that α is not dependent on anything and thus, the linear combination weight given to r_{ij} is fixed. The problem with HeuristicA is that regardless of the direct sample size of r_{ij}, the linear combination weight remains fixed. Thus, an r_{ij} with a very large direct sample size would recieve exactly α weight just as an r_{ij} with a very small direct sample size. An improvement can be made by having the linear combination weight of r_{ij} be a function of the direct sample size of r_{ij}. This means that when there is a smaller direct sample size r_{ij} will get less weight, but as direct sample size increases r_{ij} gets more weight. This is useful because as direct sample size increases, the effects of overfitting ought to decrease and thus, the original value can be weighted more heavily.

HeuristicB uses $|r_{ij}|$ to arrive at a weight for r_{ij} that is more appropriate for the specific pair of i and j. It is described in Eq. 3 where β is the linear combination weight of r_{ij} and C is a chosen constant. As with HeuristicA, $C = 0$ was chosen so that the linear combination would be skewed towards 0, the equivalent of no relationship. For $\beta = 5$ with a direct sample size of 95, HeuristicB gives r_{ij} 95% of the linear combination weight and only 5% to no relationship. If the direct sample size were only 5, r_{ij} would receive only 50% of the weight while no relationship would also get 50%. By determining a linear combination weight for r_{ij} from the direct sample size of r_{ij}, HeuristicB incorporates the conservative estimate with small amounts of data, but has little impact when there is larger amounts.

$$HeuristicB_{ij} = r_{i,j}\frac{|r_{ij}|}{|r_{ij}|+\beta} + (1 - \frac{|r_{ij}|}{|r_{ij}|+\beta})C = r_{i,j}\frac{|r_{ij}|}{|r_{ij}|+\beta} \tag{3}$$

4.2 Transitive PMCC

Our proposed Transitive PMCC algorithm (TPMCC) works to find information beyond $i \cap j$ to develop a stronger estimate for r'_{ij}. This extra information is rooted in the neighbors that are chosen to represent i's relationship with j. That is, the TPMCC algorithm takes the items most similar to i and uses their relationships with j to estimate i's relationship with j. To determine an ordering of neighbors for i by similarity, we use $abs(r_{ik})$, such that neighbors most similar will have a strong correlation. This strong linear dependence can be either positive or negative denoted by the absolute value.

The process of selecting a set of neighbors for a given pair, i and j, begins by examining all possible neighbor candidates, k. The candidates are then narrowed down, keeping only those whose $abs(r_{ik}) > \delta$ for which δ is some chosen constant. Additionally, we want to require some sufficient direct sample size on r_{ik} and r_{kj} so that we can have a degree of certainty that the neighbors themselves aren't suffering from overfitting. These constraints take the form of $|r_{ik}| \geq \gamma_{ik}$ and $|r_{kj}| \geq \gamma_{kj}$. We then take our final neighbor set N_{ij}, as the set of all ks that meet the previously stated criterion with δ, γ_{ik}, and γ_{kj}. The number of neighbors in N_{ij} will be referred to as the *transitive sample size*. The Transitive PMCC algorithm is then described in Eq. 4 where $w(i,j)$ is the weight of the actual r_{ij} and $w(i,j,k)$ is the weight of neighbor k.

$$TPMCC_{ij} = \frac{r_{ij}w(i,j) + \sum\limits_{k \in N_{ij}} r_{kj}w(i,j,k)}{w(i,j) + \sum\limits_{k \in N_{ij}} w(i,j,k)} \tag{4}$$

In order to examine all possible neighbor candidates, every unique PMCC must be computed. This step alone has an asymptotic complexity of $O(n^2)$ in running time where n is the number of vectors in the universe. The TPMCC algorithm then examines all $n-2$ neighbor candidates for each of the $O(n^2)$ unique PMCCs, making the asymptotic complexity of the Transitive PMCC algorithm $O(n^3)$. This is somewhat alleviated by being trivially executed in parallel, but the cubic complexity must be considered.

5 Experimental Methodology

The Netflix Prize data set was used to experimentally measure the performance of the heuristics and TPMCC algorithm presented in the previous section. This data set contains the rating history for 480,189 users and 17,770 movies with a total of 100,480,507 ratings. The movies data consists of a title, release year, and a unique identifier while the users consist of a unique identifier only. Lastly, the ratings data consists of a unique user identifier, a unique movie identifier, date of the rating, and a value of the rating ranging from one to five.

One hundred million ratings may appear substantial, but it only represents 1% of the total possible ratings. That is, if every user rated every movie then every possible rating would already be known, while in actuality 99% of those ratings are missing. This missing data complicates the use of PMCCs as they are based on only a subset of the possible data. Thus, the goal is to compute the PMCCs of the complete set of data using only a subset of the data. These sets could be thought of as a grading set and training set respectively. In our first set of experiments, we examine the effects of training the TPMCC algorithm on 50% of the data set (or 0.5% of the total possible ratings) to predict against the second half of the data. Subsequently, we will discuss how different amounts of training data influence the results.

For our purposes, the only points of data used were the rating's unique user identifier, the unique movie identifier, and the rating value. Using only one random half of the ratings data the PMCCs were computed for all pairs of movies. With 17,770 movies this results in 157,877,565 unique pairs of movies, each with their own PMCC. The PMCC for movies i and j in this set will be denoted $Original_{ij}$. Another set of PMCCs, *Final*, was computed using the entire set of ratings data and is used to grade the accuracy of the *Original* PMCCs and the PMCCs created by the algorithms.

The *Original* PMCCs will be used by the heuristics and TPMCC algorithm as input to provide new estimates for the *Final* PMCCs. To quantify the error between any two sets of PMCCs, we use the root mean-squared error (RMSE). The formula for RMSE is shown in Eq. 5 where a and b are sets of PMCCs of size n.

$$RMSE = \sqrt{\frac{1}{n}\sum_{x=1}^{n}(a_x - b_x)^2} \tag{5}$$

The resulting RMSE between *Original* and *Final* is 0.468. Theoretically, the worst possible RMSE could be 2.0. This would happen if, for example, the *Final* PMCCs were all 1 and all of the *Original* PMCCs were -1. However, given a distribution of data and predicting the mean yields much lower measures of error in practice. For example, the RMSE of absolutely no data, which is predicting 0 for every PMCC, yields an RMSE of 0.542. This means that using the *Original* PMCCs computed using half of the data only reduced the error of predicting 0 for all PMCCs by 13.7%.

6 Results

6.1 Heuristics

Each heuristic was run using the *Original* PMCCs as input yielding two new sets of PMCCS, *HeuristicA* and *HeuristicB*. A plot of the RMSE of *HeuristicA* is shown in Figure 1a for different values of α. In this plot it is visible that the RMSE of *HeuristicA* is minimized for $\alpha = 0.6$, which reduced the RMSE to 0.425 - a 9.1% reduction of the RMSE of *Original*. The value of α that achieved the lowest RMSE is between 0 and 1, indicating that *Original* does suffer from overfitting and benefits from the HeuristicA algorithm. If $\alpha = 0$ or $\alpha = 1$ yielded the least RMSE, it would mean that predicting 0 for all PMCCs was best or using the unmodified *HeuristicA* was best, respectively. When $\alpha = 0.6$, HeuristicA is going to scale the *Original* PMCCs down to 60% of the linear combination weight and give 40% to 0. The contrast with the effects of HeuristicA on the PMCCs produced by the TPMCC algorithm as also shown in Figure 1a will be discussed in the following section.

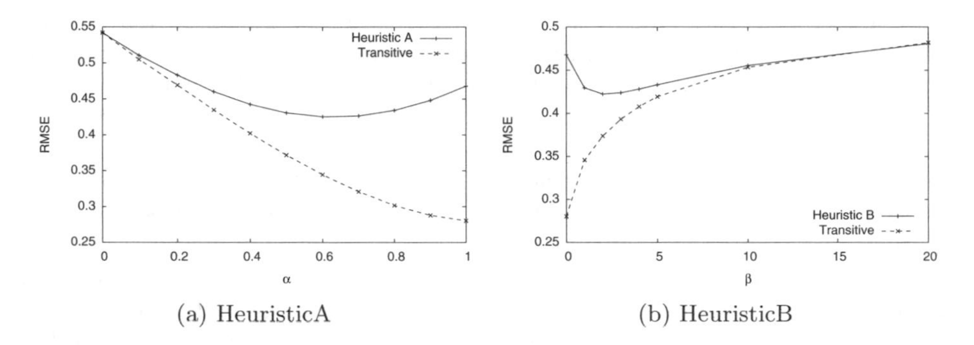

(a) HeuristicA (b) HeuristicB

Fig. 1. *RMSE of HeuristicA vs. α and HeuristicB vs. β:* HeuristicA exhibits the lowest RMSE for *Original* when α is 0.6 for a 9.1% improvement of the RMSE of *Original* while HeuristicB exhibits the lowest RMSE with *Original* when β is 2, yielding a 9.8% improvement of the RMSE over *Original*.

The RMSE of the PMCCs of the HeuristicB algorithm were computed for various values of β and displayed in Figure 1b. Note that choosing $\beta = 0$ results in no change to *Original* and $\beta = \infty$ would result in a prediction of 0 for all PMCCs. The RMSE for *HeuristicB* was minimized using $\beta = 2$, which achieved a total reduction of RMSE of nearly 9.8% over *Original*. Like HeuristicA, this demonstrates that HeuristicB does reduce the RMSE of the PMCCs indicating that *Original* does suffer from overfitting. This value of β means that PMCCs with a direct sample size of 2 were reduced to 50% of their value, while PMCCs with a direct sample size of 20 were reduced to only 90.1% of their original value. The contrast with the effects of HeuristicB on the PMCCs produced by the TPMCC algorithm as also shown in Figure 1b will be discussed in the following section.

6.2 Transitive PMCC

Multiple sets of PMCCs were computed with the proposed TPMCC algorithm using *Original* as input. The different sets were computed with different constraints on the neighbor sets. We chose a fixed $\delta = 0.9$ and $\gamma_{kj} = 1$, but used multiple values of γ_{ik} ranging from 3, 6, 12, and 24. This means that the neighbors for the $Orginal_{ij}$ were limited to ks such that $abs(Original_{ik}) \geq 0.9$, the direct sample size of $Original_{kj}$ is greater than zero and the direct sample size of $Original_{ik}$ ranged from greater than or equal to 3, 6, 12, and 24. Our implementation was in Java and the computation was performed in parallel on four machines. The machines had 4GB of RAM, 2.13GHz Intel Core 2 CPU and were running Debian GNU/linux 2.6.18. Depending on the value for γ_{ik} (which determined the size of the neighbor sets), the entire operation would take five to eight hours. In contrast, a standard PMCC calculation could be done on a single machine in less than an hour.

The RMSEs for the different values of γ_{ik} are shown in Figure 2. The second axis of the figure displays the average number of transitive and direct neighbors of i and k for each γ_{ik}. Note that with $\gamma_{ik} = 24$ it was difficult to even find a large number of direct neighbors and thus, didn't have a significant impact on the data. Both the improvements from $\gamma_{ik} = 24$ to $\gamma_{ik} = 12$ and $\gamma_{ik} = 12$ to $\gamma_{ik} = 6$ were substantial, while the change form $\gamma_{ik} = 6$ to $\gamma_{ik} = 3$ had little impact. This shows that direct sample sizes like 6 and 12 held a strong balance between attainability and usefulness. Neighbors that only have a very small direct sample size are less reliable because such a small direct sample size could easily misrepresent the complete set of data, however, they were still able to make a positive contribution to reducing the overall RMSE. The TPMCC algorithm is minimized for $\gamma_{ik} = 3$ with nearly 1300 transitive neighbors, which reduces the RMSE of *Original* to 0.28, a 40.1% reduction in RMSE. This set of PMCCs, denoted *Transitive*, will be used in subsequent comparisons to other sets of PMCCs.

Looking back at Figures 1a and 1b, both plots also display the results of each heuristic algorithm on the PMCCs produced by TPMCC. In these figures *Transitive* is minimized by the heuristic algorithms when they don't effect

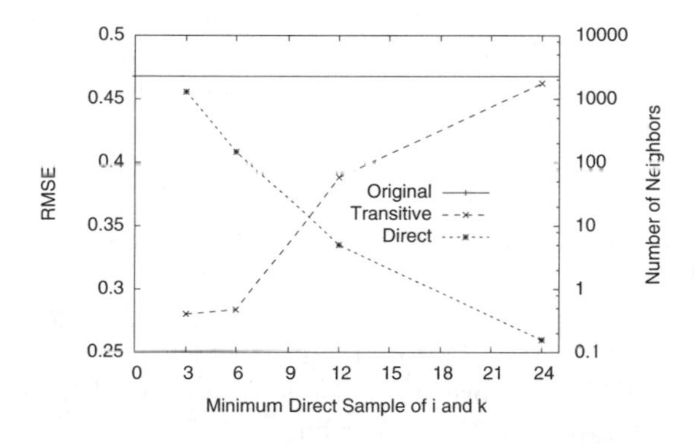

Fig. 2. RMSE of TPMCC with *Transitive* vs. γ_{ik}. The RMSE for *Original* is also displayed for comparison. The secondary Y axis is the mean number of transitive neighbors for the *Neighbors* plot. Note that *Transitive* exhibits the lowest RMSE when γ_{ik} is 3, computed with nearly 1300 neighbors, yielding a 40.1% improvement over *Original*.

them at all - namely $\alpha = 1$ and $\beta = 0$ for HeuristicA and HeuristicB respectively. As discussed in the above subsection, these values of α and β have no effect on *Transitive*, and further, the RMSE gets progressively worse as the heuristics make a larger impact. This is directly indicative that *Transitive*, unlike *Original*, already accounts for overfitting and is only made worse by the heuristics.

6.3 Error Distributions of PMCC Estimations

The distribution of error from each algorithm's PMCCs, including *Original*, are shown in Figure 3a. The plot was built by computing the absolute value of the error and counting the frequency of errors falling into each bucket. The buckets have a lower and upper threshold, all of which were chosen to have width 0.1 and range from 0 to 2. A particular error value falls into the first bucket for which the error is less than that bucket's upper threshold. We will refer to the first bucket, containing values ranging from 0 to 0.1, as the "0.1 bucket" and all subsequent buckets will be denoted by their upper threshold.

In Figure 3a, *Original* has the largest error of any other set of PMCCs. The *Transitive* PMCCs contain the most values in the 0.1 bucket with over 30% of all PMCCs falling into this category. Both heuristic PMCC sets are close behind, while *Original* has only 25%. In the next two buckets, *Original* and both heuristics differ slightly, but TPMCC has about 5% more. In addition, *Transitive* is the only set of PMCCs to have any significant effect on buckets 0.7 to 1, which each contain roughly 5% of all other sets of PMCCs. *Transitive* has much less, emphasizing the fact that it has much fewer high-error PMCCs. These buckets are likely populated by PMCCs that have a very small direct sample size which results in overfitting and high error. TPMCC's performance in this situation is indicative that the it is doing more than curing the symptoms of overfitting, but actually using the extra information to improve estimations.

To further examine the PMCCs and understand the implications of Figure 3a, a second distribution was made to show the RMSE for different direct sample sizes of i and j in Figure 3b. Like Figure 3a, this distribution was sampled using thresholds and each bucket is denoted by its upper threshold, where the first bucket contains only those PMCCs who had a direct sample size of zero. The remaining buckets have exponential widths ranging from the previous buckets upper threshold (exclusive) to its own upper threshold (inclusive).

The PMCCs for all algorithms, excluding *Transitive*, are identical for the first two buckets, 0 and 1, as they had no data on which to operate and therefore predicted 0. The TPMCC algorithm was able to produce PMCCs that reduced this error in bucket 0 by over 31.2%. For bucket 1, *Transitive* further improves and demonstrates its ability to operate with little direct data and reduces the error of all other algorithms by 41.4%. *Original* doesn't improve much in bucket 2, but both heuristics show a drastic change and reduce the RMSE of *Original* by 17%. The *Transitive* continues to improve and reduces the RMSE of *Original* for the bucket 2 by 50.4%. These contrasting results demonstrate the ability of

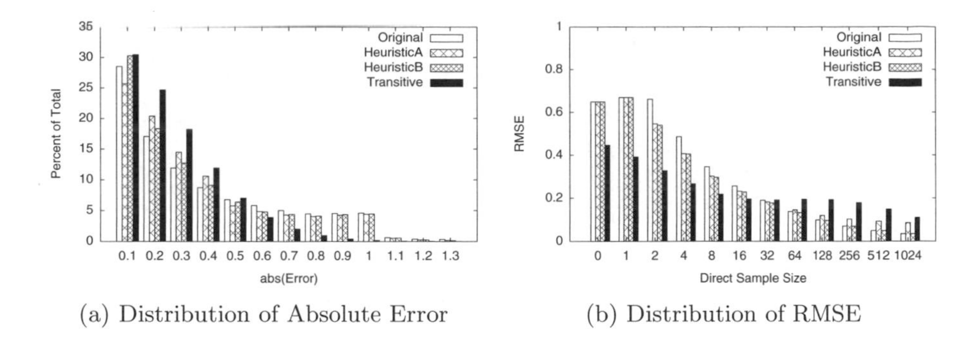

(a) Distribution of Absolute Error (b) Distribution of RMSE

Fig. 3. *Distributions of the absolute value of error and RMSEs:* In (a), the distribution of the absolute value of the error with thresholds of width 0.1 where the X axis represents upper threshold is shown. Note that *Transitive* has the most values in buckets 0.1-0.5 while the others have more in the high-error buckets. In (b), the distribution of RMSE by direct sample sizes of PMCCs where the X axis represents upper threshold (inclusive). Note that *Transitive* exhibits significantly lower RMSEs for direct sample sizes less than or equal to 8. After this point *Transitive* has higher RMSEs demonstrating that the transitive data becomes less valuable as direct sample size increases.

the TPMCC algorithm to extract indirect information from transitive neighbors and improve the accuracy of predictions with limited amounts of data.

The *Transitive* continues to outperform all other algorithms by a similarly significant margins up to bucket 8. For buckets larger than 16, a new trend develops and *Transitive* begins to have a larger RMSE than the other algorithms. This interesting behavior implies that there exists a direct sample size at which point enough direct information renders the transitive neighbor information detrimental. This is somewhat intuitive as the larger the direct sample size that is available, the more trust that can be placed on the subsequent results. Thus, when the results are sufficiently trusted, the *Transitive* uses less accurate and indirect information from transitive neighbors that actually increases the RMSE of the PMCC estimations.

To gain insight as to how significantly Figure 3b will impact the overall RMSE, a third distribution was made. This distribution is shown in Figure 4 and displays the percent of all PMCCs to fall in each of the buckets used in Figure 3b. It shows that roughly 7% of all PMCCs have a direct sample size of 0. The thresholds with direct sample sizes from 0 to 8 account for 67.9% of all data and *Transitive* was able to reduce the RMSE by 42.7%. In addition, the thresholds where *Transitive* is detrimental, direct sample sizes with thresholds 64 and greater, all combine to make up only 12.9% of all unique pairs.

6.4 Data Density

As discussed in Section 5, the Netflix Prize data set has approximately a 1% density, indicating that in the results presented above the algorithms operated

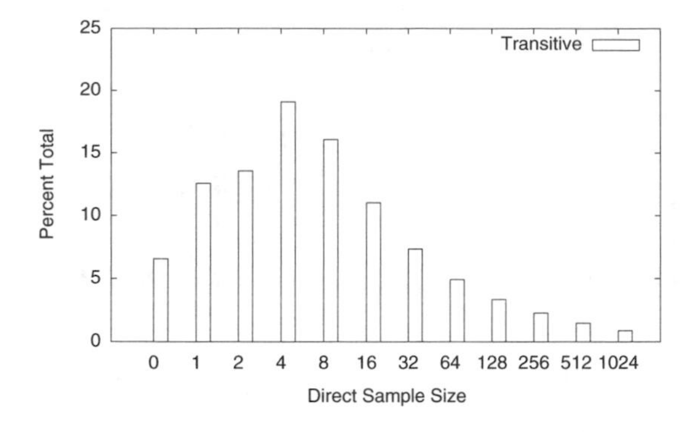

Fig. 4. Distribution of all PMCCs by direct sample sizes where X axis represents upper threshold. Note that samples sizes 0 through 8, where TPMCC performs well, account for 67.9% of all PMCCs. The TPMCC has a negative impact on buckets with direct sample sizes of 64 and greater, which combine to only 12.9%.

on only 0.5% data density since we used half of the data set. The following analysis addresses how the different algorithms perform as the amount of data is reduced.

The plot in Figure 5 shows a RMSE of different sets of PMCCs as the amount of data is varied. *OrigHeuristicA* and *OrigHeuristicB* are the PMCCs resulting from using *Original* with each heuristic. *Transitive* is the results from TPMCC algorithm while *TransHeuristicA* and *TransHeuristicB* are from the heuristics operating on *Transitive*. With 0% of the data, all algorithms produce the same set of PMCCs which amounts to predicting 0 for all PMCCs. The 2% sample points, or 0.02% data density, show little improvement because the data is still too sparse to support sufficient intersection between movies.

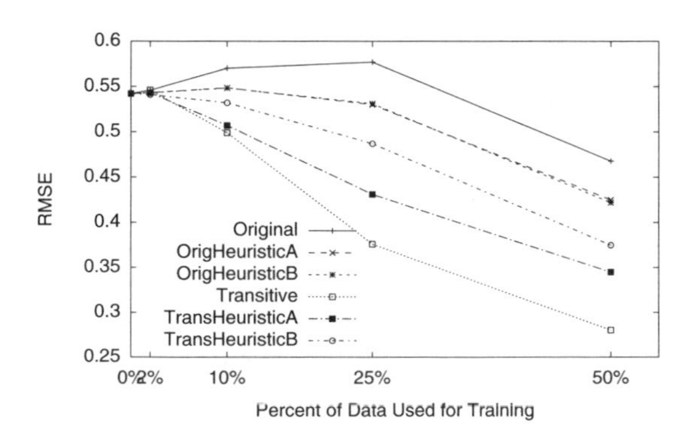

Fig. 5. RMSE vs. percent of Netflix Prize data used, *Transitive* and both heuristics for each. Note that *Original* actually increases RMSE as amount of data increases until 50% data, while *Transitive* is able to make increasingly notable improvements starting with 10% data.

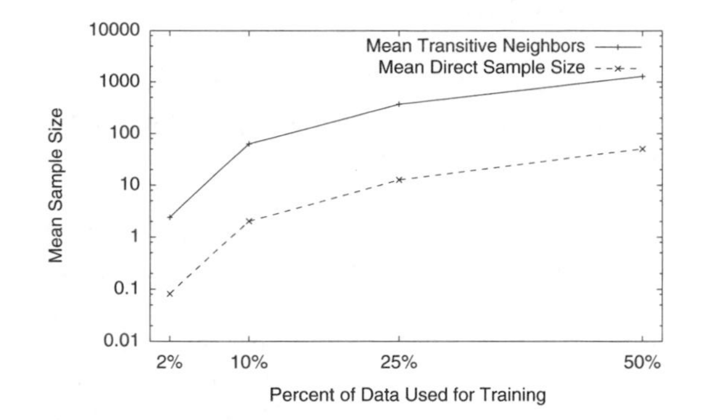

Fig. 6. Mean direct sample size and mean transitive sample size vs. percent of Netflix Prize data used. Note that neither reach usable quantities for 2% data density, but at 10% the mean number of transitive neighbors reaches 50, while mean number of direct neighbors only reaches just over 2. At 50% data there are nearly 1300 and 50 mean transitive and direct neighbors respectively.

To confirm that 0.02% data density was too low, Figure 6 displays the mean number of transitive neighbors found by TPMCC and the number of direct neighbors for normal PMCCs. For the 2% predictions, it is clear that neither transitive nor direct neighbors exist in usable quantities.

The next point at 10% (0.1% data density) shows an interesting trend. With nearly 100 transitive neighbors, *Transitive* reduces the RMSE of *Original*, by 12.5%. The heuristics for *Transitive* fail to make additional improvement. *Original* and its heuristics actually perform slightly worse than predicting 0 for all PMCCs. This demonstrates that such sparse data causes overfitting, and in this case, can actually be improved by not utilizing the data at all.

At 25% (0.25% data density) *Original* is still being out performed by predicting 0 for all PMCCs. Its heuristics do slightly better, but Transitive is able to achieve a 34.8% reduction in RMSE over *Original*. Back to 50% of the Netflix Prize data, *Original* makes a drastic improvement with a mean intersection size of 51. *Transitive* continues to improve reaching a 40.1% reduction over *Original*. Note that the heuristics operating on *Transitive* are always only made worse, demonstrating that the TPMCC algorithm has already reduced overfitting beyond the aid of those heuristics.

It is interesting to note that the Transitive PMCC algorithm is able to begin reducing RMSE with only 100 transitive neighbors as shown in the 0.1% data density point in the plots above. However, with 0.5% data density, it further benefits from over 1000 neighbors. This means that the Transitive algorithm benefits from being in a *wide* data set, or a data set that has lots of users and movies in the case of the Netflix data set. If there were only 10 movies, it would be very difficult for the Transitive algorithm to find a sufficient number of neighbors. With the Netflix Prize data, almost 20,000 movies along with half a million users exist, given plenty of opportunities to find different neighbors, transitive or not.

7 Conclusions and Future Work

The proposed nearest neighbor PMCC algorithm increases the accuracy of PMCCs estimations when dealing with sparse, sample-based data. In such sample-based data, statistical models can suffer from the lack of data and represent random error instead of underlying trends in a phenomenon known as overfitting. The results of the experiments with the Netflix Prize data demonstrate that the proposed heuristics and TPMCC algorithm are able to reduce the error in such PMCC estimations.

The PMCCs computed from the random test set reduced the error of predicting 0 for all PMCCs by only 13.7%. The heuristics reduced the error of the test set PMCCs by up to 9.8%, while our TPMCC algorithm, which took advantage of transitive relationships, was able to achieve a 40.1% reduction. For Pearson estimates with direct sample sizes of two, which account for 13.6% of the population, the TPMCC reduced the error by over 50%. Lastly, the TPMCC algorithm is able to provide comparable improvements with reduced amounts of data. This reduction in error of PMCCs will strengthen the variety of applications in which they are applied and allow statistical models to be utilized in situations where they otherwise could not. Furthermore, the abstract notion of gathering information from transitive neighbors is likely to have a positive effect in new applications.

For future work, we plan on exploring how measures of similarity other than PMCC, like the Jaccard index, Euclidean distance, and Spearman rank coefficient, could be improved by discovering transitive relationships in the data sets. The TPMCC's temporal computational complexity is $O(n^3)$, requiring an $O(n)$ operation for each of the $O(n^2)$ unique PMCCs. This running time could be reduced to $O(kn^2) = O(n^2)$ by selecting some well- chosen subset of size k to represent all possible neighbor candidates. Furthermore, if the TPMCC algorithm was computed on a subset of c PMCCs, those that are likely to benefit the most (e.g. those with a very small sample size), it could be reduced to $O(ckn) = O(n)$ which could make it much more pragmatic in real life situations.

References

1. Adomavicius, G., Tuzhilin, A.: Towards the next generation of recommender systems: A survey of the state-of-the-art and possible extensions. IEEE Transactions on Knowledge and Data Engineering 17, 634–749 (2005)
2. Ali, K., van Stam, W.: Tivo: Making show recommendations using a distributed collaborative filtering architecture. In: Proceedings of the 10th ACM International Conference on Knowledge Discovery and Data Mining, pp. 394–401 (2004)
3. Bell, R., Koren, Y.: Improved neighborhood-based collaborative filtering. In: International Conference on Knowledge Discovery and Data Mining (2007)
4. Bell, R., Koren, Y.: Lessons from the netflix prize challenge. SIGKDD Explorations 9, 75–79 (2007)
5. Bell, R., Koren, Y.: Scalable collaborative filtering with jointly derived neighborhood interpolation weights. In: IEEE International Conference on Data Mining (ICDM 2007), pp. 43–52 (2007)

6. Bell, R., Koren, Y., Volinsky, C.: The bellkor solution to the netflix prize. Tech. rep., AT&T Labs (2007)
7. Buskirk, E.V.: Winning teams join to qualify for $1 million netflix prize. Wired Magazine (2009)
8. Goldberg, D., Nichols, D., Oki, B.M., Terry, D.: Using collaborative filtering to weave an information tapestry. Communications of the ACM 35(12), 61–71 (1992)
9. Hong, T., Tsamis, D.: Use of knn for the netflix prize. Tech. rep., Stanford University (2006)
10. Koren, Y.: Factorization meets the neighborhood: a multifaceted collaborative filtering model. In: International Conference on Knowledge Discovery and Data Mining (2008)
11. Linden, G., Smith, B., York, J.: Amazon.com recommendations: Item-to-item collaborative filtering. IEEE Internet Computing 7(1), 76–80 (2003)
12. Lohr, S.: Netflix Competitors Learn the Power of Teamwork. NY Times (2009)
13. Netflix: The Netflix Prize, http://www.netflixprize.com
14. Newitz, A.: Movie Tips From Your Robot Overlords. Washington Post (2009)
15. Sarwar, B., Karypis, G., Konstan, J., Riedl, J.: Item-based collaborative filtering recommendation algorithms. In: Proc. 10th International Conference on the World Wide Web, pp. 285–295 (2001)
16. Thompson, C.: Netflix challenge to hackers: Improve our service and win big. NY Times (2007)

The NOX Framework: Native Language Queries for Business Intelligence Applications

Todd Eavis, Hiba Tabbara, and Ahmad Taleb

Concordia University, Montreal, Canada

Abstract. Over the past ten to fifteen years, Business Intelligence applications have become increasingly important and visible components of enterprize computing environments. While relational database management systems often form the backbone of the BI software stack, the unique modeling and processing requirements of BI applications often make for a relatively awkward fit with RDBMS platforms in general, and their SQL query interfaces in particular. In this paper, we present a new framework for BI/OLAP applications that directly exploits a domain specific conceptual data model. In turn, the new paradigm allows us to support native, client-side OOP querying without the need to embed an intermediate, non-OOP language such as SQL or MDX. A pre-processor essentially translates standard OOP source code into a query grammar developed specifically for BI analysis. The end result is a query facility that is far more intuitive to use, as well as being more amenable to contemporary code development tools. We provide numerous examples to illustrate the flexibility and convenience of the new framework.

1 Introduction

Over the past three decades, relational DBM systems have secured their place as the cornerstone of contemporary data management environments. During that time, logical data models and query languages have matured to the point whereby database practitioners can almost unequivocally identify common standards and best practices. With respect to operational databases, the ubiquitous relational data model and the Structured Query Language (SQL) have become synonymous with the notion of efficient storage and access of transactional data.

That being said, a number of new and important domain-specific data management applications have emerged in the past decade. At the same time, general programming languages have evolved, driven by a desire for both greater simplicity, modeling accuracy, reliability, and development efficiency. As such, opportunities to explore new data models, as well as the languages that might exploit them, have emerged.

One particular area of interest is the Business Intelligence/Online Analytical Processing (OLAP) context. Typically, such systems work in conjunction with an underlying relational data warehouse that houses an integrated, time sensitive, repository of one or more organizational data stores. At its heart, BI attempts to abstract away some of the often gory details of the large warehouses so as

T.B. Pedersen, M.K. Mohania, and A M. Tjoa (Eds.): DaWaK 2010, LNCS 6263, pp. 172–189, 2010.

to provide users with a cleaner, more intuitive view of enterprize data. Beyond trivial exploitation of BI GUI facilities, however, meaningful analysis can become quite complex and can necessitate a considerable investment of the developer's time and energy.

Of particular significance in this regard is the awkward relationship between the development language and the data itself. Given the relational model of the underlying DBMS, BI querying typically relies upon non-procedural SQL or one of its proprietary derivatives. Unlike transactional databases, however, which are often cleanly modeled by a set-based representation, the nature of BI/OLAP environments argues against the use of such languages. In particular, concepts such as cubes, dimensions, aggregation hierarchies, granularity levels, and drill down relationships map poorly at best to the standard logical model of relational systems.

A second related concern is the relative difficulty of integrating non-procedural query languages into application level source code. Larger development projects typically encounter one or more of the following limitations:

- Comprehensive compile-time type checking is often impossible. All parsing is performed at run-time by a possibly remote, often overloaded server.
- Developers must merge two fundamentally incompatible programming models (i.e., procedural OOP versus a non-procedural DBMS query language).
- There are few possibilities for the kind of code re-use afforded by OOP concepts (e.g., inheritance and polymorphism).
- The use of embedded query strings (i.e., JDBC/SQL) severely limits the developer's ability to efficiently *refactor* source code in response to changes in schema design.

In this paper, we present a comprehensive new data access framework called NOX (Native language OLAP query eXecution) that is specifically tailored to the BI/OLAP domain. Beginning with the specification of an OLAP algebra, we develop a robust query grammar that presents the developer with an Object Oriented representation of the primary OLAP structural elements. The grammar, in turn, is the foundation of a native language query interface that eliminates the reliance on an intermediate, string based embedded language. We illustrate the new design via the Java programming language, and demonstrate how developers can transparently interact with massive, remote data cubes using standard OOP principles and practices. While the underlying compilation and translation mechanism is somewhat complex, virtually all of the framework's sophistication is hidden from the developer. In short, NOX represents a significant step towards "making the OLAP DBMS disappear".

The paper is organized as follows. In Section 2, we present an overview of related work. Section 3 introduces the primary NOX components, while Section 4 discusses the underlying conceptual model. The full details of the client architecture are then presented in Section 5. Future work and final conclusions are provided in Sections 6 and 7 respectively.

2 Related Work

For more than 30 years, Structured Query Language (SQL) has been the defacto standard for data access within the relational DBMS world. Because of its *relative age*, however, numerous attempts have been made to modernize database access mechanisms. Two themes in particular are noteworthy in the current context. In the first case, Object Relational Mapping (ORM) frameworks have been used to define type-safe mappings between the DBMS and the native objects of the client applications. With respect to the Java language, industry standards such as JDO (Java Data Objects) [4], as well as the open source Hibernate framework [10] have emerged. In all cases, however, it is important to note that while the ORM frameworks do provide *transparent persistence* for individual objects, additional string-based query languages such as JDOQL (JDO), or HQL (Hibernate) are required in order to execute joins, complex selections, subqueries, etc. The result is a development environment that often seems as complex as the model it was meant to replace.

More recently, Safe Query Objects (SQO) [12] have been introduced. Rather than explicit mappings, safe queries are defined by a class containing, in its simplest form, a *filter* and *execute* method. Within the filter method, the developer encodes query logic (e.g., selection criteria) using the syntax of the native language. The compiler checks the validity of query types, relative to the objects defined in the filter. The *execute* method is then rewritten as a JDO call to the remote database. The approach is quite elegant, though it can be difficult to accurately model completely arbitrary SQL statements.

In contrast to the ORM models, a second approach extends the development languages themselves. The Ruby language [7], for example, employs *ActiveRecords* to dynamically examine method invocations against the database schema. HaskellDB [5], on the other hand, "decomposes" queries into a series of distinct algebraic operations (e.g., restrict, project) . Microsoft's LINQ extensions (C# and VisualBasic) [11] are also quite interesting in that they essentially integrate the mapping facilities of the ORM frameworks into the language itself (via the ubiquitous SELECT-FROM-WHERE format). It should be noted, however, that none of these language extensions are in any way OLAP-aware.

In terms of OLAP and BI specific design themes, most contemporary research builds in some way upon the OLAP *data cube* operator [15]. In addition to various algorithms for cube construction, including those with direct support for dimension hierarchies [21,19], researchers have identified a number of new OLAP *operators* [9,13], each designed to minimize in some way the relative difficulty of implementing core operations in "raw SQL". There has also been considerable interest in the design of supporting algebras [8,16,20]. The primary focus of this work has been to define an API that would ultimately lead to transparent, intuitive support for the underlying data cube. In a more general sense, these algebras have identified the core elements of the OLAP conceptual data model.

A somewhat orthogonal pursuit in the OLAP context has been the design of domain-specific query languages and/or extensions. SQL, for example, has been updated to include the CUBE, ROLLUP, and WINDOW clauses [18], though vendor support for these operations in DBMS platforms is inconsistent at best [14]. In addition to SQL, many commercial applications support Microsoft's MDX query language [23]. While syntactically reminiscent of SQL, MDX provides direct support for both multi-level dimension hierarchies and a *crosstab* data model. Still, MDX remains an embedded string based language with an irregular structure and is plagued by the same limitations as those discussed in Section 1.

Finally, we note that query languages such as SQL and MDX are typically encapsulated within a programmatic API that exposes methods for connection configuration, query transfer, and result handling. While relational systems utilize mature standards (e.g., JDBC, ODBC), no definitive API has emerged for OLAP. A recent attempt to do so was the ill-fated JOLAP specification, JSR-69 [3], an industry-backed initiative to define an enterprize-ready, Java-oriented meta data and query framework based upon the Common Warehouse Metamodel [2]. JOLAP proved to be exceedingly complex and, consequently, no viable JOLAP-aware applications were ever developed. At present, the most widely supported API is arguably XML for Analysis (XMLA) [1], a low-level XML/-SOAP mechanism running across HTTP. In practice, XMLA is effectively just a wrapper for MDX, though XMLA result sets are structured in an OLAP-aware format.

3 NOX: Native Language OLAP Query eXecution

To begin, we note that a fundamental design objective for any new query framework or API must be the minimization of the complexity associated with transparent persistence, as the introduction of obscure and non-intuitive design and programming patterns severely limits the likelihood of adoption. We therefore state at the outset that the NOX focus is explicitly on the OLAP/BI domain. In fact, NOX is intended to specifically support higher level analytics servers and is not expected to resolve every "ad hoc" query that might be executed against an underlying relational data warehouse. The primary motivation for this approach is the rejection of the "be all things to all people" mantra that tends to plague systems that must maintain a fully generic, lowest common denominator profile [22]. Conventional RDMSs, conceptual mapping frameworks, and even JOLAP suffer from this same "curse of generality". In the current context, the targeting of a specific application domain ultimately relieves the designer from having to manually construct a comprehensive data model, along with its constituent processing constructs.

3.1 The NOX Components

Given the preceding objective, NOX has been constructed from the ground up so as to emphasize the *transparency* in the term "transparent persistence". Doing so, of course, requires considerable infrastructure. In the remainder of the paper, we discuss the design, implementation, and use of the NOX framework, using a number of programming examples to illustrate its practical value. Before digging in to the details, however, it is useful to first provide a brief overview of the primary physical and logical elements of the framework. Keep in the mind that the following list includes elements that are both visible and invisible to the developer.

- **OLAP conceptual model.** NOX allows developers to write code directly at the conceptual level; no knowledge of the physical or even logical schema is required.
- **OLAP algebra.** Given the complexity of directly utilizing the relational algebra in the OLAP context (via SQL or MDX), we define fundamental query operations against a cube-specific OLAP algebra.
- **OLAP grammar.** Closely associated with the algebra is a DTD-encoded OLAP grammar that provides a concrete foundation for client language queries.
- **Client side libraries.** NOX provides a small suite of OOP classes corresponding to the objects of the conceptual model. Collectively, the exposed methods of the libraries form a clean programming API that can be used to instantiate OLAP queries.
- **Augmented compiler.** At its heart, NOX is a query re-writer. During a pre-processing phase, the framework's compilation tools (JavaCC/JJTree) effectively re-write source code to provide transparent model-to-DBMS query translation.
- **Cube result set.** OLAP queries essentially extract a subcube from the original space. The NOX framework exposes the result in a logical, read-only multi-dimensional array.

In short, the developer's view of the OLAP environment consists solely of the API and the Result Set. More to the point, from the developer's perspective, all OLAP data is housed in a series of cube objects housed in local memory. The fact that these repositories are not only remote, but possibly Gigabytes or even Terabytes in size, is largely irrelevant.

4 Conceptual Model

One of the great burdens associated with enterprize ORM projects is the design of accurate data models. Even when a model can be formally identified, it is often the case that the conceptual view of the data differs widely even between departments of the same organization. In the OLAP context, however, the conceptual view of the data has reached a level of maturity whereby virtually all *analytical* applications essentially support the same high level view of the data.

Briefly, we consider analytical environments to consist of one or more *data cubes*. Each cube is composed of a series of d dimensions (sometimes called *feature* attributes) and one or more *measures*. The dimensions can be visualized as delimiting a d-dimensional hyper-cube, with each axis identifying the *members*

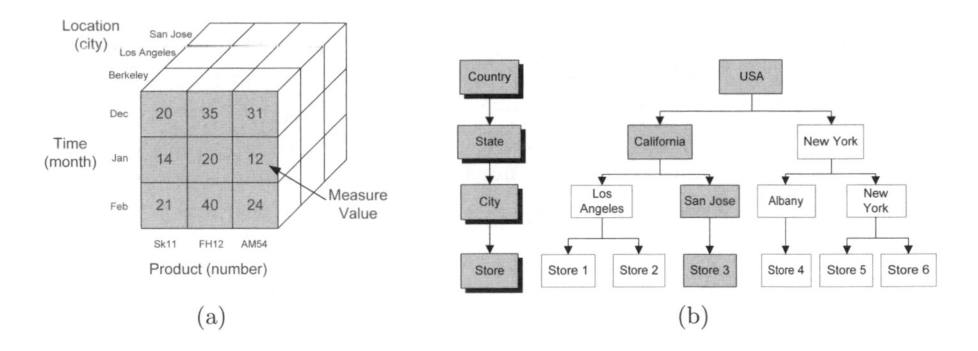

Fig. 1. (a) NOX conceptual query model (b) A simple symmetric hierarchy

of the parent dimension (e.g., the days of the year). Cell values, in turn, represent the aggregated measure (e.g., sum) of the associated members. Figure 1(a) provides an illustration of a very simple three dimensional cube. We can see, for example, that 12 units of Product AM54 were sold in the Berkeley location during the month of January (assuming a Count measure).

Beyond the basic cube, however, the conceptual OLAP model relies extensively on aggregation hierarchies provided by the dimensions themselves. In fact, hierarchy traversal is one of the more common and important elements of analytical queries. In practice, there are many variations on the form of OLAP hierarchies [17] (e.g., symmetric, ragged, non-strict). NOX supports virtually all of these, and does so by augmenting the conceptual model with the notion of an arbitrary graph-based hierarchy that may be used to *decorate* one or more cube dimensions. Figure 1(b) illustrates a simple geographic hierarchy that an organization might use to identify intuitive customer groupings.

4.1 OLAP Algebra

Given the clean, conceptual model described above, it is possible to consider the application of an OLAP algebra that directly exploits the model's structure. As noted in Section 2, a number of researchers have identified the core operations of such an algebra. We will shortly see how the exploitation of a formal algebra ultimately allows developers to program directly against the conceptual model, rather than to a far more complex physical or even logical model.

As indicated, a core set of operations common to virtually all proposed OLAP algebras has been identified. Below, we list and briefly describe these operations. Note that we do not provide a formal analysis of the semantics of the algebraic operations, nor their equivalence to the components of the relational algebra, as these issues have been extensively discussed in the original publications.

- **selection:** the identification of one or more cells from within the full d-dimensional search space.
- **projection:** the identification of presentation attributes, including *both* the measure attribute(s) and dimension members.

```
<!-- Data queries-->
<!ELEMENT DATA_QUERY (CUBE_NAME, OPERATION_LIST, FUNCTION_LIST?)>
<!ELEMENT CUBE_NAME (#PCDATA)>
<!ELEMENT OPERATION_LIST (
    SELECTION?, PROJECTION?, CHANGE_LEVEL?, CHANGE_BASE?,
    PIVOT?, DRILL_ACROSS?, UNION?, INTERSECTION?, DIFFERENCE?)>
```

Listing 1.1. Core operations of the NOX algebra

- **drill across:** the integration of two independent cubes, each possessing common dimensional axes. In effect, this is a cube "join".
- **union/intersection/difference:** basic set operations performed on two cubes sharing common dimensional axes.
- **change level:** modification of the granularity of aggregation, typically referred to as "drill down" and "roll up".
- **change base:** the addition or deletion of one or more dimensions from the current result.
- **pivot:** rotation of the cube axes to provide an alternate *perspective*.

Several explanatory notes are in order at this stage. First, the **selection** is the driving operation behind most analytical queries. In fact, if suitable defaults are available for the **projection**, many queries can be expressed with nothing more than a selection. Second, the final three operations — **change level**, **change base**, and **pivot** — are distinct from the first six in that each is only relevant as a query against an existing result set. Third, it is important to recognize that while logical data warehouse models typically require explicit joins between fact (measure) and dimension tables, there is no such requirement at the conceptual level. The result is a dramatic reduction in complexity for the developer. (Depending upon the architecture of the supporting analytics server, of course, join operations may still be performed at some point.). Finally, and perhaps most importantly, the OLAP algebra is implicitly *read only*, in that database updates are performed via distinct ETL processes.

4.2 The NOX Grammar

NOX encapsulates the algebra in a formal schema encoded by a Document Type Definition (DTD). The DTD is relatively complex as it effectively represents the foundation for an expressive, XML-based analytics language. Due to space limitations, we do not present the full schema specification here. However, key elements are presented below.

Listing 1.1 defines the core structure of a NOX query. Each query is associated with a single cube (though references to other cubes are possible), as well as a Function List and an Operations List. We do not discuss cube functions extensively in this paper but, for the sake of completeness, we can informally define a cube function as one that is logically associated with a result set, rather than a specific cell or dimension member. The common *top10* function is a simple example.

```
<!-- Selection -->
<!ELEMENT SELECTION (DIMENSION_LIST)>
<!ELEMENT DIMENSION_LIST (DIMENSION+)>
<!ELEMENT DIMENSION (DIMENSION_NAME, EXPRESSION)>
<!ELEMENT DIMENSION_NAME (#PCDATA)>

<!-- Dimension Expressions -->
<!ELEMENT EXPRESSION (RELATIONAL_EXP | COMPOUND_EXP)>
<!ELEMENT RELATIONAL_EXP (SIMPLE_EXP, COND_OP, SIMPLE_EXP)>
<!ELEMENT COMPOUND_EXP (EXPRESSION, LOGICAL_OP, EXPRESSION)>
<!ELEMENT SIMPLE_EXP (EXP_VALUE | ARITHMETIC_EXP)>
<!ELEMENT ARITHMETIC_EXP (SIMPLE_EXP, ARITHMETIC_OP, SIMPLE_EXP)>
```

Listing 1.2. Selection elements

```
<!ELEMENT UNION (DATA_QUERY)>
<!ELEMENT INTERSECTION (DATA_QUERY)>
<!ELEMENT DIFFERENCE (DATA_QUERY)>
```

Listing 1.3. Set Operations

The Operations List contains the algebraic elements of the query, and each may occur exactly zero or one time in a single query. Given the significance of the `selection` operation, we will look at it in greater detail. Listing 1.2 demonstrates that a `selection` is defined as a listing of one or more dimensions, each associated with an expression. In effect, the expression represents a query restriction on the associated dimension (this will become more clear in Section 5). Simple expressions may be combined to form compound expressions (via logical AND and OR) and can be recursively defined. In other words, as with any meaningful programming language, conditional restrictions can be almost arbitrarily complex.

Finally, in Listing 1.3, we illustrate the remarkable simplicity of the *set operation* specifications. In effect, set operations are syntactically modeled on an OOP paradigm. Consider, for example, a String equality check in a language such as Java, where we would write `myString.equals("Joe")`, rather than something like `myString == "joe"`. This same approach allows us to represent set operations simply as a nested data query, defined relative to the current query.

5 Client Side API

Within the NOX framework, the conceptual model and its associated grammar are intended to provide an abstract development environment for expressive analytical programming. In order to provide such an interface, however, supporting client side functionality is required. In a nutshell, NOX provides persistent transparency via a source code re-writing mechanism that interprets the developer's OOP query specification and decomposes it into the core operations of the OLAP algebra. These operations are given concrete form within the NOX grammar and then transparently delivered (via standard socket calls) at run-time to the back-end analytics server for processing. Results are again transparently injected back into the running application and made available through a standard OOP API.

We note at this point that we have chosen to implement the API functionality using external libraries rather than direct language modification. This is partly to encourage portability between languages, as we consider the NOX model to be broadly applicable to any modern OOP language. However, it is also due to the fact that while OLAP/BI is an immensely important commercial domain (thereby justifying this work in the first place), OLAP-specific language extensions would have virtually no relevance to the vast majority of developers working in arbitrary domains.

5.1 The NOX Preprocessor

As should be obvious, source code augmentation of this form is non-trivial. In short, NOX must identify query-specific elements of the source code and transform them as required before passing the output to the standard Java compiler. The pre-preprocessor is produced with the JavaCC parser generator and its JJTree Tree builder plug-in [6]. Briefly, JJTree is used to define parse tree building actions that are executed during the later parse process. In the NOX case, JJTree identifies query-specific code constructs (e.g., class definitions) that should be re-written. The output of JJTree is then used by JavaCC to construct a Java parser that actually "walks the parse tree" in order to locate and transform these constructs. We note that although NOX utilizes a complete Java 1.5 grammar for its parser, the pre-processor only examines and/or processes parse tree nodes defined by JJTree. In practice, this makes the pre-processing step extremely fast.

So what is the pre-processor looking for? NOX is supported by client libraries that define the relevant query components. The fundamental structure is the OlapQuery class. Listing 1.4 provides a partial listing of its contents. We make note of the following points. First, method names correspond directly to the operations of the algebra/grammar (Note: We currently do not include the **change base**, **change level**, and **pivot** methods in the OlapQuery class as we consider these operations to be manipulations of the Result Set. Their exact implementation is the subject of ongoing research). Second, method bodies have no meaningful implementation, other than a nominal return value (required for successful compilation). In fact, this is true of most client library methods, a fact that makes sense when one realizes that the only code that will actually be executed is the code eventually inserted by the pre-processor. Third, each query method has a return type unique to its own semantic abstraction (the upcoming examples will make this more clear). Fourth, the `execute` method serves as the link between the programmer's conceptual view and NOX's algebraic view. More to the point, it is the `execute` method that will be re-written to include an XML statement corresponding to the specifications of the other methods. The XML string is then "wrapped" in a message that is sent to the server when execute() is invoked in the application. Finally, the OlapQuery is declared *abstract*, though

```
public abstract class OlapQuery {
 public boolean select() {return false;}
 public Object[] project() {return null;}
 public OlapQuery drillAcross() {return null;}
 public OlapQuery union() {return null;}
 public OlapQuery intersection() {return null;}
 public OlapQuery difference() {return null;}

 public ResultSet execute() { return new ResultSet(); }
}
```

Listing 1.4. The OLAP Query class

none of its methods are abstract, a model reminiscent of Java's Adapter classes. Use of this structure allows programmers to over-ride the OlapQuery and provide only the operations necessary for the query at hand (often just selection). The remaining methods are effectively no-ops.

Figure 2 graphically illustrates the process described thus far. In the box at the left, we see the parser generation tools that produce the *translating pre-processor*. The dashed line to the pre-processor itself indicates that this association is static, and the parser building tools are not invoked directly at either compile time or run-time. In terms of the compilation process, the pre-processor take as input the original source file and then, using the parse tree constructed from this source, converts the relevant source elements into an XML decomposition of the OlapQuery. Throughout this process, various DOM utilities and services are exploited in order to generate and verify the XML. Finally, once the source has been transformed, it is run through a standard Java compiler and converted into an executable class file. We note that, in practice, the NOX translation step would be integrated into a build task (ANT, makefile, IDE script, etc.) and would be completely transparent to the programmer.

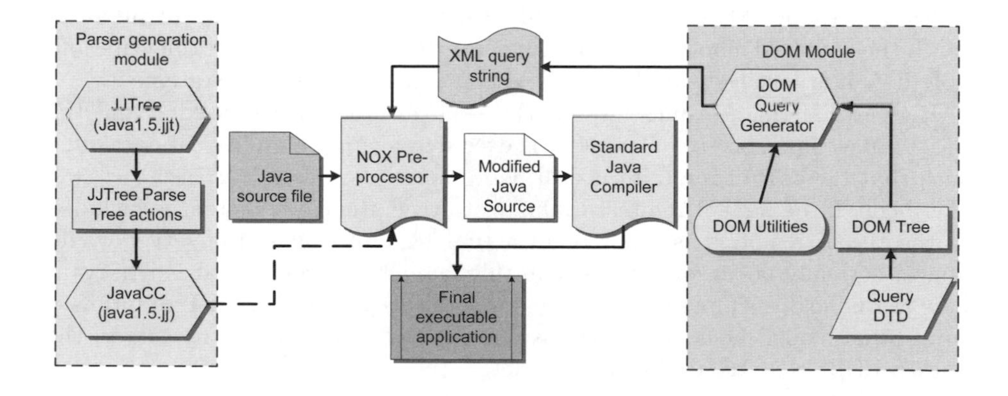

Fig. 2. The client compilation model

```
class SimpleQuery extends OlapQuery {
 public boolean select(){
  DateDimension date = new DateDimension();
  return date.getYear() == 2001;
 }
 //... projection excluded
}

public class Demo1 {
 public static void main(String[] args) {
  //...DBMS boilerplate connection
  SimpleQuery myQuery = new SimpleQuery(''SalesByDate'');
  ResultSet result = myQuery.execute();
  // ...manipulate result set
}}
```

Listing 1.5. Simple OLAP Query

5.2 Application Programming

While novel algebras, grammars, and parsing methods are interesting for their own sake, they provide little benefit unless they ultimately lead to a clean, intuitive programming experience for the developer. In this section, we provide a number of examples that demonstrate the practical use of the NOX model.

A Simple Selection. We begin with a query that specifies a simple `selection` criteria, namely that we would like to list total sales for 2001. Listing 1.5 provides the corresponding OlapQuery definition, along with a small `main` method that demonstrates how the query's `execute` method would be invoked. (For simplicity, we will ignore the projection method that would specify the measure and display attributes, as well as the "boilerplate" connection and authentication methods.) We can see that the `select` method instantiates a DateDimension and invokes its getYear() method. Because Dates are virtually universal in analytical processing, NOX provides a fully functional Date class "out of the box" (with the standard empty method bodies). In terms of the `selection` criterion, note how it is specified simply via a boolean-generating `return` statement.

It is crucial that we understand why such an approach is used. From the programmer's perspective, the query is executed against the physical data cube such that the `selection` criteria will be iteratively evaluated against *each and every* cell. If the `selection` test evaluates to true, the cell's content is included in the result; if not, it is ignored. In actual fact, of course, the server would almost certainly not resolve a query in this manner. However, that is irrelevant here as our goal is simply to allow the developer to program against an intuitive conceptual model. Once the query is decomposed and sent to the server, the backend DBMS is free to do what it likes.

```
public class CustomerDimension extends OlapDimension {
 private String name;
 private int age;
 CustHierarchy geographicHierarchy;

 public String getName() { return name; }
 public int getAge() { return age; }
 public CustHierarchy getGeographicHierarchy(){
  return geographicHierarchy;
}}
```

Listing 1.6. Simple OLAP dimension

In terms of the decomposition itself, it is of course represented in an XML string generated by the pre-processor (Due to space limitations, we do not reproduce the associated XML here; we simply note that it corresponds directly to the DTD depicted in Listing 1.2). This string is inserted — by the pre-processor — into the query's `execute` method and subsequently invoked in the `main` method. At run-time, this invocation produces a network call to the DBMS to send and receive the query and its results. Again, we stress that all of this processing is entirely invisible to the end user.

Manipulating Hierarchies. As previously noted, hierarchical queries are extremely common in OLAP environments. For this reason, much of the current NOX research focuses on extending the expressive capabilities of the framework in this context. With the example below, we give the reader a sense of the NOX philosophy with respect to hierarchical navigation.

Let us assume that we would like to find sales data for older customers from California cities who purchased products in the first half of 2007. Because we now have an arbitrarily defined dimension to restrict (as opposed to the built-in Date dimension), we need a mechanism to statically type-check the relevant dimension attributes so that we can ensure at compile time that all query element are being used appropriately (e.g, integers compared with integers). (Again, we do NOT want to rely on embedded strings like SQL/MDX since type validity could then only be assessed at run-time.) To do this, the programmer simply sub-classes the library-provided OlapDimension class and adds the relevant attributes/types and *getter* methods (NOX can strip the "get" from the getters to obtain case insensitive attribute names). Both dimension attributes and hierarchies can be specified in this manner. Listing 1.6 illustrates this simple approach. Note that CustHierarchy is a simple extension of the NOX OlapHierarchy class.

Now, given this simple Customer class, and a geographic hierarchy corresponding to that of Figure 1(b), we can now discuss the hierarchical query of Listing 1.7. Here, conditions are expressed on both Date and Customer. We can see how the NOX `Path` object is used to identify the elements of a partial hierarchy path. (Note that the path strings refer to *raw* cube data, NOT typed-checked meta data). Furthermore, we see the use of the built-in `includes` method to constrain the hierarchy condition. How does one interpret

```
public boolean select(){
   DateDimension date = new DateDimension();
   CustomerDimension customer = new CustomerDimension();

   CustHierarchy hierarchy = customer.getGeographicHierarchy();
   OlapPath path= new OlapPath(''USA'', ''California'');

   return ( customer.getAge() > 65 && hierarchy.includes(path))
       &&
      ( date.getYear() == 2007 && date.getMonth() <= 6);
}
```

Listing 1.7. Manipulating hierarchies

```
class OuterQuery extends OlapQuery{
 public boolean select(){
  CustomerDimension customer = new CustomerDimension();
  ProductDimension product = new ProductDimension();
  return ((customer.getAge() < 30) && (product.getWeight() >
      10.0));
 }

 public OlapQuery intersection(){ return new InnerQuery();}
}
```

Listing 1.8. Set operations

the expression `hierarchy.includes(path)`? Again, all selection criterion are defined relative to the current cube cell. Logically, this condition simply asks "Is this partial path consistent with the hierarchy members of this cell?" We note that while there are many variations on hierarchy traversal, NOX always uses this same simple approach.

Several additional points are worth noting. First, our NOX objects are fully amenable to standard IDE refactoring methods. For example, should the DB administrator modify the customer `name` field to `cname`, we can directly refactor the attribute name/schema without relying on a tedious and error-prone "find and replace". Second, pre-computation of the query verifies its sematic validity. In other words, while we cannot guarantee that the user's specific selection criteria will actually match any cells in the remote database, we *can* guarantee at compile-time that the query is structurally sound in terms of its use of dimensions, hierarchies, members, etc. Finally, by decomposing the query into its constituent algebraic elements at compile time, we relieve the server of the computational overhead that would normally be done at run-time. Embedded query string APIs — while superficially appealing for trivial queries — simply cannot provide this functionality.

Set Operations. Previously, we showed that *set* operations are defined quite simply in the NOX grammar. As it turns out, their specification in the native language is just as straightforward. Listing 1.8 provides a simple illustration.

```
class OldQuery extends OlapQuery{
 // ... select method definition
 // ... project method definition
}

class NewQuery extends OldQuery {
 public Object[] project() {
  CustomerDimension customer = new CustomerDimension();
  ProductDimension product = new ProductDimension();
  SalesMeasure measure =  new SalesMeasure();

  Object[] projections = {measure.getCount(), customer.getName(),
      product.getLabel()};
  return projections;
}}
```

Listing 1.9. Over-riding query classes

Here, the programmer defines the "outer" query using the standard `selection` method (and possibly others). In the *intersection* method, the "inner" query (previously defined) is specified merely by returning a reference to the relevant query object. Using this info, the NOX pre-parser can combine both queries into a single XML string corresponding to the nested style of the grammar.

Query Inheritance. One of the reasons that we represent algebraic operations in separate methods is simply because most operations are semantically unique, making it very difficult to combine them into a single native language method (with a single return type). However, a second rationale is just as important. Namely, we feel that it is extremely valuable to allow for the re-use of previous, often very complex, queries. We saw a simple example of this with the "inner" query above. A more powerful opportunity would be to allow programmers to re-use portions of already defined queries. Perhaps the most obvious example would be to re-define the `projection` method to simply identify a different measure or display attribute. With virtually all current approaches, this would involve cutting and pasting previous chunks of source code, each of which would have to be independently located and updated in the future.

With NOX's distinct query methods, we now have a great deal more latitude in this regard. Listing 1.9 demonstrates how a "new" query extends an "old" one by providing a new `projection` method. Because NOX obeys *inheritance chaining*, it sees that a new projection has been specified, and creates a new query that consists of the `selection` method of the "old" query and the projection method of the "new" query. Any subsequent changes to the source of OldQuery will be automatically integrated into the NewQuery upon re-compilation.

As a final point, this listing also demonstrates the use of the `projection` method itself. Note that its `return` argument is an array of Objects, indicative of its purpose to identify *measure* and *display* attributes (strings, ints, floats, etc). Measures extend the OlapMeasure class and are defined in a manner similar to dimension classes; that is, a list of measures and their associated getters.

5.3 Result Sets

One of the great advantages of ORM systems is that they allow data to be more or less transparently mapped back into client applications. NOX offers the same functionality in the context of multi-dimensional cube results. Specifically, the framework retrieves results from the server and transforms them into a multi-dimensional array object that can be directly accessed via the OlapResultSet reference. The format of the result is again defined by a DTD and is essentially structured as a combination of *meta data* and *cell data*. The meta data consists of the relevant dimensions, along with those dimension members actually included in the query result. The cell data, on the other hand, is listed in a row format that maps cell values to the corresponding axis coordinates. For example, a meta data element defined in the DTD as `(MEMBER_NAME, MEMBER_ID)` would associate a member — say the customer John Smith — with an integer representing the axis offset – say 4. In the *cell data* section of the XML document, this ID would then be embedded within a record of the form `<4,1,2,345.24>`. Assuming a Sales measure and a Customer–Product–Location cube, the row `<4,1,2,345.24>` would indicate that John Smith has purchased $345.24 of Product 1 at Location 2.

Once the XML result is received at the client, it is immediately transformed into a multi-dimensional object. The XML is parsed using the same DOM facilities used to create the original query (albeit with a different DTD). The aforementioned MEMBER_ID values are directly utilized as cube axes coordinates, thereby allowing a linear time population of the Result Set object. Meta data is inserted into a series of lookup data structures (i.e., maps and dictionaries) that not only allow efficient searches, but also permit transparent mapping between "user friendly" member names and the server generated member IDs that are meaningless to the end user.

The Result Set API then exposes a series of methods that allow for the simple manipulation of the cube results. Individual cell values can be retrieved merely by specifying the appropriate coordinates, either by axis value or member value. More sophisticated access can also be layered on top of the simpler access primitives. For example, Listing 1.10 shows how one might produce a simple report of all cells in a simple Customer-Product cube, assuming that the `execute` method has already been invoked and an OlapResultSet created. One merely has to retrieve the member values for each dimension and then, with a set of nested `FOR` loops, combine the relevant coordinates for each cell. It should be clear that this is really quite trivial relative to the alternatives (e.g., a JDBC ResultSet model).

```
// ... retrieve lists of dimension members from result object
for (String CustMember: CustList){
 for (String ProdMember: ProdList){
   coordinates = new LinkedList<CubeCoordinate>();
   coordinates.add(new CubeCoordinate(CustDimension,
       CustMember));
   coordinates.add(new CubeCoordinate(ProdDimension,
       ProdMember));
   System.out.println( result.getCellValue(coordinates));
}}}
```

Listing 1.10. Trivial report method

6 Future Work

NOX is already a very large system and is currently the subject of a great deal of ongoing research. Of particular importance at the present time are the following challenges:

- The expansion of the facilities for hierarchical navigation to include more flexible and varied traversal options.
- The enhancement of the ResultSet model to include transparent **change base**, **change level**, and **pivot** operations, as well as paged retrieval of result sets that are either too big or too sparse to be fully encapsulated inside a local array.
- Support for run-time parameterization of query values (i.e., user-defined query parameters). This will likely be done via query constructors, with "stubs" identifying the location for run-time XML augmentation.
- Full integration with the OLAP DBMS. While NOX includes a simple server that validates and parses the final XML query (including all examples in this paper), a parallel project is currently developing an optimized OLAP server that natively understands the algebra of the NOX model. However, even in the absence of such a server, we note that it is entirely possible to convert the NOX output to MDX and deliver it to an XMLA-compliant server.

7 Conclusions

In this paper we have provided a relatively thorough presentation of NOX, the Native language OLAP query eXecution framework. The current version of NOX represents a comprehensive implementation of the native language query model. In building upon the notion of a consistent OLAP conceptual model, we have been able to provide almost fully transparent cube persistence functionality that allows the programmer to view remote, possibly very large, analytical repositories merely as local objects. In addition to the ability to program against the conceptual model, our framework also provides compile-time type checking, clean re-factoring opportunities, and direct Object-Oriented manipulation of Results

Sets. While we chose to target Java in this initial implementation, the fundamental concepts are language agnostic and could easily be applied to other modern OOP languages. Given the awkward, loosely standardized nature of the current OLAP application marketplace, we believe that NOX offers exciting possibilities for those building and utilizing products and services in this extremely important area.

References

1. XML for Analysis Specification v1.1. (2002), http://www.xmla.org/index.htm
2. CWM, Common Warehouse Metamodel (2003), http://www.cwmforum.org/
3. JSR-69 JavaTM OLAP Interface (JOLAP), JSR-69 (JOLAP) Expert Group (2003), http://jcp.org/aboutJava/communityprocess/first/jsr069/index.html
4. JSR 243: Java Data Objects 2.0 - An Extension to the JDO specification (2008), http://java.sun.com/products/jdo/
5. HaskellDB (2010), http://www.haskell.org/haskellDB/
6. JavaCC, the Java Compiler Compiler (2010), https://javacc.dev.java.net/
7. Ruby programming language (2010), http://www.ruby-lang.org/en/
8. Agrawal, R., Gupta, A., Sarawagi, S.: Modeling multidimensional databases. In: International Conference on Data Engineering (ICDE), Washington, DC, USA, pp. 232–243. IEEE Computer Society, Los Alamitos (1997)
9. Akinde, M.O., Bohlen, M.H.: Efficient computation of subqueries in complex OLAP. In: International Conference on Data Engineering (ICDE), pp. 163–174 (2003)
10. Bauer, C., King, G.: Java Persistence with Hibernate. Manning Publications Co., Greenwich (2006)
11. Blakeley, J.A., Rao, V., Kunen, I., Prout, A., Henaire, M., Kleinerman, C.: .NET database programmability and extensibility in Microsoft SQL Server. In: ACM SIGMOD International Conference on Management of Data, pp. 1087–1098. ACM, New York (2008)
12. Cook, W.R., Rai, S.: Safe query objects: statically typed objects as remotely executable queries. In: International Conference on Software Engineering (ICSE), pp. 97–106 (2005)
13. Cunningham, C., Graefe, G., Galindo-Legaria, C.A.: PIVOT and UNPIVOT: Optimization and execution strategies in an RDBMS. In: International Conference on Very Large Data Bases (VLDB), pp. 998–1009 (2004)
14. Dittrich, J.-P., Kossmann, D., Kreutz, A.: Bridging the gap between OLAP and SQL. In: International Conference on Very Large Data Bases (VLDB), pp. 1031–1042 (2005)
15. Gray, J., Bosworth, A., Layman, A., Pirahesh, H.: Data Cube: A relational aggregation operator generalizing group-by, cross-tab, and sub-total. In: International Conference on Data Engineering (ICDE), Washington, DC, USA, pp. 152–159. IEEE Computer Society, Los Alamitos (1996)
16. Gyssens, M., Lakshmanan, L.V.S.: A foundation for multi-dimensional databases. In: International Conference on Very Large Data Bases (VLDB), pp. 106–115. Morgan Kaufmann Publishers Inc., San Francisco (1997)
17. Malinowski, E., Zimanyi, E.: Hierarchies in a multidimensional model: From conceptual modeling to logical representation. Data Knowl. Eng. 59(2), 348–377 (2006)

18. Melton, J.: Advanced SQL 1999: Understanding Object-Relational, and Other Advanced Features. Elsevier Science Inc., New York (2002)
19. Morfonios, K., Ioannidis, Y.: CURE for cubes: cubing using a ROLAP engine. In: International Conference on Very Large Data Bases (VLDB), pp. 379–390. VLDB Endowment (2006)
20. Romero, O., Abelló, A.: On the need of a reference algebra for OLAP. In: Song, I.-Y., Eder, J., Nguyen, T.M. (eds.) DaWaK 2007. LNCS, vol. 4654, pp. 99–110. Springer, Heidelberg (2007)
21. Sismanis, Y., Deligiannakis, A., Kotidis, Y., Roussopoulos, N.: Hierarchical dwarfs for the rollup cube. In: International Workshop on Data Warehousing and OLAP (DOLAP), pp. 17–24. ACM, New York (2003)
22. Stonebraker, M., Madden, S., Abadi, D.J., Harizopoulos, S., Hachem, N., Helland, P.: The end of an architectural era (it's time for a complete rewrite). In: International Conference on Very Large Data Bases (VLDB), pp. 1150–1160 (2007)
23. Whitehorn, M., Zare, R., Pasumansky, M.: Fast Track to MDX. Springer, New York (2005)

Experience in Extending Query Engine for Continuous Analytics

Qiming Chen and Meichun Hsu

HP Labs
Palo Alto, California, USA
Hewlett Packard Co.
{qiming.chen,meichun.hsu}@hp.com

Abstract. Combining data warehousing and stream processing technologies has great potential in offering low-latency data-intensive analytics. Unfortunately, such convergence has not been properly addressed so far. The current generation of stream processing systems is in general built separately from the data warehouse and query engine, which can cause significant overhead in data access and data movement, and is unable to take advantage of the functionalities already offered by the existing data warehouse systems.

In this work we tackle some hard problems not properly addressed previously in integrating stream analytics capability into the existing query engine. We define an extended SQL query model that unifies queries over both static relations and dynamic streaming data, and develop techniques to extend query engines to support the unified model. We propose the ***cut-and-rewind*** query execution model to allow a query with full SQL expressive power to be applied to stream data by converting the latter into a sequence of "chunks", and executing the query over each chunk sequentially, but without shutting the query instance down between chunks for continuously maintaining the application context across the execution cycles as required by sliding-window operators. We also propose the ***cycle-based transaction model*** to support Continuous Querying with Continuous Persisting (CQCP) with cycle-based isolation and visibility.

We have prototyped our approach by extending the PostgreSQL. This work has resulted in a new kind of tightly integrated, highly efficient system with the advanced stream processing capability as well as the full DBMS functionality. We demonstrate it with the popular Linear Road benchmark, and report the performance. By leveraging the matured code base of a query engine to the maximal extent, we can significantly reduce the engineering investment needed for developing the streaming technology. Providing this capability on proprietary parallel analytics engine is work in progress.

1 Introduction

Streaming analytics is a data-intensive computation chain from event streams to analysis results. In response to the rapidly growing data volume and the pressing need for lower latency, Data Stream Management Systems (DSMSs) provide a paradigm shift from the load-first analyze-later mode of data warehousing [8,13,15].

T.B. Pedersen, M.K. Mohania, and A M. Tjoa (Eds.): DaWaK 2010, LNCS 6263, pp. 190–202, 2010.

1.1 The Problem

However, the current generation of DSMS is in general built separately from the data warehouse query engine, due to the difference in handling stream data and static data; as a result, the data transfer overhead between the two has become a performance and scalability bottleneck [4,6,10]. The standalone DSMS's also lack the full SQL expressive power and DBMS functionalities of managing persistent data. It does not have the appropriate transaction support for continuously persisting and sharing results along with continuous querying. As stream processing evolves from simple to complex, these functionalities are likely to be redeveloped.

1.2 The Prior Art

Since a stream query is defined on unbounded data and in general limited to non-transactional event processing, the current generation of DSMSs is mostly built from scratch independently of the database engine. Big players along this direction include System S (IBM) [12], STREAM (Stanford) [3], TelegraphCQ (Berkeley) [5], as well as Aurora, Borealis, etc [1,2,7,11]. Managing data-intensive stream processing outside of the query engine causes the data copying and moving overhead, and fails to leverage the full SQL and DBMS functionality. Two recently reported systems, the TruSQL engine [13] developed by Truviso Inc, USA, and the DataCell engine [15] developed by CWI, Netherlands, do leverage database technology but are characterized by providing a workflow like service for launching a SQL query for each chunk of the stream data during stream processing. Oracle currently offers a "continued query" feature but it is based on automatic view updates and is not the same feature as stream processing.

Managing data-intensive stream processing outside of the query engine causes the data copying and moving overhead, and fails to leverage the full SQL and DBMS functionality. To the best of our knowledge, none of the existing approaches has solved the difficulty of processing stream in terms of truly continued SQL query with chunk-wise semantics but continuously tracked application context, by leveraging the query engine without introducing an additional loosely-coupled "middleware" layer.

1.3 The Solution

We view a query engine essentially as a streaming engine, although this potential has not been thoroughly explored. With this vision, we advocate an extended SQL model that unifies queries over both streaming and static relational data, and a new architecture for integrating stream processing and DBMS to support continuous, "just-in-time" analytics with window-based operators and transaction semantics.

Our proposed stream model is defined as follows: given a query Q over a set of relations $R_1,..,R_n$ and an infinite stream of relation tuples S with a criterion C for cutting S into an unbounded sequence of chunks, e.g. by every 1-minute time window,

$$<S_{C0}, S_{C1}, \ldots, S_{Ci}, \ldots>$$

where S_{Ci} denotes the *i-th* "chunk" of the stream according to the chunking-criterion C. S_{Ci} can be interpreted as a relation. The semantics of applying the query Q to the unbounded stream S plus the bounded relations $R_1,..,R_n$ lies in

$$Q\,(S, \mathrm{R}_1,..,\mathrm{R}_n) \rightarrow <Q\,(S_{C0}, \mathrm{R}_1,..,\mathrm{R}_n), \ldots\, Q\,(S_{Ci}, \mathrm{R}_1,..,\mathrm{R}_n), \ldots >$$

which continuously generates an unbounded sequence of query results, one on each *chunk* of the stream data.

Our goal is to support the above semantics using a continuous query that runs cycle by cycle for processing the stream data chunks, each data chunk to be processed in each cycle, in a single, long-standing query instance. In this sense we also refer to the *data chunking criterion* C as the *query cycle specification*. The cycle specification can be based on time or a number of tuples, which can amount to as small as a single tuple, and as large as billions of tuples per cycle. The stream query may be terminated based on specification in the query (e.g. run for 300 cycles), user intervention, or a special end-of-stream signal received from the stream source. Specifically, our solutions include the following.

- We start with providing unbounded relation data to feed queries continuously, by using function-scan instead of table-scan, to turn captured events into unbounded sequence of relation tuples without first storing them on disk.
- We develop UDF shells [9] to deliver operators with stream semantics (e.g. moving average, notification) that are not available in conventional SQL.
- We propose the *cut-and-rewind* query model, namely, cutting a query execution based on some granule ("chunk") of the stream data (e.g. in a time window), and then rewinding the state of the query without shutting it down, for processing the next chunk of stream data. This mechanism, on one hand, allows applying a query continuously to the stream data chunks falling in consecutive time windows, within a single, long-standing query; on the other hand, allows retaining the application context (e.g. data buffered with UDFs) continuously across the execution cycles to perform sliding-window oriented, history sensitive operations.
- To support *Continuous Querying with Continuous Persisting* (CQCP), we introduce the cycle-based transaction model with the *cycle-based isolation* mechanism.

The proposed *cut-and-rewind* approach enables us to support truly continuous query in a completely different way from other DSMSs, and seamlessly integrate the stream processing capability into a full-functional database system, creating a powerful and flexible system that can run SQL over tables, streams (tuple by tuple or chunk by chunk), and the combination of the two.

We report our experience in leveraging the PostgreSQL engine for supporting stream processing, and demonstrate the merit of our platform using the popular Linear Road (LR) stream processing benchmark. Providing this capability on a proprietary parallel database engine is currently being explored.

The rest of this paper is organized as follows: Section 2 reports our approach in handling stream source and stream analytic functions by extending a DBMS with new source functions and UDFs for stream operations; Section 3 proposes the *cut-and-rewind* approach; Section 4 deals with the transaction issues in cycle-based stream processing; Section 5 shows how the proposed approach is applied to the LR benchmark, and discusses the experiment results; Section 6 concludes the paper.

2 Stream Processing as Continuous Querying

There exist some fundamental differences between the conventional query processing and the stream processing. First, a query is defined on bounded relations but stream data are unbounded; next, stream processing adopts window-based semantics, i.e. processing the incoming data chunk by chunk falling in consecutive time windows; however, the SQL operators are either based on one tuple (such as filter operators) or the entire relation; Further, stream processing is also required to handle sliding window operations continuously across chunk based data processing; and finally, endless stream analytics results must be continuously accessible along their production, under specific transaction semantics.

The above stream processing challenges can be indicated by the widely-accepted LR benchmark [14]. This benchmark models the traffic on multiple express ways divided by directions and segments. It requires computing the segment aggregates, i.e. the number of cars and their average speed from each data chunk falling in every 1-minute time window, as well as computing the past 5-minute moving average speed continuously atop of the 1-minute average speed but across the 1-minute query cycles.

2.1 Stream Source Function

We start with providing unbounded relation data to fuel queries continuously. The first step is to replace the database table, which contains a set of tuples on disk, by the special kind of table function, called Stream Source Function (SSF) that returns a sequence of tuples to feed queries without first storing on disk. A SSF can listen or read data/events sequence and generate stream elements tuple by tuple continuously. A SSF is called multiple, up to infinite, times during the execution of a continuous query, each call returns one tuple. When the end-of-cycle event or condition is seen, the SSF signals the query engine to terminate the current query execution cycle.

We rely on SSF and query engine for continuous querying on the basis that "as far as data do not end, the query does not end", rather than employing an extra scheduler to launch a sequence of one-time query instances. The SSF scan is supported at two levels, the SSF level and the query executor level. A data structure containing function call information, *hFC*, bridges these two levels. *hFC* is initiated by the query executor and passed in/out the SSF for exchanging function invocation related information. We use this mechanism for minimizing the code change, but maximize the extensibility, of the query engine.

2.2 Stream Analytics through UDF

One important characteristics of stream processing is the use of stream-oriented history-sensitive analytic operators such as moving average or change point detection. This represents a different requirement from the regular query processing that only cares about the current state. While the standard SQL engine contains a number of built-in analytic operators, stream history-sensitive operators are not supported. Using UDFs is the generally accepted mechanism to extend query operators in a DBMS. A UDF can be provided with a data buffer in its function closure, and for caching stream processing state (synopsis). Furthermore, it is also used to support one or more *emitters* for delivering the analytics results to interested clients in the middle of a cycle, which is critical in satisfying stream applications with low latency requirement.

We use UDFs to add window operators and other history sensitive operators, buffering required raw data or intermediate results within the UDF closures. A scalar UDF is called multiple times on the per-tuple basis, following the typical FIRST_CALL, NORMAL_CALL, FINAL_CALL skeleton. The data buffer structures are initiated in the FIRST_CALL and used in each NORMAL_CALL. A window function defined as a scalar UDF incrementally buffers the stream data, and manipulates the buffered data chunk for the required window operation. Since the query instance remains alive, as supported by our *cut-and-rewind* model, the UDF buffer is retained between cycles of execution and the data states are traceable continuously (we see otherwise if the stream query is made of multiple one-time instances, the buffered data cannot be traced continuously across cycle boundaries). As a further optimization, the static data retrieved from the database can be loaded in a window operation initially and then retained in the entire long-standing query, which removes much of the data access cost as seen in the multi-query-instances based stream processing.

3 Cycle Based Continuous Query

We propose to run a SQL query cycle by cycle for deriving a sequence of data-chunk based results, but never shutting down the query instance in order to have the per-tuple based data processing history continuous tractable. To support the cycle based execution of stream queries, we propose the *cut-and-rewind* query execution model, namely, cut a query execution based on the cycle specification (e.g. by time), and then rewind the state of the query without shutting it down, for processing the next chunk of stream data in the next cycle.

Under this *cut-and-rewind* mechanism, a stream query execution is divided into a sequence of *cycles*, each for processing a chunk of data only; it, on one hand, allows applying a SQL query to unbounded stream data chunk by chunk within a single, long-standing query instance; on the other hand, allows the application context (e.g. data buffered within a User Defined Function (UDF)) to be retained continuously across the execution cycles, which is required for supporting sliding-window oriented, history sensitive operations. Bringing these two capabilities together is the key in our approach.

Cut. *Cutting* stream data into chunks is originated in the SSF at the bottom of the query tree. Upon detection of end-of-cycle condition, the SSF signals *end-of-data* to the query engine through setting a flag on the function call handle, that, after being interpreted by the query engine, results in the termination of the current query execution cycle.

If the cut condition is detected by testing the newly received stream element, the *end-of-data* event of the current cycle would be captured upon receipt of the first tuple of the next cycle; in this case, that tuple will not be returned by the SSF in the current cycle, but buffered within the SSF and returned as the first tuple of the next cycle. Since the query instance is kept alive, that tuple can be kept across the cycle boundary.

Rewind. Upon termination of an execution cycle, the query engine does not shut down the query instance but *rewinds* it for processing the next chunk of stream data.

Rewinding a query is a top-down process along the query plan instance tree, with specific treatment on each node type. In general, the intermediate results of the standard SQL operators (associated with the current chunk of data) are discarded but the application context kept in UDFs (e.g. for handling sliding windows) are retained. The query will not be re-parsed, re-planned or re-initiated.

Note that rewinding the query plan instance aims to process the next chunk of data, rather than re-deliver the current query result; therefore it is different from "rewinding a query cursor" for re-delivering the current result set from the beginning.

As mentioned above, the proposed *cut-and-rewind* approach has the ability to keep the continuity of the query instance over the entire stream while dividing it to a sequence of execution cycles. This is significant in supporting history sensitive stream analytic operations, as discussed in the previous section.

4 Continuous Querying with Continuous Persisting (CQCP)

One problem of the current generation of DSMSs is that they do not support transactions. Intuitively, as stream data are unbounded and the query for processing these data may never end, the conventional notion of transaction boundary is hard to apply. In fact, transaction notions have not been appropriately defined for stream processing, and the existing DSMSs typically make application specific, informal guarantees of correctness.

However, to allow a hybrid system where stream queries can refer to static data stored in a database, or to allow the stream analysis results (whether intermediate or final) to persist and be visible to other concurrent queries in the system in a timely manner, a transaction model which allows the stream processing to periodically "commit" its results and makes them visible is needed.

Note that if a stream application does not use static data in the database, or does not need to persist results and make them visible to other concurrent applications, then transaction semantics are not needed. In our design, the transaction semantics is used, and thus transaction management overhead is incurred, only when a stream application requires persistent data management.

4.1 Query Cycle Based Transaction Model

Conventionally a query is placed in a transaction boundary. In general, the query result and the possible update effect are made visible only after the commitment of the transaction (although weaker transaction semantics do exist). In order to allow the result of a long-running stream query to be incrementally accessible, we introduce the cycle-based transaction model incorporated with the *cut-and-rewind* query model, which we call *continuous querying with continuous persisting*, (CQCP). Under CQCP, a stream query is committed one cycle at a time in a sequence of "micro-transactions". The transaction boundaries are consistent with the query cycles, thus synchronized with the chunk-wise stream processing. The per-cycle stream processing results are made visible as soon as the cycle ends. The isolation level is Cycle based Read Committed (CRC). To allow the cycle results to be continuously visible to external world, regardless of the table is under the subsequent cycle-based transactions, we enforce record level locking.

We extended both SELECT INTO and INSERT INTO facilities of the PostgreSQL to support CQCP. We also added an option to force the data to stay in memory, and an automatic space reclaiming utility should the data be written to the disk.

4.2 Continuous Persisting

In a regular database system, the queries with SPJ (Select, Project, Join) operations and those with the update (Insert, Delete, Update) operations are different in the flow of resulting data. In a SPJ query, the destination of results is a query receiver connected to the client. In a data update query, such as insert, the results are emitted to, or synched to, the database.

In stream processing, such separation would be impractical. The analytic results must be streaming to the client continuously as well as being stored in the database if needed for other applications to access. Therefore, we extended the query engine to have query evaluation and results persisting integrated and expressed in a single query. This two-receiver approach makes it possible to have the results both persisted and streamed out externally.

Certain intermediate stream processing results can be deposited into the database from UDFs. To do so the UDF must be relaxed from the read-only mode, and employ the database internal query facility to form, parse, plan and execute queries efficiently. In our prototype, the PostgreSQL SPI (Server Program Interface) is used.

5 Example and Experiments

5.1 Modeling the Linear Road Benchmark

We use the widely-accepted Linear-Road (LR) benchmark [14] to demonstrate our extended query engine. The LR benchmark models the traffic on express ways for the 3-hour duration; each express way has two directions and 100 segments. Cars may enter and exit any segment. The position of each car is read every 30 seconds and each reading constitutes an event, or stream element, for the system. A car position report has attributes *vid* (vehicle ID), *time* (seconds), *speed* (mph), *xway* (express way), *dir* (direction), *seg* (segment), etc. The benchmark requires computing the traffic statistics for each highway segment, i.e. the number of active cars, their average speed per minute, and the past 5-minute moving average of vehicle speed. Based on these per-minute per-segment statistics, the application computes the tolls to be charged to a vehicle entering a segment any time during the next minute, and notifies the toll in real time (notification is to be sent to a vehicle within 5 seconds upon entering the segment). The application also includes accident detection; an accident occurring in one segment will impact the toll computation of that segment as well as a few downstream segments. An accident is flagged when multiple cars are found to have stopped in the same location. The graphical representation of our implementation of the LR stream processing requirement is shown in Fig. 1 together with its corresponding stream query.

```
INSERT INTO toll_table SELECT minute, xway, dir, seg, lr_toll(r.traffic_ok, r.cars_volume)
FROM (
     SELECT minute, xway, dir, seg, cars_volume,
               lr_moving_avg(xway, dir, seg, minute, avg_speed) as mv_avg, traffic_ok
```

```
        FROM (
            SELECT floor(time/60)::integer AS minute, xway, dir, seg,
                   AVG(speed) AS avg_speed, COUNT(distinct Vid)-1) AS cars_volume,
                   MIN(trffic_ok) AS traffic_ok
            FROM (
                SELECT xway, dir, seg, time, speed, vid,
                       lr_acc_affected(vid,speed,xway,dir,seg,pos) AS traffic_ok
                FROM STREAM_CYCLE_lr_data(60, 180)
                WHERE lr_notify_toll(vid, xway, dir, seg, time)>=0
        ) s
            GROUP BY minute, xway, dir, seg
    ) p
) r
WHERE r.mv_avg > 0 AND r.mv_avg < 40;
```

This query provides the following major functions.

Stream Source Function. The streaming tuples are generated by the SSF *STREAM_CYCLE_lr_data(time, cycles)*, from the LR data source file with timestamps, where parameter "*time*" is the time-window size in seconds; "*cycles*" is the number of cycles the query is supposed to run. For example, *STREAM_CYCLE_lr_data(60, 180)* delivers the position reports one-by-one until it detects the end of a cycle (60 seconds), and performs a "cut", then onto the next cycle, for a total of 180 cycles (for 3 hours).

Segment statistics and toll generation. As illustrated by the left hand side of Fig. 1, the tolls are derived from the segment statistics, i.e. the number of active cars, average speed, and the 5-minute moving average speed, as well as from detected accidents, and dimensioned by express way, direction and segment. We leveraged the *minimum, average* and *count-distinct* aggregate-groupby operators built into the SQL engine, and provided the *moving average* (lr_moving_avg) operator and the *accident detection* (lr_accident) operator in UDFs.

Toll persisting. Required by the LR benchmark, the segment tolls of minute *m* should be generated within 5 seconds after *m*. The toll of a segment calculated in the past minute is applied to the cars currently entering into that segment. The generated tolls are inserted into a *segment toll table* (SegToll) with the transaction committed per cycle (i.e., per minute). Therefore the tolls generated in the past minutes are visible to the current minute.

Toll notification. As shown on the right side of Fig. 1, the per-car toll notification is provided by the UDF *lr_notify_toll()* appearing in the following phrase

```
WHERE lr_notify_toll(vid, xway, dir, seg, time) >= 0
```

This UDF keeps enough information about active cars so as to detect the event of a car entering a new segment; and for each car entering a new segment, it emits a toll notification while persisting the toll to a table (carAccount table) for future account balance queries. This UDF reads the segment tolls of the previous minute within the

FIRST_CALL part of the UDF (represented by the dash line), enabling it to use the information produced by the previous cycle of the stream query. Since this UDF is a *per-tuple* UDF (i.e., the NORMAL_CALL part of the UDF is invoked per input tuple), the toll notification is emitted immediately after the position report is received from the source stream, and does not wait for the current cycle (minute) to terminate. This UDF also persists the toll into the car account table. While the toll is notified immediately upon receiving the car position report, persisting the toll is committed once per cycle, in accordance to our CPCQ model.

Multiple features of our cycle-based stream processing approach are illustrated in this query:

Cut-and-Rewind. This query repeatedly applies to each data chunks falling in 1-minute time-window as an execution cycle, and rewinds 180 times in the single query instance; the sub-query with alias *p* uses the standard SQL aggregate-groupby function to yield the number of active cars and their average speed for every minute dimensioned by segment, direction and express way. The SQL aggregate functions are computed for each cycle with no context carried over from one cycle to the next.

Sliding Window Function (per-tuple history sensitive). The sliding window function *lr_moving_avg()* buffers the up to 5 per-minute average speed for accumulating the dimensioned 5-minute moving average; since the query is only rewound but not shut down, this buffer is retained continuously across query cycles – this is a critical advantage of cut/rewind over shutdown/restart.

Continuous Querying with Continuous Persisting. The top-level construct of the LR query is actually the INSERT-SELECT phrase; with our engine extension, it

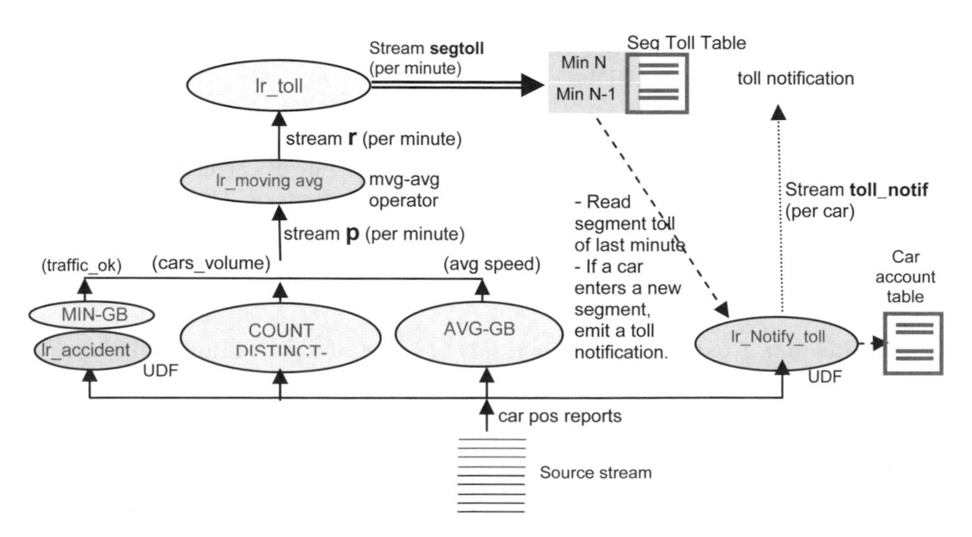

Fig. 1. Cycle based stream query for LR benchmark, for both the generation of per-minute, per cycle tolls common to all cars, and the per car based retrieval of resulting tolls

persists the result stream returned from the SELECT query (r) to the toll table on the per-cycle basis. The transactional LR query commits per cycle to make the cycle based result accessible to subsequent cycles or other concurrent queries after the cycle ends. This cycle-based isolation level is supported with the appropriate locking mechanism.

Self-Referencing. The per-car toll notification is generated by the UDF *lr_notify_toll()*. It efficiently accesses the segment toll in the *last minute* directly from the toll table. This kind of self-referencing provides a handshake mechanism for the *producer* part and the *consumer* part of the same query to rely on the query engine to synchronize, to perform history sensitive stream analytics, and to gain extremely high performance due to their seamless integration. We believe that such self-referencing represents a common paradigm in stream processing.

5.2 Experimental Setup

The experimental results are measured on HP xw8600 with 2 x Intel Xeon E54102 2.33 Ghz CPUs and 4 GB RAM, running Windows XP (x86_32) and PostgreSQL 8.4. The input data are downloaded from the benchmark's home page. The "L=1" setting was chosen for our experiment which means that the benchmark consists of 1 express way (with 100 segments in each direction). The event arrival rate ranges from a few per second to peak at about 1,700 events per second towards the end of the 3-hour duration. Fig. 3(Left) shows the distribution of data volume per minute, i.e. the per-minute throughput. The LR data can be supplied in the following two modes:

- Stress test mode: the data are read by the SSF from a file continuously without following the real-time intervals (continuous input)
- Real-time input: the data are received from a data driver outside of the query engine with real-time intervals.

We report our experimental results in these 2 different modes.

5.3 Performance under Stress Test Mode

Time for computing segment tolls. Calculating the segment statistics and tolls has been recognized as the computation bottleneck of the benchmark in the literature. The LR benchmark requires the segment toll to be calculated based on the segment statistics and traffic status (whether affected by accidents) every minute.

We took the left-hand-side of our LR model in Fig 1 and ran that branch of the query up until the toll is computed, under the stress test mode. The total computation time with L=1 setting is shown in Fig. 2 (Left). It shows that our system is able to generate the per-minute per-segments tolls for the total 3 hours of LR data (approx. 12 Million tuples) in a little over 2 minutes.

Performance of Persisting with Heap-Insert. Unlike other reported DSMSs where the stream processing results are persisted by connecting to a separate database and issuing queries, with the proposed cycle-based CQCP approach, the continuous, minute-cycle based query results are stored through efficient heap-insert.

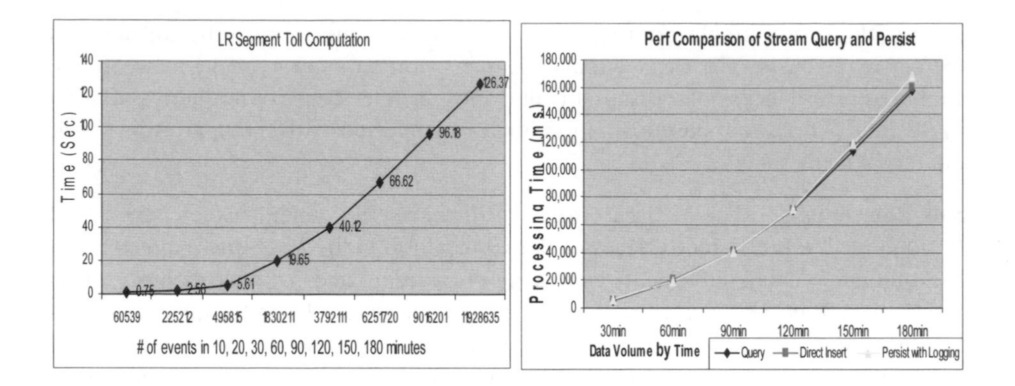

Fig. 2. (Left) Total time of toll computation. **(Right)** Performance comparison of querying-only and query+persisting (with continuous input)

From Fig. 2 (Right) we can see that persisting the cycle based stream processing results either by inserting with logging (using INSERT INTO with extended support by the query engine) or by direct inserting (using SELECT INTO with extended support by the query engine – not shown in this query), does not add significant performance overhead compared to querying only. This is because we completely push stream processing down to the query engine and handle it in a long running query instance with direct heap operations, with negligible overhead for data movement and for setting up update queries.

Post Cut Elapsed Time. In cycle-based stream processing, the remaining time of query evaluation after the input data chunk is cut, called Post Cut Elapsed Time (PCET), is particularly important since it directly affects the delta time for the results to be accessible after the last tuple of the data chunk in the cycle has been received.

Fig 3 (Left) shows the input data volume over 1-minute time windows (i.e., the stream workload). Fig. 3 (Right) shows the query time, as well as the PCET, for processing each 1-minute data chunk. It can be seen that the PCET (the blue line) is well controlled around 0.2 second, meaning that the maximal response time for the segment toll results, as measured from the time a cycle (a minute) ends, is around 0.2 second.

5.4 Performance under the Real-Time Input Mode

With real-time input, the events (car position reports) are delivered by a data driver in real-time with additional system-assigned timestamps. The query runs cycle by cycle on each one-minute data chunk. Fig 4 shows the maximal toll notification response time in each of the 180 1-minute windows.

The maximal response time of toll notification really depends on the PCET measure introduced above, i.e. it is essentially the delay after a cycle is "cut" in completing the segment toll part of the query of that cycle. This is because in the beginning of each cycle, the toll notification cannot be emitted until the segment toll generation of the last cycle completes. In the first 2 hours the toll notification response time is

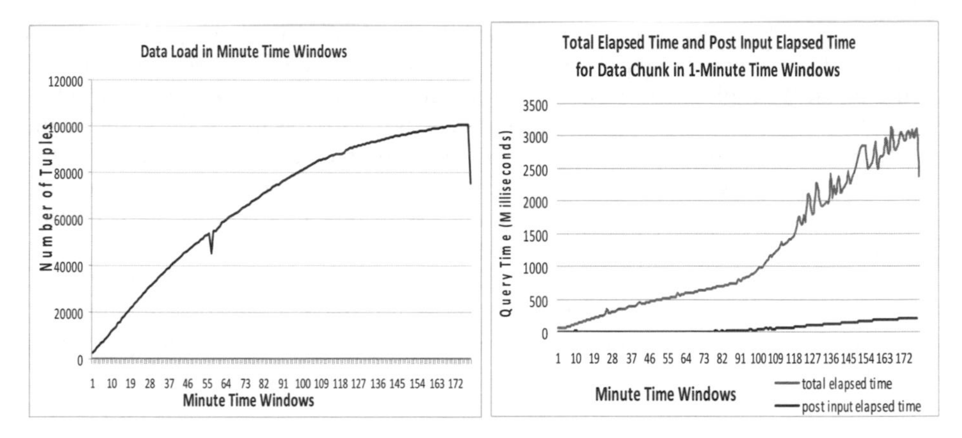

Fig. 3. (Left) Data load distribution over minute time windows **(Right)** Query time as well as PCET on the data chunk falling in each minute time window

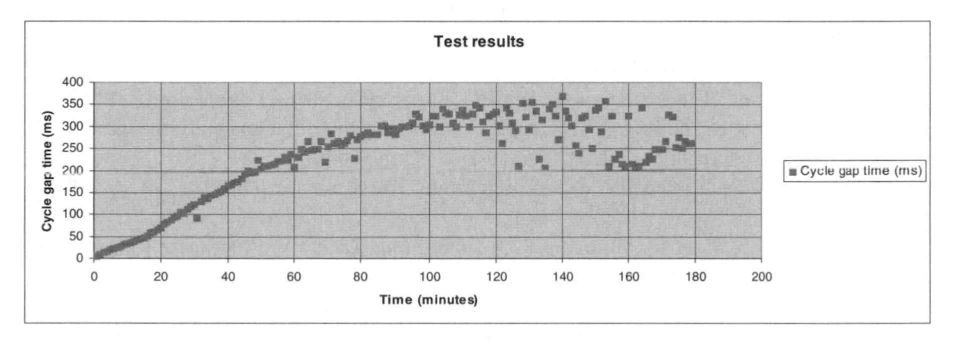

Fig. 4. Maximal toll notification response time in consecutive one-minute time windows

rather small, and with the increased data load in the last hour, it reaches the maximal value of about 0.3 second, which is still well within the 5-second latency requirement of the benchmark. Note that the maximal notification latency is not the average response time of notification. On the average, the notification response time is near zero, as the ones after the beginning of each cycle are not measurable by millisecond.

The experimental results indicate that our approach is highly competitive to any reported one. This is because we completely pushed stream processing down to the query engine with negligible data movement overhead and with efficient direct heap-insert. We eliminated the middleware layer, as provided by all other systems, for scheduling time-window based querying.

6 Conclusions

Due to the growing data volume and the low-latency requirement, the *platform separation* of analytics and data management has become the performance bottleneck, and their integration offers great potential in real-time, data-intensive analytics.

In this paper we reported our experience in leveraging the DBMS for continuous stream analytics. We tackled the key technical issues for integrating stream analytics capability into the existing query engine, and built an integrated, efficient and robust system with stream processing capability while retaining the full DBMS functionality, giving the query engine a new role. We proposed the *cut-and-rewind* query execution model for chunk-wise continuous stream processing with the full SQL power, while enabling history-sensitive stream operations. We provided advanced stream processing capability by extending the existing query engine directly without introducing separate executor or additional "middleware". With this approach we have bridged SQL and stream processing in a single engine.

The proposed approach has been implemented on the PostgreSQL engine. We are currently refining our unified query and transaction model, and investigating into using HP SeaQuest for providing a MPP-based, data-intensive streaming analytics platform.

References

1. Abadi, D., Carney, D., Cetintemel, U., Cherniack, M., Convey, C., Lee, S., Stonebraker, M., Tatbul, N., Zdonik, S.: A New Model and Architecture for Data Stream Management. VLDB J. 2(12), 120–139 (2003)
2. Abadi, D.J., et al.: The Design of the Borealis Stream Processing Engine. In: CIDR (2005)
3. Arasu, A., Babu, S., Widom, J.: The CQL Continuous Query Language: Semantic Foundations and Query Execution. VLDB Journal 2(15) (June 2006)
4. Bryant, R.E.: Data-Intensive Supercomputing: The case for DISC, CMU-CS-07-128 (2007)
5. Chandrasekaran, S., et al.: TelegraphCQ: Continuous Dataflow Processing for an Uncertain World. In: CIDR 2003 (2003)
6. Chaiken, R., Jenkins, B., Larson, P.-Å., Ramsey, B., Shakib, D., Weaver, S., Zhou, J.: SCOPE: Easy and Efficient Parallel Processing of Massive Data Sets. In: VLDB 2008 (2008)
7. Chen, J., et al.: NiagaraCQ: A Scalable Continuous Query System for Internet Databases. In: SIGMOD (2000)
8. Chen, Q., Hsu, M.: Cooperating SQL Dataflow Processes for In-DB Analytics. In: Proc. CoopIS 2009 (2009)
9. Chen, Q., Hsu, M., Liu, R.: Extend UDF Technology for Integrated Analytics. In: Pedersen, T.B., Mohania, M.K., Tjoa, A.M. (eds.) DAWAK 2009. LNCS, vol. 5691, pp. 256–270. Springer, Heidelberg (2009)
10. Cooper, B.F., et al.: PNUTS: Yahoo!'s Hosted Data Serving Platform. In: VLDB 2008 (2008)
11. Cranor, C.D., et al.: Gigascope: A Stream Database for Network Applications. In: SIGMOD 2003 (2003)
12. Gedik, B., Andrade, H., Wu, K.-L., Yu, P.S., Doo, M.C.: SPADE: The System S Declarative Stream Processing Engine. In: ACM SIGMOD 2008 (2008)
13. Franklin, M.J., et al.: Continuous Analytics: Rethinking Query Processing in a NetworkEffect World. In: CIDR 2009 (2009)
14. Jain, N., et al.: Design, Implementation, and Evaluation of the Linear Road Benchmark on the Stream Processing Core. In: SIGMOD (2006)
15. Liarou, E., et al.: Exploiting the Power of Relational Databases for Efficient Stream Processing. In: EDBT 2009 (2009)

Development of a Business Intelligence Environment for e-Gov Using Open Source Technologies

Eduardo Zanoni Marques, Rodrigo Sanches Miani,
Everton Luiz de Almeida Gago Júnior, and Leonardo de Souza Mendes

Department of Communication, School of Electrical and Computer Engineering,
University of Campinas
Campinas, SP, Brazil
{emarques,rsmiani,elagj,lmendes}@decom.fee.unicamp.br

Abstract. It has become common for modern organizations to use advanced information systems for helping their daily operational task. However, there is still a large demand for software solutions that enable straightforward data analysis from these systems. Aiming to solve this problem, Business Intelligence (BI) environments were created. Electronic Government (e-Gov) systems, which typically work with governmental operational data, can take great benefits from a BI environment. Therefore, in e-Gov systems BI tools can be used, among others, to pursue the following goals: enhance the relationship between city and state government and the citizen; help administrating public resources; monitor the impacts of public policies upon the society. This paper presents a proposal for creating a BI environment for Electronic Government systems, using open source technologies with a special application to Social Welfare, developed for the city of Campinas, SP, Brazil.

Keywords: business intelligence, e-government, open source, software architecture.

1 Introduction

With the technological progress of our society, demands for software solutions have been growing in all sorts of areas, both in public and private organizations. Generally speaking, information systems are focused on data storage and processing, which help organizations' operational sector execute their daily tasks, such as register commercial transactions, calculate payroll, register employee and client personal data and so on. However, there is an eminent demand from organizations management sectors for software solutions to help in the processing, analysis and interpretation of their data. This can provide opportunities for better monitoring of their sectors, making projections and finding business deficiencies and/or opportunities. To accomplish this goal, several tools, technologies and solutions were created and aggregated under the concept of Business Intelligence (BI) [1].

The creation of a BI environment is a very challenging task, since it requires the implementation and concise integration of such technologies as Data Warehouse (DW), Data Mining and Online Analytical Processing (OLAP). Thus, it is fundamental the

T.B. Pedersen, M.K. Mohania, and A M. Tjoa (Eds.): DaWaK 2010, LNCS 6263, pp. 203–214, 2010.

usage of a methodology and a reference architecture to support the creation of this kind of environment.

Trying to define a generic architecture for this environment, Moss [2] presents a three-layered architecture, shown in Fig. 1. In the bottom of the figure are the data sources from pre-existing systems. These data are transformed by an extraction, transformation and loading (ETL) process, executed by the bottom layer of the architecture, and sent to the middle layer, where the data is stored. Then, data views are exposed to the upper layer, where are the BI applications that allow users to visualize and manipulate data. These applications are provided to users by an interface, like a web site or a web service. The communication between layers is always mediated by a middleware.

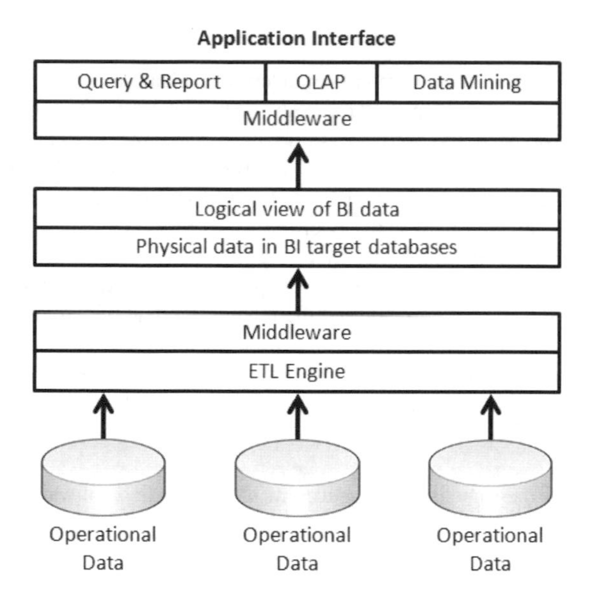

Fig. 1. Generic architecture for a Business Intelligence environment [2]

Although the creation of a BI environment is typical of private organizations, e-Gov systems can also benefit from their applications.

This paper presents a proposal for the creation of a BI environment for e-Gov, utilizing open source technologies. As proof of concept, we present the use of this proposal to develop a BI environment to manage Social Welfare for the city of Campinas, SP, Brazil.

The paper is organized as follow: Section 2 presents a discussion about related works, Section 3 brings the technical background of this paper, Section 4 presents the proposal of this paper, Section 5 describes the case study that is being conducted and, closing the paper, Section 6 brings final considerations.

2 Related Work

One can find several works describing the construction of a BI environment in various areas.

In [3] a BI environment is created for a supermarket by building a DW using Microsoft Analysis Services. The work applied clustering techniques to define clients profile and discover what types of merchandise most influenced their purchases. In the paper, no information is given on how the BI data will be made available to the users.

In [4] it is presented the elaboration of a BI environment for a life insurance company. This environment has an architecture similar to the one defined in [2], with the adoption of a DW for data storage. The paper does not inform either how the environment was built or the technologies used.

In [5] it is presented the creation of a BI environment for a Taiwan Internet Service Providers (ISP). The paper defines its own methodology for the BI construction, with an architecture similar to the one defined in [2]. In this environment, an ETL process is applied to the users' Internet traffic data, with results stored in a DW, using support tools from MS-SQL 2000. Then, clustering techniques are applied to the data to define the Internet users' usage profile. This data was made available to the users through a Web Site developed for this purpose using Java e Flash technologies.

In Brazil, [6] discusses the creation of a BI environment for the judiciary system. Here, a DW is chosen to store data generated from the ETL process. The DW was built using the methodology proposed in [7]. However, this paper presents neither the environment architecture nor the technologies used.

Regarding DW construction, which is the most common option for data storage in BI environments, there are two works, [8] and [9], treating of health related systems. The first used SAS Data Warehouse Administrator, while the second followed its own methodology using the DBMS MySQL. In [10] the creation of DW for Geographical Information Systems (GIS) and its challenges are discussed. Because all papers focused only in the DW creation, none discuss how data will be provided to the users. In the Brazilian scenario, Mussi [11] presents how the National Agency of Sanitary Surveillance (ANVISA) created its DW by adopting the methodology of [7] and using the DBMS Oracle to store the DW data and the OLAP tool MicroStrategy to make data available to end users.

This paper differs from those mentioned above by presenting: i) all phases for the creation of the BI environment, covering since the planning of the data storage until providing this data to end users; ii) the architecture of the environment; iii) the technologies and tools used in the BI implementation and their integration, highlighting the fact that all technologies and tools used are open source.

3 Technical Background

3.1 Business Intelligence

Created in 1989 by Howard Dresner [12], the Business Intelligence concept defines an architecture and a set of operational, decision support and database systems that, integrated, aims at offering to business community easy access to business information [2].

As can be seen in Fig. 1, BI can be divided in three main areas: extract, transform and load process (ETL), data storage and tools provided to end users visualize and manipulate data.

The ETL process focus on loading data from the operational data sources to BI data storage. It is developed in three well defined steps [13]:

- extract: in the first step, operational data from the organization, which can be stored in many forms (like in relational, hierarchical or multidimensional DBMS, spreadsheets or emails) is loaded and sent to the next step;
- transform: in this step the data is corrected and suited for the BI through the application of different operations, like misspelling correction, domain standardization and purging unneeded fields;
- load: final step, where the resulting data from the transform step is stored in the BI data storage.

The most common option for BI data storage is Data Warehouse (DW). The DW was first defined for Inmon [14] as "a subject-oriented, integrated, nonvolatile, and time-variant collection of data in support of management's decisions". Although not mandatory, its utilization in a BI is so usual that Almeida [15] and Biere [16] consider BI as the evolution of DW.

Addressing the applications provided to end users, Biere [16] classify them in three groups: traditional query and report, OLAP and Data Mining.

Query and Report tools help common users browse files and DBMS using friendly graphical interfaces, enabling the creation of reports from these data [16].

Defined by Codd [17], OLAP tools offers, through a well defined set of operations, support to visualization and analysis of multidimensional databases. These databases are composed by data cubes, with a set of data cubes being defined as a data mart, that have two basic entities: **facts**, which are interest items to the organization, like the sales of a product holding related numerical data, like its value and/or discount, and **dimensions**, where the data that gives context to the fact is stored hierarchically, like date of a sales [13]. The main operations of an OLAP tool are [18]:

- rollup: where the data from the fact is aggregated to an upper level of a defined hierarchy, decreasing data details;
- drill-down: where data from the fact is decomposed to the lower level of a defined hierarchy, increasing data details;
- slice and dice: which enables the application of filters to the resulting data;
- pivot: that enables re-ordering data.

These two types of tools previously presented focused on simplifying access to BI data. To complement these, Data Mining tools aims at generating intrinsic information from data stored in the BI. This kind of tools used techniques divided in two groups [19]:

- descriptive: characterize general properties of the data, finding patterns and relations, using techniques like clustering and market basket analysis;
- predictive: composed of techniques like classification, regression and time series analysis, make inferences on the current data to predict future data.

3.2 Eletronic Government

An Electronic Government (e-Gov) system is defined by the use of information and communication technologies to develop and offer governmental services and information through electronic media (usually the Internet) for citizens, enterprises and government employees agencies [20].

The construction of e-Gov systems presents particular challenges, depending on the target users of these systems. For this reason, they are divided into five classes [21]:

- Government to Business (G2B): systems developed focusing the improvement of communication between government agencies and enterprises, in such relations as service providing and regulation;
- Government to Citizens (G2C): these systems seek to improve communication between government agencies and the citizens, using initiatives like creating information portals, where citizens can both find governmental information and express their opinions;
- Government to Government (G2G): these systems focus on improving communication between government agencies, enabling them to work together in a simpler and efficient way;
- Government Internal Efficiency and Effectiveness (IEE): these systems seek to improve the internal operation of government agencies;
- Overarching Infrastructure (Cross-Cutting): systems used to facilitate the integration between the systems from all other classes, applying software and hardware integration techniques.

There are diverse papers discussing the challenges of creating e-Gov systems, like [22], [23] and [24].

4 Proposal

This paper proposes a technological composition to create a BI environment to e-Gov systems using open source technologies. The architecture used here is based on the one described in [2], being divided in three layers: the ETL layer; the data storage and view providing layer; and the end users applications layer, which are accessed through a web site. A Fig. 2 presents the technological composition proposed.

4.1 ETL Layer

The Talend Open Studio [25], which is a specialized tool for data integration and migration, was adopted to implement the ETL process over the ETL Layer. Through the configuration and composition of its components, it is possible to read data from diverse formats, like Excel, CSV and XML. The tool also enables connecting to relational, hierarchical and multidimensional DBMS. Different types of transformations, like data normalization and denormalization, data joining using key match and string processing can be applied over the data read. After applying the selected transformations, data can be stored in diverse formats, like the ones from the reading step.

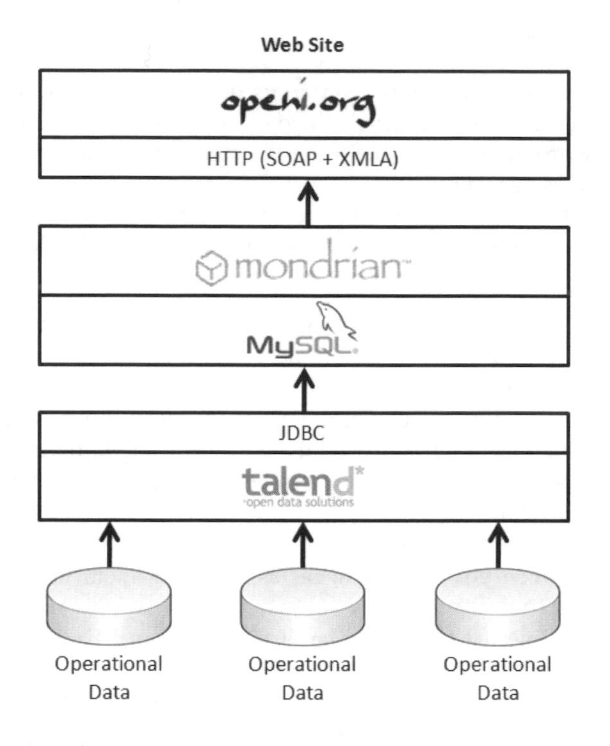

Fig. 2. Realization of the architecture proposed in [2] using open source technologies

4.2 Data Storage and View Providing Layer

In the Data Storage and View Providing Layer, the data storage is implemented using relational DBMS MySQL [26]. This DBMS was chosen based on its support to various types of indexes (B-Tree and Hash, for an instance) and fast data load and access, by using MyISAM tables. The communication between Talend and MySQL is done through JDBC middleware, which is a standard to connect Java applications to relational DBMS.

To implement the data view, Mondrian OLAP Server [27] was selected. Mondrian is an OLAP Server written in Java that allows the execution of multidimensional queries, written in Multi Dimensional Expressions (MDX), on relational databases, presenting the results in multidimensional format. It is done by creating a multidimensional mapping over the relational data, defining data cubes, where the queries are executed.

The MDX format is a standard defined by Microsoft to execute queries on data cubes, providing an easier and intuitive form to access data from multidimensional databases. In a certain way, MDX is to multidimensional databases what SQL is to relational databases, except that MDX queries return data cubes, while SQL queries return tables. Also, SQL has a Data Definition Language (DDL), which is not included in current MDX specification. Besides Mondrian, other tools support MDX queries, like MicroStrategy, SAS e Oracle Essbase [28].

Using data views brings extra benefits to BI, considering it enables creating different views to different users and providing a clearer presentation of data to end users.

4.3 End Users Applications Layer

To provide data access to end users, the OpenI [29] application was chosen. OpenI is a web application for executing queries in multidimensional databases. Using OpenI, users can select a data cube and, through a graphical interface, browse data using the main operations of an OLAP client. It is also possible to browse data using MDX to more advanced queries. The resulting data can be displayed in multidimensional tables or charts, like bar and pie. These queries can be saved in the application for further visualization and/or edition.

Another important OpenI feature is the dashboard creation, where previously saved queries are put together for fast analysis of important data. Thus, it is possible to choose different types of data display for each query result.

The communication between OpenI and Mondrian OLAP Server is done through HTTP, using Simple Object Access Protocol (SOAP) and XML for Analysis (XMLA) protocols.

The SOAP protocol structures information exchange through HTTP. Using XML, this protocol is composed by an envelope which has two tags: **header**, where information pertinent to data processing is allocated, and **body**, that contains the data from the message being transferred [30].

Using SOAP, XMLA protocol defines a couple of XML messages for communication between an OLAP client and server, establishing an interoperable client-server communication channel where diverse tools from different enterprises can work along. The messages defined in this protocol are [28]:

- discover: used to query the OLAP server properties, like a list of available data sources;
- execute: used to execute queries to a specific data source, using MDX language.

5 Case Study

This section describes the application of the proposal to create a BI environment to analyze the operational data from the Social Welfare Department of the city government of Campinas, SP, Brazil.

The Social Welfare Department has the mission of "rescuing citizenship to people that are in a social vulnerability situation caused by poverty and exclusion". Thus, diverse social programs have been developed aiming to offer to citizens in vulnerable state new possibilities to enable their financial emancipation. These social programs include different areas, like offering microcredit, digital inclusion and training courses [31].

To manage data from this area, the Social Welfare module from Integrated System for Municipal e-Gov (SIGM) was adopted. The SIGM is an ERP-like system for managing "all services, citizen records, processes management and relevant data for a city's administration" [24]. In the Social Welfare module, any data from the social programs developed by the city government can be stored and manipulated as needed.

As the first step to construct the BI environment, the utilization of a DW to data storage was chosen. Then, a data-oriented analysis, as proposed in [32], was conducted. In this, only operational data sources are considered to define the DW data. This analysis resulted in various candidate data cubes to compose the BI database. From these cubes, the Inclusion in Social Program data cube was selected to be the first offered by the BI.

With the data cube selected, the next step was defining what data would compose this data cube. It was done in two steps. First, a conceptual model was generated (Fig. 3), having the fact, the dimensions, the dimensions hierarchical levels and fields. This model was created in UML using the profile proposed in [33]. Based in this model, the logical and physical modeling was conducted. In this step, the star model proposed in [7] was adopted, where the fact is the central element surrounded by its dimensions. The resulting physical model is illustrated in Fig. 4.

After composing the models, the ETL process was configured using Talend tool. Given the number of treatments needed to be applied in the data before it was inserted on the BI data storage, this step, as predicted in [13], was the one that took the largest development effort in the case study.

Parallel to the ETL step, the DW data view exposed by Mondrian was configured. It was done using the Schema Workbench, an auxiliary tool from Mondrian. In this tool, data cubes are created by defining a fact and the dimensions related to this fact. It is also configured both fact and dimensions fields, which can be from data consolidated on the DW or calculated at runtime. The resulting data view is similar to the conceptual model (Fig. 3), which provides a clearer representation of the BI data to its users.

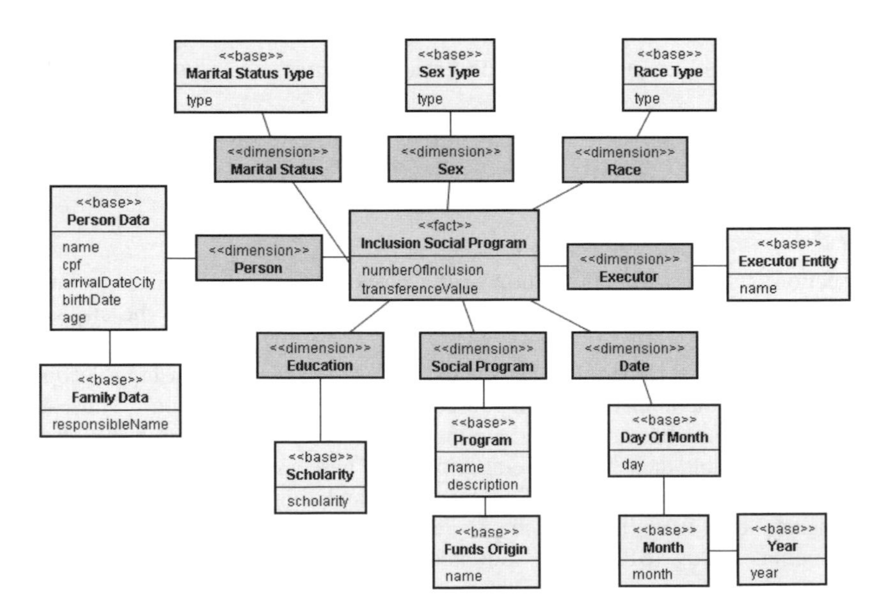

Fig. 3. The Inclusion in Social Program data cube conceptual model

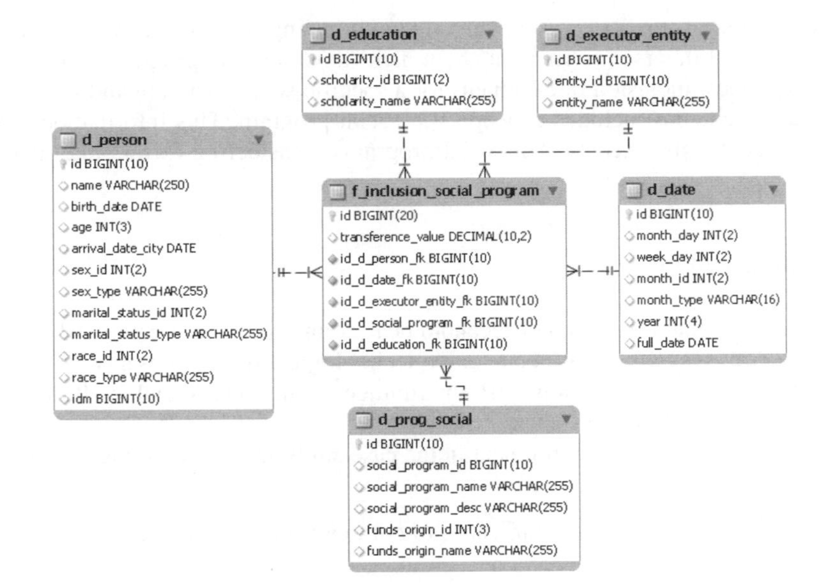

Fig. 4. The Inclusion in Social Program data cube physical model

Date	Sex	numberOfInclusion
-2006	+All Sexes	94
+January	+All Sexes	1
+March	+All Sexes	2
-April	+All Sexes	14
1	+All Sexes	8
19	+All Sexes	1
26	+All Sexes	5
+May	+All Sexes	5
+June	+All Sexes	8
+July	+All Sexes	7
+August	+All Sexes	11
+September	+All Sexes	4
[illegible]	[illegible]	[illegible]
+November	+All Sexes	19
+December	+All Sexes	13

Fig. 6. The OpenI OLAP Client interface, with a query resulting data displayed in a dimensional table.

Once both ETL and data view steps were concluded, the next step was to configure OpenI to access the data through the data view exposed in Mondrian. This configuration is done using its administrator interface, where the data sources are registered. After registering the data source, it was already possible to access the BI data using OLAP queries, which were created using OpenI OLAP Client interface, as can be seen in Fig. 6. It is important to say that the data exhibited in the figure does not correspond to the real data from the BI constructed.

The creation of this BI environment, although having only one data cube, enabled the discovery of diverse information about inclusions in social programs, like periods when the highest inclusion rate happens for a specific social program and major characteristics of citizens included in a specific social program. This information is very useful to analyze the execution of social programs, considering factors like its effectiveness.

6 Conclusion

This paper presented a proposed technological composition to create a BI environment to e-Gov systems using open source technologies. It also presents the application of this proposal to develop a BI environment to manage Social Welfare for the city of Campinas, SP, Brazil.

The creation of the BI environment in the case study was done through the following steps:

i. analysis of candidate data cubes to compose the DW and selection of one to be implemented;
ii. conceptual, logical and physical modeling of the chosen data cube;
iii. configuration of the ETL process in Talend tool;
iv. creation of the data view in Mondrian OLAP Server;
v. configuration of OpenI to access the data view exposed in Mondrian, providing users access to data.

The case study is in its beginning, having still a few requirements not being considered, like data access security and queries performance. Nevertheless, analyzing the data already provided to end users, we could conclude that the case study validates the proposal.

As future work, besides the mitigation about the requirements not considered and the inclusion of new data cubes in the BI, will be conducted a study analyzing data mining techniques that can be applied in this scenario. It is also planned a deeper analysis of the impacts of applying BI environments to e-Gov.

Acknowledgments. Rodrigo Sanches Miani's work is supported by the State of São Paulo Research Foundation (FAPESP).

References

1. Xu, L., Zeng, L., Shi, Z., He, Q., Wang, M.: Research on Business Intelligence in enterprise computing environment. In: ISIC. IEEE International Conference on Systems, Man and Cybernetics, pp. 3270–3275 (2007)
2. Moss, L.T., Atre, S.: Business Intelligence Roadmap: The Complete Project Lifecycle for Decision-Support Applications. Addison-Wesley Longman Publishing Co., Inc., Amsterdam (2003)
3. Hong, X., Zai-wen, L., Hai-yang, M.: Study and Realization of Supermarket BI System Based on Data Warehouse and Web Technique. In: CSSE 2008: Proceedings of the 2008 International Conference on Computer Science and Software Engineering, pp. 482–485. IEEE Computer Society, Washington (2008)

4. Xu, Z., Zhang, M., Jiang, X.: Business Intelligence - A Case Study in Life Insurance Industry. In: ICEBE 2005: Proceedings of the IEEE International Conference on e-Business Engineering (2005)
5. Li, S., Shue, L., Lee, S.: Business intelligence approach to supporting strategy-making of ISP service management. J. Expert Syst. Appl. 35, 739–754 (2008)
6. Ruschel, A.J.: Governo eletrônico: Business Intelligence para a modernização do Judiciário (2008)
7. Kimball, R.: The data warehouse toolkit: practical techniques for building dimensional data warehouses. John Wiley & Sons, Inc., Chichester (1996)
8. Sahama, T.R., Croll, P.R.: A data warehouse architecture for clinical data warehousing. In: ACSW 2007: Proceedings of the Fifth Australasian Symposium on ACSW Frontiers, pp. 227–232. Australian Computer Society, Inc., Australia (2007)
9. Wah, T.Y., Sim, O.S.: Development of a data warehouse for Lymphoma cancer diagnosis and treatment decision support. J. WSEAS Trans. Info. Sci. and App. 6, 530–543 (2009)
10. Rifaie, M., Blas, E.J., Muhsen, A.R.M., Mok, T., Kianmehr, K., Alhajj, R., Ridley, M.J.: Data warehouse architecture for GIS applications. In: iiWAS 2008: Proceedings of the 10th International Conference on Information Integration and Web-based Applications & Services, pp. 178–185. ACM, New York (2008)
11. Mussi, C., Murahovschi, D., Bettni, G., Kratz, L.G.: Data Warehouse - A Experiência da ANVISA. In: IX CBIS - Congresso Brasileiro de Informática em Saúde (2004)
12. Business intelligence at age 17, `http://www.computerworld.com/s/article/266298/BI_at_age_17`
13. Kimball, R., Reeves, L., Ross, M., Thornwaite, W.: The Data Warehouse Lifecycle Toolkit: Expert Methods for Designing, Developing and Deploying Data Warehouses. John Wiley & Sons, Inc., Chichester (1998)
14. Inmon, W.H.: Building the Data Warehouse. John Wiley & Sons, Inc., Chichester (1992)
15. Almeida, M.S., Ishikawa, M., Reinschmidt, J., Roeber, T.: Getting Started with Data Warehouse and Business Intelligence. IBM Redbooks (1999)
16. Biere, M.: Business intelligence for the enterprise. IBM Press (2003)
17. Codd, E.F., Codd, S.B., Salley, C.T.: Providing OLAP to User-Analysts: An IT Mandate (1993)
18. Chaudhuri, S., Dayal, U.: An overview of data warehousing and OLAP technology. J. SIGMOD Rec. 26, 65–74 (1997)
19. Dunham, M.H.: Data Mining: Introductory and Advanced Topics. Prentice Hall PTR, Englewood Cliffs (2002)
20. Yu, C.: Role-Based and Service-Oriented Security Management in the E-Government Environment. In: EGOV 2009: Proceedings of the 8th International Conference on Electronic Government, pp. 364–375. Springer, Berlin (2009)
21. Lee, S.M., Tan, X., Trimi, S.: Current practices of leading e-government countries. J. Commun. ACM 48, 99–104 (2005)
22. Velsen, L., Geest, T., Hedde, M., Derks, W.: Engineering User Requirements for e-Government Services: A Dutch Case Study. In: Wimmer, M.A., Scholl, H.J., Ferro, E. (eds.) EGOV 2008. LNCS, vol. 5184, pp. 243–254. Springer, Heidelberg (2008)
23. Al-Omari, H.: E-Government Architecture in Jordan: A Comparative Analysis. J. of Computer Science 2(11), 846–852 (2006)
24. Tilli, M., Panhan, A.M., Lima, O., Mendes, L.S.: A Web-based Architecture for e-Gov. Application Development. In: ICE-B: Proceedings of the International Conference on e-Business (2008)
25. Talend: Open Source ETL and Data Integration Software, `http://www.talend.com`

26. MySQL, http://www.mysql.com
27. Mondrian, http://mondrian.pentaho.org
28. XML for Analysis, http://www.xmla.org
29. OpenI: Open Source Business Intelligence for On-Demand Deployments, http://openi.org
30. Snell, J., Tidwell, D., Kulchenko, P.: Programming Web services with SOAP. O'Reilly & Associates, Inc., Sebastopol (2002)
31. Prefeitura Municipal de Campinas, http://2009.campinas.sp.gov.br/trabalho/
32. Inmon, W.H.: Building the Data Warehouse, 4th edn. John Wiley & Sons, Inc., Chichester (2005)
33. Luján-Mora, S., Trujillo, J., Song, I.: A UML profile for multidimensional modeling in data warehouses. J. Data Knowl. Eng. 59, 725–769 (2006)

A Fast Randomized Method for Local Density-Based Outlier Detection in High Dimensional Data

Minh Quoc Nguyen, Edward Omiecinski, Leo Mark, and Danesh Irani

College of Computing,
Georgia Institute of Technology,
Atlanta, GA 30332, USA
{quocminh,edwardo,leomark,danesh}@cc.gatech.edu

Abstract. Local density-based outlier (LOF) is a useful method to detect outliers because of its model free and locally based property. However, the method is very slow for high dimensional datasets. In this paper, we introduce a randomization method that can computer LOF very efficiently for high dimensional datasets. Based on a consistency property of outliers, random points are selected to partition a data set to compute outlier candidates locally. Since the probability of a point to be isolated from its neighbors is small, we apply multiple iterations with random partitions to prune false outliers. The experiments on a variety of real and synthetic datasets show that the randomization is effective in computing LOF. The experiments also show that our method can compute LOF very efficiently with high dimensional data.

1 Motivation

Recently, several different methods for outlier detection [6] have been presented. We can roughly categorize the methods into parametric and nonparametric methods. The nonparametric methods have a great advantage over the parametric methods is that they do not require prior knowledge of the processes that produce the events (e.g. data distribution). These methods can further be categorized into globally based and locally based. The globally based methods [8] identify the observations that are considered to be the top outliers with respect to distance for the entire dataset. Breunig et al [5] introduced a local density-based method (LOF) to detect local outliers with respect to its neighbors. The concept of local outliers show to be very useful [6] [15] [12] [13] for two reasons. First, it is because, in practice, an observation is an outlier due to its deviation from its locally similar observations rather than the entire dataset. Second, it can detect outlier without requiring any statistical model assumption.

However, the k-nearest neighbors need to be computed for the LOF method. The time complexity is $O(N^2)$ for a data set of size N. This is costly. In very low dimensional data, one may use indexing methods to speedup the nearest neighbor searches, namely R*-tree [3], X-tree [4], Kd-tree [9], etc. The main idea

T.B. Pedersen, M.K. Mohania, and A M. Tjoa (Eds.): DaWaK 2010, LNCS 6263, pp. 215–226, 2010.

of the indexes is to create a hierarchical tree of the point boundaries. The index trees scale well with n but the number of boundaries exponentially increases with the number of dimensions. In fact, Bay and Schwabacher [2] showed that the performance of index trees is worse than the brute force search for more than 20 dimensions. Therefore, even though LOF is very useful method, it is challenging to compute LOF for large and high dimensional datasets [6] [15].

In this paper, we present a randomization method to compute LOF efficiently for datasets with more than 50 dimensions. The method is made possible by our observation of the outlier consistency property of local outliers, which will be discussed in the formalism section. As a result, we can employ a randomization method to compute LOF. From now on, we will refer to the original version of LOF with full k-nearest neighbor search as the **nonrandomized version** of LOF. In the following sections, we will formally define the randomized method. In the experiment section, we will evaluate the effectiveness and efficiency of the **randomized version** against the nonrandomized version of LOF.

2 Related Work

Outliers have been studied extensively in the field of statistics [10] by computing the probability of an event against its underlying distribution. However, this method requires prior knowledge about the process that produces the events, which is usually unknown. Knorr et al [8] introduce a distance-based method to identify the outliers. The outliers are those points whose distance to other observations are the largest. Their method can detect global outliers. The advantage of this method is that no prior knowledge about the underlying distribution is required. Breunig et al [5] introduce a local density based method for outlier detection. An outlier is a point that deviates from its neighbors. The local outlier factor is measured by the ratio of its distance to its neighbors and the local density. Spiros et al [18] introduce the method to detect outliers by using multi-granularity deviation factor (MDEF). The authors then propose an approximate version to speed up the method. The method is based on the modification of an approximate nearest neighbor search algorithm (quad-tree) in order to avoid the cost of computing the MDEF scores for all the points in the dataset. Thus, the method depends on the performance of the index tree. Recently, Kriegel et al [12] propose an angle-based method that computes outlier scores based on the angles of the points with respect to other points. The method aims to provide more accurate rankings of the outliers in high dimensions. However, the method can not detect outliers surrounded by other points. The naive implementation of the algorithm runs in $O(n^3)$. Bay and Schwabacher [2] introduce a randomize method to detect distance-based outlier. However, their method can not be used for density-based outlier.

3 Generalized Local Density-Based Outlier

We revisit the concept of local density-based outlier introduced by Breunig et al [5]. The local density-based outlier is based on the k-nearest neighbor distance, the local reachability and local outlier factor. The local outlier factor (LOF) of p is the ratio between the average local reachability of its neighbors and its local reachability. If p is in a deep cluster, the local outlier factor is close to 1. If p is outside the clusters, it is greater than 1. The local outlier factor measures the degree of local deviation of p with respect to its neighbors. We observe that the main idea of the local outlier factor is in fact similar to computing of the ratio between the distance from p to its nearest points and the density of its local subspace, in order to identify local outliers. Breunig et al measure the local density by using the average k-distance of the nearest neighbors of p. This metric, however, can be generalized to other local density functions without affecting the meaning of local density-based outlier. A reasonable choice can be a kernel density function. We say that S is approximately uniform if the following two conditions hold. The variance of the k-nearest distances is less than a small ϵ and there is no k-nearest distance is larger than the average k-nearest distance with ϵ unit for some k. In this study and in the following theorems, we measure the local density by the average closest distance between the points in S ($density(S)$). We also observe that if the distance of p to its nearest points is much greater than the density of any subset in D ($dist(d, S)$) that is approximately uniform, then p is not in any cluster. Clearly, p is an outlier in D. On the contrary, if there is a subset S' such that the difference is small, then p is likely to be generated from the same distribution of S'. We can not conclude that p is an outlier, so p is considered to be normal. These two observations lead to the conclusion that the ratio between the distance and the density must be high for all the subsets in the dataset for a point to be an outlier. Thus, we can define a local density-based outlier as follows:

Definition 1. *Given a point p, a dataset D, and for any subset S of D such that S_i is approximately uniform, then p is an outlier with respect to D iff* $\frac{dist(p,S)}{density(S)} \gg 1$.

Figure 1 illustrates two outliers p_1 and p_2 based on this definition. In the figure, p_1 is not only a local outlier for the cluster containing S_1, but p_1 is also an outlier with respect to S_2, S_3, and S_4. Similarly, p_2 is also an outlier with respect to S_1.

By this definition, we observe that if we take a random hyperplane to partition a data set into two subsets, then in most cases, the local outlier factors will not change dramatically. Hence, we can recursively partition the subsets into smaller subsets. We can partition the data set until the subsets are small enough for us to compute the local outlier factors efficiently. As we see, we do not need to perform the nearest neighbor computation for the entire dataset in order to detect the local outliers.

Figure 1 illustrates an example of partition. L_1 and L_2 partition the dataset. $S_1 \ldots S_4$ are unchanged after the partitions. L_2 cuts S_3 into two subsets S_3' and S_3''. We see that S_3' and S_3'' are still approximately uniform after the partition. The points p_1 and p_2 remain to be outliers in the new subsets partitioned by L_1 and L_2.

The procedure to detect local outliers assumes that a partition does not affect the density for the partitioned sets. There are two cases where it can go wrong. The first case is when a point q is on a cluster boundary and the partition isolates

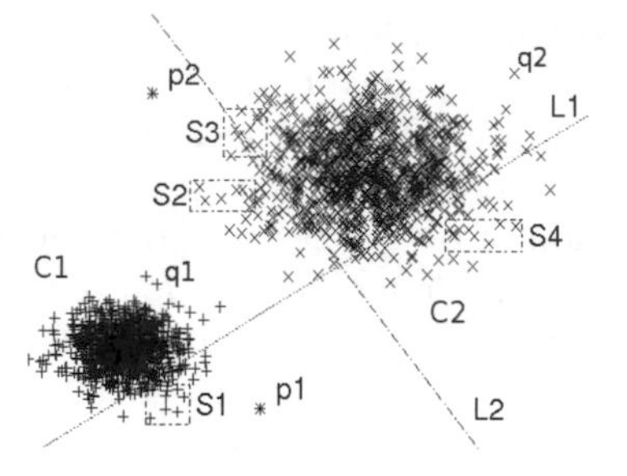

Fig. 1. Outliers with respect to their local subspaces

it from the cluster it belongs to. If the distance between the local subset of q and the new cluster in the subset it belongs to is large, then q is incorrectly identified as an outlier with respect to the new clusters. The second case is when there are many points like q that are separated from their clusters. It may make an outlier p to be normal in the new subset contains only these points.

These problem in fact can be avoided if during the separation, the new subsets contain enough neighbors of these points. Fortunately, it can be shown that the probability of partitions that separate a normal point from all of their neighbors is small. This is due to the fact that if a set C which contains q (on the cluster boundary) is large, then the probability of drawing a hyperplane cutting C such that it only contains q is small.

Theorem 1. *Given a point p in a set of size N, the probability of selecting k nearest neighbors of p or less is k/N.*

Proof. The probability to choose a value k is $1/N$. Thus, the probability to choose up to k nearest neighbors is $\sum_{i=1}^{k} \frac{1}{N} = \frac{k}{N}$.

If p is not an outlier, it should belong to a cluster. This implies that $k \ll N$. The theorem shows that the probability of a point p being on the boundary to be separated from its cluster is small. This is an important observation because we can detect local outliers effectively using randomization. If we randomly partition the dataset multiple times, in most partitions, q will appear to be normal. Thus, if a point appears to be normal in most partitions, we can flag it as normal with high confidence. We can illustrate this using figure 1. It will be rare for the small group of points S_1 to be always separated from its cluster using random partitioning. The observations above are the principles of the randomized method for computing outliers by randomly partitioning a dataset and running the algorithm multiple times so that the false outliers can be ruled out. The discussions above are based on the assumption that the local subsets are approximately uniform. Practically, data sets do not usually contain uniform subsets. However, this does not affect the randomization method. In the section above, we discuss the definition based the density of local set but there is no requirement about the size of the set. The subsets do not have to be large for the definition to be correct. In fact, S can be any size of at least two and the definition is still applicable. Therefore, we can consider any data set as a set of approximately uniform subsets.

4 Algorithm

The randomized algorithm is described in Algorithm 1. In this algorithm, PARTITION takes a dataset D as input. Then, it will call SPLIT to split the dataset into two subsets S_1 and S_2 in the following way. SPLIT randomly selects two points p_1 and p_2 in D. For every point in D, SPLIT computes the distance from it to p_1 and p_2. D will be split into S_1 and S_2 where S_1, S_2 contain all the points that are closer to p_1, p_2 respectively. This SPLIT is equivalent to choosing a hyperplane P to partition the dataset. Then, for $S \in \{S_1, S_2\}$, if the size of S is still greater than a threshold M_θ, PARTITION will be applied to S. This recursive PARTITION will be performed until the size of the result sets are smaller than a chosen size of M_θ. At this point, the LOF for all the points in S will be computed with respect to S. M_θ should be greater than the parameter k of LOF. Other than that, it can be any value that allows the outlier detection to be computed efficiently. In the end, we will have all the outlier scores for D. As discussed in section 3, the result set of outliers may contain false outliers due to isolated points. PARTITION is run multiple times to rule out false outliers. The final LOF for each point will be its minimum score over all the iterations. We use the parameter N_{iter} to set the number of iterations of the algorithm. According to the experiments, the output tends to be stable with $N_{iter} = 10$. We can speed up the algorithm by filtering points with low scores that are less than a threshold δ_{out}. The points with the scores computed in the first few iteration less than δ_{out} will not be considered in the next iterations.

It is expected that there will always be some small differences in the rankings between the original method and the randomized method. Even for the original LOF method, the ranking depends on k. The choice of k is subjective. A small change in k will lead to a change in the ranking by LOF. Therefore, it is acceptable for the ranking to be slightly different. In the case that a more similar LOF ranking is desired, we can recompute the outlier scores for the top N outliers by using the original nonrandomized version. It will give the exact scores for these points. The number of top outliers is small, thus the computation time is fast. We call this version the **recompute version** of the randomized method, while we call the earlier one the **naive version**.

We can also run the algorithm multiple times with the new final scores being the average of all the runs. We call it the (**merge version**). We notice that even though the recompute version can produce a ranking which is nearly the same as the ranking of the nonrandomized version, it is limited to the top N outliers. On the other hand, the output of the merge version is less similar for the top outliers, but the similarity can be improved for all the points. Thus, we first produce the average outlier scores using the merge version, then we recompute the score of the top outliers (**hybrid version**). Finally, we have a ranking similar to that of the nonrandomized method for all the points.

In the algorithm, the parameter M_θ is the stop condition for the partition step. A partition will stop if there is less than M_θ points in the partition. As discussed earlier, the value for M_θ should not affect the quality of the algorithm as long as it is large enough. In our experiments, the scores are not affected when we increase M_θ.

Algorithm 1. PARTITION(Set D)

```
SPLIT(D, S1, S2)
if |S1| > Mθ then
    PARTITION(S1)
else
    COMPUTECANDIDATES(S1)
end if
if |S2| > Mθ then
    PARTITION(S2)
else
    COMPUTECANDIDATES(S2)
end if
```

4.1 Query Time of New Point

The partition can actually be treated as the creation of a binary tree with two-key nodes. Each key represents a new subset. The keys are two selected points (called split points) for the partition. Each key has a pointer to the child node. The structure is then recursively created. A leaf node is a node which represents a subset which will not be further partitioned. The leaf node contains all points of the subset. The keys of the root node are the first two randomly selected points. To traverse the tree, we start with the root node. We compare a query point p to the keys of the parent node and choose the key which is closest to p. Then, we traverse the tree to the child node referred by this key. We repeat this process until we reach a leaf node where we will compute the outlier score for p with respect to the leaf node.

As we discussed earlier, we will maintain multiple trees for ruling out false outliers. The number of trees corresponds to the number of iterations. The score of a point will be the minimum score computed from all the trees. The time complexity of a query is $O(h + f(M_\theta))$, where h is the height of the tree and $f(M_\theta)$ is the time required to compute the outlier scores for the subset. If the trees are balanced, the number of steps to reach the leaf nodes is $O(\log n)$.

4.2 Time Complexity Analysis

We use the tree structure discussed in section 4.1 to analyze the time complexity of the algorithm. The algorithm consists of three main steps: partition, outlier score computation for local sets, and merge.

The *partition step* is fastest when the tree is perfectly balanced. Multiple partitions are required until the subsets are less than M_θ. For each level h, there are 2^h subsets, the size of each set is $\frac{n}{2^h}$, thus the partition cost at this level is $O(2^h \times \frac{n}{2^h}) = O(n)$. The total time for all levels is $O(H \times n)$, where H is the height of the tree. If the tree is balanced, $H \approx \log n$. The total time will be $O(n \log n)$. In the *outlier score computation step*, we consider it a constant $O(c)$ because the sizes of the subsets are very small. The maximum number of subsets is n, the worst time complexity to compute the scores is $O(n)$. In the worst case, the *merging process* for different runs and iterations can be done in $O(n)$.

In total, the upper bound for a balanced tree is $0(n \log n)$. In practice, we may not have a balanced tree, however, if we assume that most of the time the ratio of the sizes of subsets after a split is a reasonable value, the time complexity can be roughly approximated as in the balanced tree. It is possible that a partition may result in two completely unbalanced subsets where one set contains most of the points. Therefore, the key is to ensure that the probability of completely unbalanced subsets is small. If a tree is completely unbalanced, the data set is always divided into two groups such that one of them contains most of the data. However, theorem 1 shows that the probability of isolating a point from its neighbors is small. Therefore, the probability of always isolating a point from its neighbors is small. In other words, the probability for the algorithm to approach $O(n^2)$ is small. The speed is guaranteed under this assumption; however, in practice, it is showed that the algorithm can yield fast performance.

5 Experiments

5.1 2D Example

We use a two dimensional dataset to show that the randomized method can detect local outliers correctly. We generate two Gaussians with different means and standard deviations. We then generate two local outliers p_1 and p_2 for clusters C_1 and C_2. The dataset is illustrated in figure 1.First, we compute the outlier scores using the nonrandomized version. The LOF method detects two outliers p_2 (2.75) and p_1 (2.4) as two top outliers. In addition, it returns two other outliers q_2 (2.2) and q_1 (2.1). These outliers are synthetically generated by the Gaussians. Then, we compute the scores using the merge version. We set $M_\theta = 100$ and $N_{run} = 6$. The points p_2 and p_1 are consistently detected as the top two outliers for all the different runs. Their final scores are 2.65 and 2.35 respectively, which are very close to the original scores. In contrast with p_1 and p_2, the rankings for q_2 and q_1 are not consistent. However, when using the merge version, they were ranked correctly. The scores are 2.1 and 1.9 respectively. The experiment shows that the randomized method is as good as the original method using full nearest neighbor computation. In some cases, some outliers may be ranked differently but on the average the output of the randomized method converges to the original method.

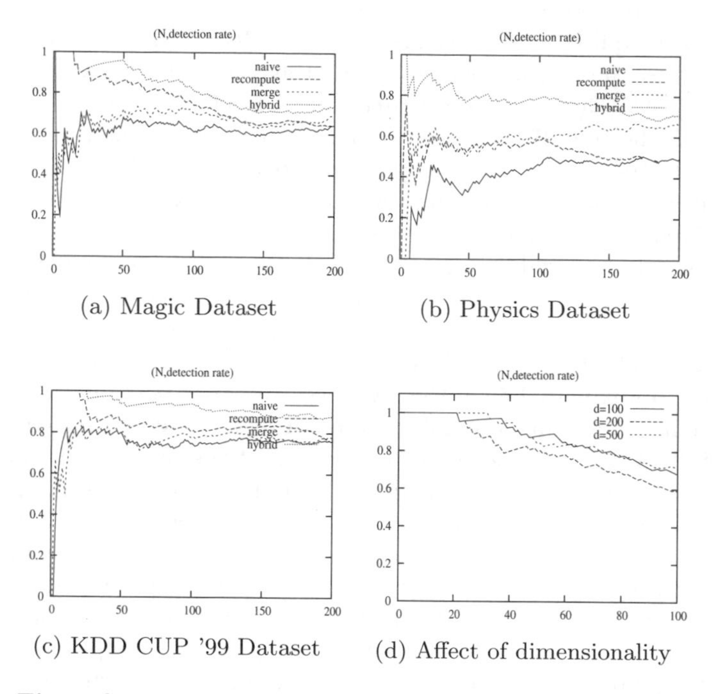

Fig. 2. Similarity in ranking for randomized outlier detection

5.2 Real Datasets

Dataset Description. We evaluate the performance of our method against the original LOF method with three different datasets: MAGIC Gamma Telescope [17], Physics [7], and KDD Cup '99 Network Intrusion [17]. The number of attributes and the size of the Magic, Physics, and KDD Cup after normalization and after removing nonnumerical attributes are (10, 19K), (69, 50K) and (34, 80K) respectively.

Evaluation Metrics. Before proceeding with the experiments, we first discuss the metrics for evaluating the effectiveness of the randomized method. The LOF method returns two values which are the local outlier factor (which we call score in our method) and the ranking of the points according to the local outlier factor. We observe that LOF method is sensitive to the parameter k. With the same LOF method, a small change in k can lead to changes in the ranking and the scores. Except for very strong outliers where the scores are distinct, the ranking is sensitive to the scores. It is even more sensitive for points with low outlier scores. For an example, there is not much difference between the rankings of 200 and 203 for the outliers with the scores of 1.90 and 1.85 due to the statistical variation. Therefore, the objective is not to have the exact same scores and rankings between the original and randomized versions. Instead, the main objective is to have similar scores with some acceptable statistical variation.

Since we are interested in the top outliers, we try to preserve the ranking for these outliers. This preservation is important if there are strong and distinct outliers in the dataset. Therefore, we evaluate the method using the following metrics:

We use the "detection rate" $\frac{N-N_{lof}}{N}$ to evaluate the top outliers, where N_{lof} is the number of outliers in the top N outliers in our method that also appear in the top N nonrandomized outliers. According to the experiments, the scores drop quickly when the points are outside the top 100 outliers, which makes the ranking sensitive to small changes of the scores. Thus, we vary N up to 200 outliers. For the weak outliers, we compute the mean and standard deviation of the ratios of the absolute differences between the methods for every point. If they are small, the two methods produce similar results.

Effectiveness of the Randomized Method. We then evaluate the effectiveness of the randomized method as follows:

First, we run the nonrandomized LOF on the datasets to compute the outlier scores ($k = 20$). Then, we run the randomized method on the datasets ($M_\theta = 500$, $N_{iter} = 20$, and $N_{run} = 6$). The results are shown in Figure 2a, 2b, and 2c. In all the figures, the naive version performs worst in comparison with the others. Nonetheless, in all the experiments, it still guarantees a detection rate of 40% for $N = 25$. It means that at least the top ten outliers are detected. The method performs best for the KDD dataset where the top 20 outliers are identified. The merge version produces slightly better results for the Magic and KDD datasets. At least 50% of the top 50 outliers are detected. The performance of the merge version is more stable compared with the naive version when N increases. As expected, the recompute version boosts the performance for all the datasets. In the figures, all top five outliers are correctly detected. At least 80% of the top 50 outliers are detected in the Magic and KDD datasets. However, the differences in the rankings start to increase when N increases. By using the hybrid approach, the performance of the randomized version becomes stable with high accuracy. As we can see, this approach is the best in all the experiments.

By manually examining the results, we found that the KDD dataset contained many strong outliers. The outlier scores for the KDD dataset are high while those in the Magic and Physics datasets are low. It can be explained by the fact that the KDD dataset contains many intrusion attack connections. This makes the distinction between the outlier scores in the KDD dataset more obvious. Therefore, the results in the KDD dataset are more stable than those in the other two datasets.

For the weak outliers, we compute the mean and standard deviation as mentioned earlier. We found that the top 180, 167, and 151 outliers had the exact same scores with the outliers computed by the original LOF in the Magic, Physics, and KDD datasets respectively. The statistics imply that our method and the original LOF method produce similar results.

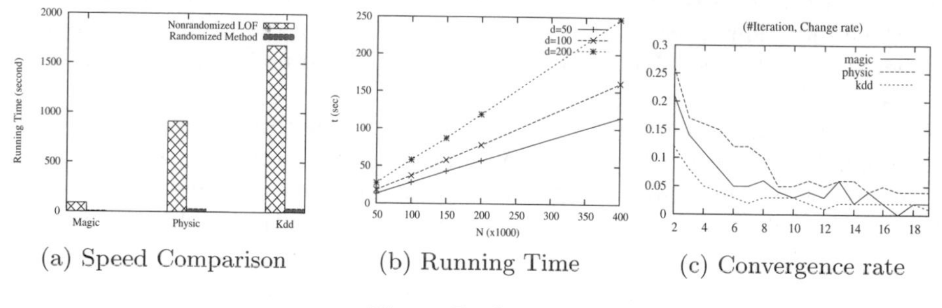

(a) Speed Comparison (b) Running Time (c) Convergence rate

Fig. 3. Performance

5.3 Performance

Dimensionality. We want to answer the question whether the effectiveness of the randomized method will be reduced by the "curse of dimensionality" as is the case for index trees. We generate synthetic datasets with the dimensionality up to 500. We run the experiments with d = 100, 200 and 500. The datasets consist of the Gaussians with randomly generated means and standard deviations. We also inject ten randomly generated outliers into the datasets. According to figure 2d, the ten injected outliers are correctly identified and the top 20 outliers are correctly identified in all the experiments. We notice that there is a slight decrease in the detection rate when d increases. When we examine the outliers manually, we find that it is due to the fact that the scores of the outliers become closer when d increases which makes the ranking fluctuate. This experiment shows that the randomized method is still viable in very high dimensions.

Speed Comparison. We evaluate the running time of the randomized method against the nonrandomized version of LOF using the Magic, Physics, and Kdd Cup '99 datasets. In these datasets, Magic is the smallest (19K points) while Kdd is the largest (80K points). In Figure 3a, the running time of the nonrandomized version grows quickly when the size of the datasets increase from 19K to 80K. However, the running time of the randomized method grows slower. In addition to Magic, Physics, and KDD Cup '99 datasets, we use a synthetic dataset with 200 dimensions and 100K points. The synthetic dataset contains five randomly generated Gaussians and ten random outliers. According to the experiment, the randomized method is consistently faster than the original LOF method.

Running Time. We randomly generate the Gaussian clusters with different means and standard deviations for the sizes from 50K to 400K. We randomly injects the top 10 outliers in the datasets. We generate the datasets for d = 50, 100 and 200. According to the results, all the generated outliers are detected as the top outliers. Figure 3b shows the running time for different datasets. The vertical axis shows the running time in seconds. In the figure, the running time is linear with the size of the dataset for different dimensions. The experiments show that the algorithm can scale well with high dimensionality.

Convergence Rate. The method relies on multiple iterations in order to rule out false outliers. We will evaluate how the iterations affect the effectiveness of the method. We observe that in the first iteration there will be many false outliers. However, when the number of iterations (N_{iter}) increases, these outliers will be ruled out in subsequent iterations. The quality of detected outliers will become at some iteration. We will evaluate it based on the changes in the scores. This experiment aims to identify a good value of N_{iter} in practice. Figure 3c shows the rate of change in the size of outliers for the Magic, Physics, and KDD dataset (after filtering out the low score outliers). As expected, the figure shows that the number of outliers changes rapidly in the first few iterations and the rate of change becomes stable when N_{iter} approaches 10. The rate of change is insignificant when $N_{iter} > 10$. We perform multiple runs with the datasets and found that in general, $N_{iter} = 10$ is a reasonable choice for the randomized algorithm.

6 Conclusion

We have shown that it is unnecessary to perform the KNN computation for the entire dataset in order to identify local density-based outliers. We introduced a randomized method to compute the local outlier scores very fast with high probability without finding KNN for all data points by exploiting the outlier consistency property of local outliers. We also introduced a hybrid version for the randomized method by recomputing the scores and combining the scores using multiple runs of the algorithm to improve its accuracy and stability. The parameters can be selected intuitively. We have evaluated the performance of our method on a variety of real and synthetic datasets. The experiments have shown that the scores computed by the randomized method and the original LOF are similar. The experiments also confirm that the randomized method is fast and scalable for very high dimensional data.

References

1. Aggarwal, C.C., Yu, P.S.: Outlier detection for high dimensional data. In: SIGMOD 2001: Proceedings of the 2001 ACM SIGMOD International Conference on Management of Data, pp. 37–46. ACM, New York (2001)
2. Bay, S.D., Schwabacher, M.: Mining distance-based outliers in near linear time with randomization and a simple pruning rule. In: KDD 2003: Proceedings of the Ninth ACM SIGKDD International Conference on Knowledge Discovery and Data Mining, pp. 29–38. ACM, New York (2003)
3. Beckmann, N., Kriegel, H.-P., Schneider, R., Seeger, B.: The r*-tree: an efficient and robust access method for points and rectangles. SIGMOD Rec. 19(2), 322–331 (1990)
4. Berchtold, S., Keim, D.A., Kriegel, H.-P.: The x-tree: An index structure for high-dimensional data. In: VLDB 1996: Proceedings of the 22th International Conference on Very Large Data Bases, pp. 28–39. Morgan Kaufmann Publishers Inc., San Francisco (1996)

5. Breunig, M.M., Kriegel, H.-P., Ng, R.T., Sander, J.: LOF: identifying density-based local outliers. SIGMOD Rec. 29(2), 93–104 (2000)
6. Chandola, V., Banerjee, A., Kumar, V.: Outlier detection: A survey. ACM Computing Surveys, 1–72 (September 2009)
7. Young, C., et al.: KDD Cup 2004: Quantum physics dataset (2004)
8. Knorr, E.M., Ng, R.T.: Algorithms for mining distance-based outliers in large datasets. In: VLDB 1998: Proceedings of the 24rd International Conference on Very Large Data Bases, pp. 392–403. Morgan Kaufmann Publishers Inc., San Francisco (1998)
9. Freidman, J.H., Bentley, J.L., Finkel, R.A.: An algorithm for finding best matches in logarithmic expected time. ACM Trans. Math. Softw. 3(3), 209–226 (1977)
10. Hawkins, D.: Identification of outliers. Chapman and Hall, London (1980)
11. Korn, F., Pagel, B.-U., Faloutsos, C.: On the 'dimensionality curse' and the 'self-similarity blessing'. IEEE Transactions on Knowledge and Data Engineering 13(1), 96–111 (2001)
12. Kriegel, H.-P., Hubert, M.S., Zimek, A.: Angle-based outlier detection in high-dimensional data. In: KDD 2008: Proceeding of the 14th ACM SIGKDD International Conference on Knowledge Discovery and Data Mining, pp. 444–452. ACM, New York (2008)
13. Lazarevic, A., Kumar, V.: Feature bagging for outlier detection. In: KDD 2005: Proceeding of the Eleventh ACM SIGKDD International Conference on Knowledge Discovery in Data Mining, pp. 157–166. ACM, New York (2005)
14. Mannila, H., Pavlov, D., Smyth, P.: Prediction with local patterns using cross-entropy. In: KDD 1999: Proceedings of the Fifth ACM SIGKDD International Conference on Knowledge Discovery and Data Mining, pp. 357–361. ACM, New York (1999)
15. Kamber, M., Han, J.: Data Mining: Concepts and Techniques, 2nd edn. Morgan Kaufmann Publishers, San Francisco (March 2006)
16. Nguyen, M.Q., Omiecinski, E., Mark, L.: A Fast Feature-based Method to Detect Unusual Patterns in Multidimensional Data. In: 11th International Conference on Data Warehousing and Knowledge Discovery (August 2009)
17. Newman, C.B.D., Merz, C.: UCI repository of machine learning databases (1998), `http://archive.ics.uci.edu/ml/`
18. Papadimitriou, S., Kitagawa, H., Gibbons, P.B., Faloutsos, C.: LOCI: Fast outlier detection using the local correlation integral. In: Proceedings of the 19th International Conference on Data Engineering: 2003, pp. 315–326. IEEE Computer Society Press, Los Alamitos (March 2003)

Specialty Mining

Hanuma Kumar, Rohit Paravastu, and Vikram Pudi

International Institute of Information Technology, Hyderabad 500032, India
{hanuma,prohit}@research.iiit.ac.in, vikram@iiit.ac.in

Abstract. In this paper, we consider the problem of mining the special properties of a given record in a relational dataset. In our formulation, a property is a combination of multiple attribute-value pairs. The support of a property is the number of records that satisfy it. We consider a property as special if its support occurs to us as a *shock* and the measure of this shock factor is more than a user defined threshold η. We provide a way to define this notion of shock based on entropy. We also output the shock factor for records in the dataset in a convenient, easily-interpretable manner. An illustrated example is provided on how users can interpret the results. Experiments on real and synthetic data sets reveal interesting properties of data records that cannot be mined using traditional approaches.

1 Introduction

In this paper, we consider the problem of mining special properties of a given record in a relational data set. Our goal is to discover special properties of *any* given record that distinguish it from most other records. There are many scenarios in which this kind of mining can be useful:

1. A student comes to a faculty member in a university asking for a project. It would be nice for the faculty to know how the student is different from other students in terms of marks in various subjects.
2. A shopper can select from a range of products by analyzing the special properties of each product in comparison with the other products. For example, there are many varieties of mobile phones in the market and it would be convenient for the customer to know how each mobile phone differs from the remaining group.

Intuitively, we say that a property is *special* if it is present in only a few objects. However, we argue that if *similar* properties are present in a large number of objects, then even such a property shouldn't be considered special. For example, in a database of 20 people if only 2-3 persons know English, they are special with respect to the language attribute. However, if for every language in the database, there are only 2-3 people who know that language, then it means that *all* people are special with respect to the language attribute.

This is absurd – if everyone is special, no one is truly special. To ensure that the definition of specialty captures this intuition in an elegant fashion, we use concepts from information theory such as surprisal and entropy. Our specialty mining technique uses ideas from clustering [1,2], frequent itemset mining [3] and concepts from information theory.

T.B. Pedersen, M.K. Mohania, and A M. Tjoa (Eds.): DaWaK 2010, LNCS 6263, pp. 227–238, 2010.

We experimentally evaluate our approach on two datasets. The first is a synthetic dataset which contains the information about students in a university. The data set is described over seven attributes and contains ten thousand records. The second is a cricket dataset which contains the statistics of 2527 players described over 25 attributes. We also showed the scalability of our approach over huge number of data points and attributes generated synthetically.

The remainder of the paper is organized as follows: In Section 2 we formulate the specialty mining problem. In Section 3 we present our algorithm and experimentally evaluate it in Section 4. Related work that seems connected to the problem definition or the algorithm is reviewed in Section 5. Finally, in Section 6, we summarize the conclusions of our study.

2 The Specialty Mining Problem

In this section we formulate the specialty mining problem.

Definition 1. *Candidate Property: A boolean property P that is computable for all objects in a dataset D is a candidate property if at least σ objects in D satisfy P.*

Candidate properties form the search space for mining special properties. Generally we fix $\sigma = 1$. The reason for this will be clarified later.

In a relational database D, the candidate properties are defined on the attributes of D. Without loss of generality, we treat attributes as being either categorical or numeric. Example candidate properties in a student database could be: "Major of this student = Computer Science" (categorical) or "Marks in Data Structures is in the range (70, 80)" (numeric). We also allow properties to be combinations of several categorical and numeric attributes along with their corresponding values and ranges.

Definition 2. *Support of a Property: The support of a boolean property P is the fraction of records in D that satisfy P, i.e.* $\mid \{y\colon y \in D \wedge P(y) = True\} \mid / \mid D \mid$.

Definition 3. *Generalization of a Property: The generalization G of a property P is the set of attributes of P with their values left unspecified.*

The generalization of the property "Is the CGPA of the student in the range $(6,7)$ and Major = Computers?" would be "CGPA = ? and Major = ?".

Definition 4. *Specialization of a Property: The specialization of a property P is achieved by adding additional attributes to P.*

For example consider a property "Major of the student = Computer Science?". Specializations of this property include: "Major = Computer Science and Range of CGPA = (6,7)", "Major = Computer Science and Range of CGPA =(6,7) and Grade = B", etc.

Definition 5. *Instance of a Generalization: An instance of a generalization G is achieved by specifying values to all attributes in G.*

Specifying values to attributes of G results in boolean properties. For example consider a generalization "Major = ?". Instances of this property include: "Major = Computer Science?", "Major = Electronics?", and so on for all the major degrees that the university is offering to the students.

Lemma 1. *Any record X satisfies exactly one instance among the instances of a generalization G.*

Lemma 2. *All instances of a generalization are not candidate properties.*

There may be some instances of a generalization which may not be satisfied by at least σ data points(σ=1). One possible instance of this generalization may be CGPA in the range(5.5 - 6) and Grade='A'. This property may not be satisfied by a single data point. For mining specialty of data points, we are not interested in properties that are not actually present in the data set and for this reason, we keep $\sigma = 1$.

Definition 6. *Sibling Property: The set of all possible candidate instances of the generalization of a property P are said to be sibling properties of P.*

For example consider a property: "Major of the student = Computer Science?". The generalization for this property is "Major of the student = ?". So its sibling properties will be the set of all the major degrees that the university is offering to the students i.e. "Major of the student = Electronics ", " Major of the student = Civil " etc.

Definition 7. *Uncertainty of a Generalization: The uncertainty $H(G)$ of a generalization G is the entropy of G calculated over its candidate properties P_i, i.e. $H(P) = -\sum_i p(P_i)\log_2 p(P_i)$*

The intuition behind this definition is as follows: If we take a record at random and we are to guess the property that it satisfies, the uncertainty in that guess is given by the above quantity, which is the uncertainty of the generalization.

Note that the logarithm is taken to the base 2 and so the outcome is in *bits*. We chose to measure the uncertainty in *bits* because then it can be interpreted in an intuitive manner – the user can get a feel for how much information (in bits) is uncertain. This ease of interpretation is important in decision support scenarios where the end-users may not be tech-savvy data mining specialists.

The ease of interpretation is illustrated in the following example: The uncertainty of a coin toss experiment is equal to 1 bit. The uncertainty in guessing the correct answer for a multiple choice question containing four options is equal to 2 bits.

Definition 8. *Surprisal of a Property: The surprisal of a property P is the measure of the information content associated with P w.r.t. its generalization, i.e. $-\log_2 p(P)$*

By definition, the amount of surprisal contained in a property P depends only on its probability (support). The smaller its probability, the larger the surprisal associated with receiving the information that a record indeed satisfied that property. Consider a scenario where you need to guess the outcome of a toss of an unbiased coin. Suppose you guessed heads and the coin indeed landed with heads up, the amount of surprise you feel corresponds to one bit. Two bits of surprise corresponds to guessing two heads when two coins are tossed simultaneously.

Intuitively, the uncertainty of a generalization can be thought of as the average or expected surprisal of all its instances.

Definition 9. *Shock Factor of a Property: The shock factor of a property P is the difference between its surprisal and the uncertainty of its generalization, i.e.*

$$\{-\log_2 p(P)\} - \{-\sum_i p(P_i)\log_2 p(P_i)\}$$

Intuitively, we compare the surprisal contained in a property with the total uncertainty present in the property's generalization. If the surprisal contained in a property is greater than the uncertainty in its generalization then it is more surprising than its average sibling.

Example: In the coin toss example described above, the uncertainty present in guessing the coin toss is 1 bit for both heads and tails. Thus, the average uncertainty for a coin toss event is also 1 bit. The surprise associated with an event of 'heads up' is the same as the average surprise of all coin toss events. Thus, there is no "shock" in observing a 'heads up' event. The corresponding shock factor turns out to be zero.

Definition 10. *Special Property: A property P is special with respect to a threshold η if its shock factor is greater than or equal to η.*

NOTE: If the user is not interested in giving any shock factor threshold as input then all the properties whose shock factor is greater than 0 are special properties.

Theorem 1. *If E is the entropy of a generalization G, and suppose that all the candidates of G are equiprobable, then the number of candidates of G is equal to $\lfloor 2^E \rfloor$.*

Definition 11. *Infimum of a Generalization: The number of equiprobable properties that are possible for a generalization with entropy E is called infimum for that generalization and is denoted by P_{inf}.*

Definition 12. *Bias of a Generalization: Bias of a generalization B_G is the ratio of infimum of a generalization, P_{inf} to the total number of sibling properties P_{sib} for that generalization, i.e*

$$B_G = P_{inf}/P_{sib}$$

Theorem 2. *The infimum of a generalization is always less than or equal to the total number of sibling properties of the generalization.*

Lemma 3. *The bias of a generalization ranges between 0 and 1, i.e., $0 < B_G \leq 1$.*

A generalization is unbiased if $B_G = 1$ and highly biased when the value of B_G nears zero but it is never equal to zero. The bias of generalization increases as the value of B_G approaches zero while the randomness in the generalization increases as the value approaches one.

Our intention is to prune away generalizations that are highly random and uninteresting. We cannot do this by only looking at the bias of a generalization G, because the specializations of these generalizations may be interesting.

Definition 13. *Chaotic Generalization: If $B_G > \theta$ and if the number of candidate properties in a generalization G is greater than or equal to $\delta\%$ times the size of the dataset D, then the generalization is a chaotic generalization.*

Generally we keep $\theta = 0.5$, because the generalization should be random enough in order to become a chaotic generalization. After the number of candidate properties cross the δ threshold, the data points are expected to be distributed among the candidate properties just by chance and not by virtue of the property. At this point, the dataset size is too small to infer anything for this number of candidate properties in G. The specializations of G contains even more number of candidate properties than G and can be ignored. Therefore, if G is a chaotic generalization we can neither have any special properties from this generalization nor from the generalizations containing the specializations of properties of G. The value of the δ threshold depends upon the type of dataset. If the dataset is a dense dataset, we should have a low δ threshold and if it is a sparse dataset we should have a little bit higher δ threshold as the number of candidate properties is expected to be higher.

3 The Specialty Mining Algorithm

In this section we describe our algorithm for the specialty mining problem that was formulated in the previous section. The Specialty Mining algorithm consists of many steps and is shown in Figure 2.

FormBaseProperties(D, θ, δ, O)
Input:Dataset D,Threshold θ, δ, Ontology O
1. For each categorical attribute A:
2.   if Ontology O is provided for A:
3.   For each level in the O:
4. $A' - generalize(A, O, D)$
5.   Add A' to the list of attributes
6.   Compute and store frequency of values in A'
7.   if $A \neq ChaoticGeneralization$:
8.   compute and store frequency of values in A
9.   else:
10. $Prune(A)$//ignore A from now on
11. For each numerical attribute A:
12.   cluster the values of attribute A
13.   Compute and store the frequency of each cluster.

Fig. 1. Forming Base Properties

3.1 Forming Base Properties

The pseudo code for mining base properties is shown in Figure 1.

Numeric Attributes: Consider a dataset of 10,000 students having CGPA as one of the attributes. Suppose a student has a CGPA of 7.75. Even if only this one student has a CGPA of 7.75, it does not imply he is special because all the 10,000 students may have unique CGPA's as CGPA is a continuous valued numeric attribute.

The solution to this problem is to divide the whole continuous range of CGPA into a discrete set of ranges. These ranges form the base properties for specialty mining. There is a wide range of methods for discretization such as binning by equal depth [4], or clustering. Partitional clustering algorithms find well separated clusters and these clusters serve our purpose well as we want to differentiate between data points.

3.2 Mining Candidate Properties

In addition to base properties, we also allow properties to be combinations of several categorical and numeric attributes along with their corresponding values and ranges. The total space of possible attribute combinations is exponential.

If we observe carefully, candidate properties follow the *downward closure property* which was first used in the Apriori algorithm [3] for mining frequent itemsets – All subsets of a frequent itemset are frequent. In our case, we use this principle to prune away all specializations of a non-candidate property.

Candidate Property Generation: The specialty mining algorithm utilizes the candidate property generation procedure shown in figure 3 to mine candidate properties efficiently. The candidate generation procedure generates new candidate properties of length k, using the candidate properties of length $k - 1$. This method is similar to the one in apriori [3]. However, because of the different nature of our problem from apriori, other optimizations can be made which increases the overall efficiency of the algorithm.

Exploiting nature of Relational Datasets

Lemma 4. *A candidate property P can have at most one occurrence of each relational attribute.*

We can further prune the search space by utilizing the above lemma. For example, no record in a relational dataset can contain two CGPA values. Thus when forming *candidate* properties in our approach, we ensure that only one occurrence of each attribute is generated. Also, the maximum length of a candidate property set will be equal to the number of attributes/dimensions present in the relational dataset. This is much more efficient than the generic apriori that is exponential in the number of base properties.

The above lemma is implemented in the candidate property generation procedure by introducing the condition $Gen(L_1[k-1])! = Gen(L_2[k-1])$, which is shown in figure 3 in line 9. The condition ensures that two candidate properties of length $k-1$ are joined if and only if the $(k-1)^{th}$ base properties of both belong to different generalizations.

```
Input: D, Relational database
       Threshold θ, δ, Ontology O
Output: S=All special properties
Algorithm:
1.    S={}
2.    L1 = FormBaseProperties(D,θ,δ,O)
3.    Convert(D,L1) //Convert values in all records to their
4.                   respective base properties.
5.    for(k=2;L(k-1) ≠ φ,k++)
6.        Lk={}
7.        Ck = Candidate_Gen(L(k-1))
8.        Gk = {} // A hash map whose values are also hash maps
9.        for each record rϵD :
10.           Ct = subset(Ck,r)
11.           for each candidate cϵCt :
12.               hash_increment(c,Gk)
13.               //i.e hash to its respective generalization
14.                 and increment its count.
15.       for each k-length generalization gϵGk :
16.           Eg = Entropy(g)
17.       if ChaoticGeneralization(Eg):
18.           for each candidate cϵEg :
19.               Delete(c,Ck)
20.       for each candidate cϵCk:
21.           if count(c) > 0:
22.               Add(c,Lk)
23.               if shock_factor(c) > 0 :
24.                   Add(S,c,shock_factor(c))
```

Fig. 2. Specialty Mining Algorithm

3.3 Chaotic Generalization

The formulation of Chaotic generalization 13 helps us to further prune the search space.

Lemma 5. *All properties of a chaotic generalization are non-candidate properties.*

From definition 13 it is clear that we can neither have any special properties from a chaotic generalization G nor from the generalizations containing specializations of properties of G. So it is unnecessary to create specializations of properties of chaotic generalizations and hence these properties are treated as non-candidate properties. As a result, all the properties that belong to a chaotic generalization are removed from the set of candidate properties which is reflected in the specialty mining algorithm shown in figure 2.

3.4 Mining Special Properties

The candidate properties whose support count is greater than zero and whose shock factors are positive become special properties. When the user provides a query record

```
Candidate_Gen(L_{k-1})

1.   if k - 1 == len_of_a_record(r):
2.       C_k = φ
3.       return C_k
4.   for each property l_1 ε L_{k-1}
5.       for each property l_2 ε L_{k-1}
7.           if l_1[1] = l_2[1] and l_1[2] = l_2[2] and .....and
8.             l_1[k-2] = l_1[k-2] and l_1[k-1] = l_2[k-1] and
9.             Gen(l_1[k-1])! = Gen(l_2[k-1]) :
10.                C = l_1 ⋈ l_2
11.                if NonCandidateSubset(c_1 L_{k-1}) :
12.                    delete c
13.                else:
14.                    add c to C_k
15.  return C_k
```

Fig. 3. Candidate Property Generation

and threshold η, we traverse through all the stored special properties and output those whose shock factor is greater than or equal to η. These properties put together determine the ways in which that record is special relative to other records in the dataset.

3.5 Scalability

Our algorithm for mining candidate properties is linear with respect to the size of dataset. Without any optimizations for mining candidate properties we have to do an exhaustive search of the attribute combinations which is exponential. The lemma 5 in addition to the optimizations described for mining candidate properties help us to deal with this exponential complexity effectively. As the complexity of a generalization increases, the number of candidate properties possessed by the generalization also increases and thereby increasing the chance to become chaotic generalization. This nature of relational datasets helps us to handle large number of attributes efficiently.

4 Performance Study

In this section, we evaluated the specialty mining algorithm discussed earlier using synthetic and real datasets. In these experiments we attempt to demonstrate that the algorithm and framework are useful in mining special properties of records.

4.1 Experiment 1

We generated a synthetic dataset containing 1,00,000 records replicating student information. The attributes on which the student database is built are Name, Gender(M, F), Major (Computers, Electronics, Mechanical,...), Date of Birth, Telephone, City of residence, CGPA, Major Course Grade(A, A-, B, B-, C, C-, D, F). A total of 8 major degrees are issued and the domain for City of residence contains 10 values.

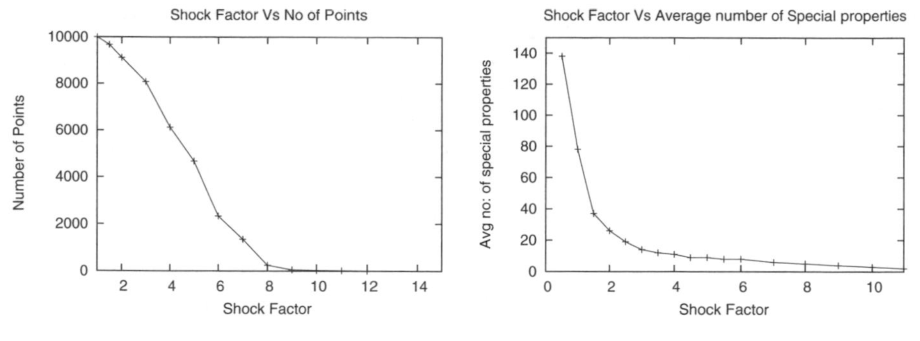

Fig. 4. Student Dataset

Fig. 5. Shock Factor Vs Avg no of properties

While forming base properties the attributes Name, Phone and Date of Birth were pruned as they are chaotic generalizations and have too many distinct values. The numerical attribute CGPA was clustered using the K-means partitional clustering algorithm. The parameter K should be carefully selected as the quality of the results depend upon the quality of the clusters formed. Run the k-means algorithm with different parameters of K and select the K for which the squared error is minimum. For this experiment, we have $\sigma = 1, \delta = 20\%$.

Table 1. Sample Output

Property	# Records	Shock Factor
{Hyderabad, India, 6.03-6.28}	19	7.284
{A, 6.03-6.28}	12	10.92
{Female, A, 6.03-6.28}	7	7.12

The number of records for which we can extract at least one special property for a shock factor η is shown in Figure 4. Notice in this figure 4 the number of records which have at least one special property decrease as the η increases. This is because it becomes increasingly difficult for the record to possess special properties which occur to us as a shock at this threshold. With this figure 4 we assume that the user gets a clear picture of the dataset and this can be used to set the η threshold appropriately.

We demonstrate the utility of the specialty mining approach with the following typical trial-run yielding special properties that cannot be mined using traditional approaches. Consider a record in the database:

Sita, Female, Electronics, Hyderabad, CGPA=6.23, 'A', 6 Sept. 1986, 9885445278

The output given by our algorithm for Sita when η=7 bits is shown in Table 1. In this manner the special properties of any student are generated and the user can know how special a student is relative to all other students in the data set.

4.2 Experiment 2

In this experiment we evaluated our algorithm on a real life cricket dataset. It contains statistics of 2,527 players described over 25 attributes. This data is collected by parsing the website cricinfo.com.

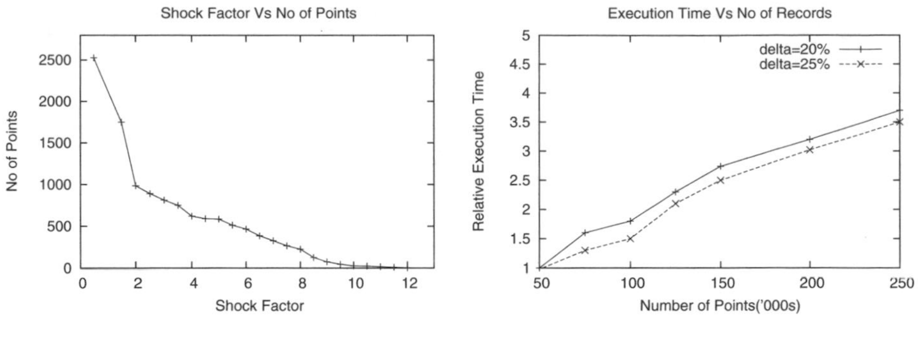

Fig. 6. Cricket Dataset

Fig. 7. Execution Time Vs # Records

Name of the Player becomes a chaotic generalization and hence removed. We used K-means to find clusters that describe individual attributes. For this experiment we have, $\sigma = 1$ and $\delta = 20\%$. The number of records for which we can extract atleast one special property for a shock factor η is shown in Figure 6.

From the graph in Figure 6 it is clear that all the players are special with respect to at least one property upto a threshold of η=1.5 bits. From η=1.5 bits to 2 bits there is a drastic decrease in the number of players because majority of the players have all their special properties in that range of threshold.

Table 2. Top 3 players for a threshold η=2.5 bits

Player	No of Spl Properties	Avg. Shock Factor
Sachin Tendulkar	72	8.64
Jacques Kallis	55	7.2
Ricky Ponting	47	8.1

The graph in figure 5 shows the average number of properties by which a player is special with respect to a shock factor threshold η. We also calculated the top k-players i.e sorted w.r.t number of special properties and their average shock factors. As an example, the top 3 players mined are shown table2. Suppose a user who is not too familiar with cricket wants to know about Sachin Tendulkar, who is one of the greatest players in world cricket. By directly looking at the statistics of Sachin, the user will not be able to comprehend whether those statistics are really special. By applying our specialty mining algorithm and for a threshold η=2.5 bits, Sachin Tendulkar is special with respect to 72 properties which is far greater than the average, which is 21 properties. So, the result of the specialty mining algorithm coupled with the actual data gives a better idea than providing only the actual data to the user.

4.3 Scalability and Accuracy

The student data set described in Experiment 1 is a synthetic data set and we increased the number of records, attributes to test the scalability of our algorithm. The running time for the algorithm can be divided into two parts.

1. **Counting Support for the Properties:** The time for this is linear as the underlying apriori based approach is expected to scale linearly w.r.t number of records. This is confirmed by figure 7 which shows the relative execution time as we increase the number of records from 50,000 to 2,50,000. The times are normalized with respect to the times for 50,000 records. The experiment is conducted for 25 attributes with thresholds $\sigma = 1$ and $\delta = 20\%$ and $\delta = 25\%$.

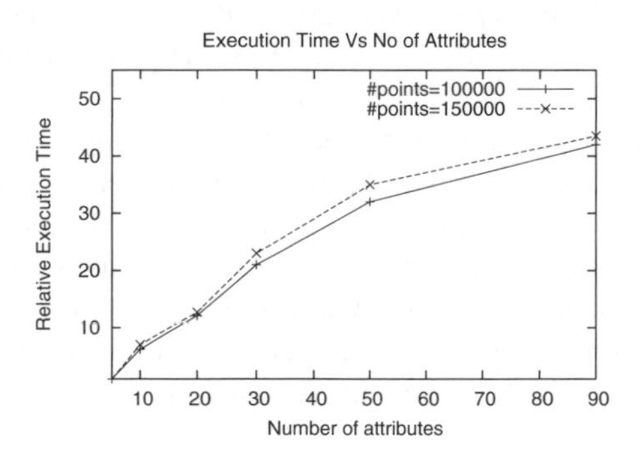

Fig. 8. Execution Time Vs # Attributes

2. **Mining Candidate Properties:** The time for this depends more on the number of attributes present in the dataset. Our algorithm is scalable with respect to number of attributes present in the dataset. This is explained clearly in the subsection 3.5 and is reflected in the figure 8. The times are normalized with respect to the times for 5 attributes. The experiment is conducted with a threshold $\delta = 20\%$ and for 1,00,000, 1,50,000 points, respectively.
3. **Accuracy of the Results:** We purposefully put some records with some combination of attribute values which are very rare in the dataset. Our algorithm is able to identify those combinations as properties with very good shock factors. However, the choice of δ threshold effects the nature of results. From our experiments, $\delta = 20\%$ provided good results for dense datasets, while $\delta = 30\%$ gave good results for sparse datasets respectively.

5 Related Work

The specialty mining problem introduced in this paper is related to several other areas such as subspace clustering [5], outlier mining [6,7] and frequent itemset mining [3]. Our problem is significantly different from the normal outlier mining algorithms as it considers each and every subspace to be important and looks for records with outlier properties in that space without applying any complex algorithms on that subspace. Also, distance functions work well when the data is in pure numerical form. But most of the real world data is a mixture of both numerical and categorical values. In this scenario, outlier detection algorithms that are based on pure distance functions do not serve our purpose well. Frequent itemsets when used effectively is a good tool to exploit relations between data in the presence of both numerical and categorical values. We used frequent itemset mining ideas effectively in our algorithm.

Our problem may also initially appear as a version of subspace clustering. However, upon analysis, the two problems are quite different. Subspace clustering algorithms such as Clique [5] seek to find clusters as *dense contiguous regions* that occur in any subspace. In our approach, we only form clusters in *single dimensions* which represent natural groupings in those single dimensions. These single-dimensional clusters are then used to form candidate *properties* in higher dimensions – these properties do not represent clusters in those higher dimensions. Thus, the problem addressed in this paper is different from the task of mining clusters in sub-spaces.

For example, in this paper, a property such as {A in Programming, CGPA in 6.03-6.28} could represent a candidate special property. However, this property may only be part of a much larger contiguous cluster in that sub-space. Moreover, this larger cluster may be of arbitrary shape and so it may not be possible to represent it concisely in terms of simple range constraints on attributes. Our task is not to mine such clusters. Instead, we are interested in identifying succinct properties that are easily interpretable by the user and can be represented in terms of simple constraints on attributes.

6 Conclusions

In this paper we introduced the problem of specialty mining on relational datasets. We are interested in determining what makes a given record special or different from majority of the records. To this end, we used ideas from information theory to define notions such as shock factor and special properties. We discussed several applications where this kind of mining is useful. Our solution uses ideas from clustering, attribute-oriented induction and frequent itemset mining. We demonstrated the merit and scalability of our approach using synthetic and real datasets.

References

1. Kanungo, T., Mount, D.M., Netanyahu, N.S., Paitko, C.D., Silverman, R.: An efficient k-means clustering algorithm algorithm: Analysis and implementation. IEEE Trans. Pattern Analysis and Machine Intelligence (2002)
2. Ester, M., Kriegel, H., Sander, J., Xu, X.: A density-based algorithm for discovering clusters in large spatial databases with noise. In: Intl. Conf. on Knowledge Discovery and Data Mining, KDD (1996)
3. Agrawal, R., Srikant, R.: Fast algorithms for mining association rules. In: Proc. of Intl. Conf. on Very Large Databases (VLDB) (September 1994)
4. Srikant, R., Agrawal, R.: Mining quantitative association rules in large relational tables. In: Proc. of ACM SIGMOD Intl. Conf. on Management of Data (June 1996)
5. Agrawal, R., Gehrke, J., Gunopulos, D., Raghavan, P.: Automatic subspace clustering of high dimensional data for data mining applications. In: Proc. of ACM SIGMOD Intl. Conf. on Management of Data (1998)
6. Agrawal, C., Yu, P.: Outlier detection for high dimensional data. In: Proc. of ACM SIGMOD Intl. Conf. on Management of Data (2001)
7. Ng, R., Breunig, M., Kriegel, H., Sander, J.: Identifying density based local outliers. In: Proc. of ACM SIGMOD Intl. Conf. on Management of Data (2000)

Region of Interest Based Image Categorization

Ashraf Elsayed[1], Frans Coenen[1], Marta García-Fiñana[2], and Vanessa Sluming[3]

[1] Department of Computer Science, University of Liverpool,
Ashton Building, Ashton Street, Liverpool L69 3BX, United Kingdom
[2] Centre for Medical Statistics and Health Evaluation, University of Liverpool,
Shelley's Cottage, Brownlow Street, Liverpool L69 3GS, United Kingdom
[3] School of Health Sciences, University of Liverpool,
Thompson Yates Building, The Quadrangle, Brownlow Hill,
Liverpool L69 3GB, United Kingdom
{a.el-sayed,coenen,m.garciafinana,vanessa.sluming}@liv.ac.uk

Abstract. Region Of Interest Based Image Classification (ROIBIC) is a mechanism for categorising images according to some specific component or object that features across a given image set. This paper describes and compares two such approaches. The first is founded on a weighted graph mining technique whereby the ROI is represented using a tree structure which allows the application of a weighted graph mining technique to identify features of interest, which can then be used as the foundation with which to build a classifier. The second approach is founded on a time series analysis technique whereby the ROI are represented as time series which can then be used as the foundation for a Case Based Reasoner. The presented evaluation focuses on MRI brain scan data where the classification is focused on the *corpus callosum*, a distinctive region in MRI brain scan data. Two scenarios are considered: distinguishing between musicians and non-musicians and epilepsy patient screening.

Keywords: Image mining, Image categorisation.

1 Introduction

Image categorisation is concerned with the labelling of images into one or more predefined classes. The principal challenge of image categorisation is the capture of the significant features within images that facilitate the desired classification. Edge detection, segmentation and registration all have a significant part to play in this process. One method of simplifying the image categorisation process is to focus on some particular feature or Region Of Interest (ROI) within the image set. The advantage offered is that the remainder of the image can be ignored and thus computational advantages gained. Alternatively, the representation can be more detailed. We refer to this approach is ROI Based Image Categorisation (or ROIBIC). Of course ROIBIC is not suited to every image categorisation application; not all such applications include a significant and identifiable ROI the appears across the data set. The most appropriate applications for ROIBIC are those where the data set includes a common feature whose size and shape

T.B. Pedersen, M.K. Mohania, and A M. Tjoa (Eds.): DaWaK 2010, LNCS 6263, pp. 239–250, 2010.

strongly influences the categorisation, i.e. a set of images that can be categorised according to the shape of some object that consistently appears across the image set. The focus of the work described in this paper is the categorisation of Magnetic Resonance Imaging (MRI) brain scan data according to a specific feature within the data called the *corpus callosum*.

This paper describes and compares two approaches to ROIBOC: graph based and time series based. Both approaches, although operating in very different manners, are essentially supervised learning mechanisms where by a pre labelled training set is used to build a "classifier" which can be applied to unseen data. The first approach uses a tree based representation for the common feature, one tree per image. A graph mining technique is then applied to identify frequently occurring sub-graphs (sub-trees). The identified set of trees are then used to describe the image set in terms of a set of attributes each of which equates to a frequently occurring sub-tree. A classification algorithm is then applied to this attribute set to build a classifier to be applied to "unseen" data. The second approach is founded on a time series representation coupled with a Case Based Reasoning (CBR) technique. The features of interest, when identified, are represented as time series, one per image. These time series are then stored in a Case Base (CB) which can be used to categorise unseen data. The unseen data is compared with the categorisation on the CB using a Dynamic Time Warping (DTW) based similarity checking mechanism, the categorisation associated with the most similar time series (case) in the CB is then adopted as the categorisation for the unseen data. Both approaches require the application of a registration process and segmentation, and this will entail the established difficulties encountered when conducting this process (i.e. poor image contrast, intensity inhomogeneities and partial-volume effects). Both approaches are intended to preserve the size and shape of the feature of interest (ROI).

The rest of this paper is structured as follows. In section 2 a brief overview of related previous work is presented, followed in section 3 with an overview of the application that acts as the focus for this paper. The advocated approaches are described in Sections 4 and 5 respectively. A complete evaluation of these approaches is reported in Section 6, followed by some conclusions in Section 7.

2 Previous Work

Current image categorisation techniques can be divided into two groups according to the image features used for the classification: global approaches and ROI based approaches. The first use features that reflect all the information contained within image sets. One such technique is the use of colour histograms to represent images. For example in [20] a k-nearest neighbour classifier is applied to colour histogram represented images to discriminate between "indoor" and "outdoor" images, in [12] a time series technique is applied to classify histogram represented retina images, and in [21] a Bayesian classifiers is applied to edge direction histograms to categorise city and landscape images. Support Vector Machines (SVMs) built on colour histograms were applied to classify images containing a generic set of objects in [3]. Although the global features can usually be computed with little cost and are effective for certain classification tasks, a significant drawback is that structural and relative spatial information is lost. Furthermore, for many applications (such as medical applications) image attributes such as colour and intensity have limited discriminative power.

A number of ROI-based approaches have been proposed to maintain local and spatial properties of an image using the concept of *regions* or *blocks*. In [9] city-scape images are divided into 16 non-overlapping equal-sized blocks. Dominant orientations are then computed for each block, and the images classified as city or suburb as determined by the majority orientations of blocks. In [23] graph and photograph images are divided into blocks and each block assigned to one of two categories. If the percentage of blocks classified as photograph is higher than a threshold, the image is marked as a photograph; otherwise, the image is marked as a graph. A disadvantage of this approach is that a rigid partition of an image into fixed-size blocks often breaks an object into several blocks or puts different objects into a single block. Thus visual information about objects may be destroyed by a rigid partition. In [19] an alternative approach is described where images are classifying according to spatial orderings of regions where each region is represented by a symbol corresponding to an entry in a pattern library. Image segmentation is one way to extract object information whereby an image is decomposes into a collection of regions, each corresponding to an object. There are many examples where image segmentation has been applied successfully to the image categorisation problem. ROIBIC, as described in this paper, advocates an approach where the focus is on a single ROI common across the image set.

3 Application Domain

Although generally applicable, the ROIBIC approaches described in this paper are directed at MRI brain scan data, more specifically the categorisation of MRI data according to a specific object contained in these images called the corpus callosum. The corpus callosum is a highly visible structure in MRI scans whose function is to connect the left and right hemispheres of the brain, and to provide

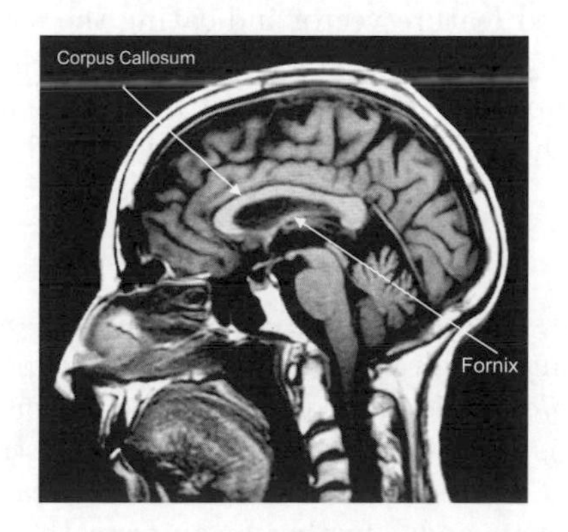

Fig. 1. Corpus callosum in a midsagittal brain MR image

the communication conduit between these two hemispheres. Figure 1 gives an example MRI scan, the corpus callosum is located in the centre of the image. A related structure, the *fornix* is also indicated. The fornix often "blurs" into the corpus callosum and thuds presents a particular challenge in the context of the segmentation of these images so as to isolate the corpus callosum ROI.

The corpus callosum is of interest to medical researchers for a number of reasons. The size and shape of the corpus callosum has been shown to be correlated to sex, age, neurodegenerative diseases and various lateralized behaviour in people. It is also conjectured that the size and shape of the corpus callosum reflects certain human characteristics (such as a mathematical or musical ability). Several studies indicate that the size and shape of the corpus callosum, in humans, is correlated to sex [1,6,18], age [18,24], brain growth and degeneration [11,15], handedness [5], epilepsy [4,17,22] and brain disfunction [7,13].

4 Graph Based Approach

In this and the following section the two proposed techniques, graph bases ROIBIC and time series based ROIBIC, are described commencing with graph based ROIBIC. A schematic of the graph based process is given in Figure 2. The process commences with segmentation and registration to isolate the ROI. The details of the identified ROI are then acquired by tessellating the images into homogeneous sub-regions, according to (say) colour or intensity, and then storing the result in a *quad-tree data structure*. A *weighted sub-graph mining* approach is then applied to the tree represented image set (one tree per image) to identify frequent sub-graphs. The identified sub-trees (graphs) then form the fundamental elements of a *feature space*, i.e. a set of attributes with which to describe the image set. Experiments conducted by the authors have revealed that, for many image sets, the graph mining process can identify a great many frequent sub-graphs; more than required for the desired categorisation. Therefore a feature selection strategy is applied so that only those sub-graphs that serve as the best discriminators are retained. Each image is then described in terms of a binary-valued feature vector indicating the selected attributes (sub-graphs) that appear in each image. Once the image set has been represented in this manner any appropriate classifier generator may be applied; for the corpus callosum application Quinlan's C4.5 algorithm was used [16].

4.1 Tessellation

The tessellation process comprises the recursive decomposition of the identified ROI, for each image, into quadrants. The tessellation proceeds until either sufficiently homogeneous *tiles* are identified or some user specified level of granularity is reached. The result is then stored in a quadtree data structure such that each roots node represents a tile in the tessellation. Nodes nearer the root of the tree represent larger tiles than nodes further away from the root. Thus the tree is "unbalanced" in that some root nodes cover larger areas of the ROI than others.

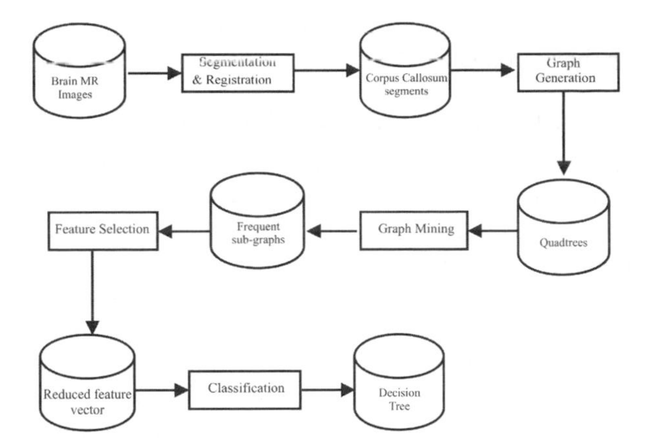

Fig. 2. Framework of graph mining ROIBIC

Note also that the relative location of the each tile is maintained in the structure of the tree. Thus the advantage of the representation is thus that information about the relative location and size of groups of pixels is maintained.

The tessellation can be conducted according to a variety of image features such as colour or intensity. With respect to the corpus callosum application a binary encoding was used, the tiles included in the corpus callosum were allocated a "1" (black) and tiles not included a "0" (white). Sufficiently homogeneous was defined as a tile that was 95% black or white. The research team experimented with a number of "granularity" settings as reported in the evaluation described in Section 6. Interested readers may like to refer to a previously published work by the authors [8] regarding further details of the tessellation process.

4.2 Weighted Graph Mining

As noted above, in the quad-tree representation nodes nearer the root can be considered to be more significant than others (because they cover a larger area). A weighted frequent sub-graph mining algorithm was therefore developed to identify commonly occurring features across the tree represented image set. The weightings were calculated according to the proximity of individual nodes to the root node in each tree. This weighting concept was built into a variation of the well known gSpan algorithm [25]. The algorithm operates in a depth first search manner, level by level, following a "generate, calculate support, prune" loop. Candidate sub-graphs are pruned if their *support* (frequency of occurrence across the graph set) is below a user defined "support threshold". The lower the threshold the greater the number of frequent sub graphs that will be identified. Space restrictions preclude further discussion of this weighted sub-graph mining algorithm here, however, interested readers are referred to [14].

Experimentation with respect to the Corpus Callosum application indicated that, to capture the necessary level of detail, a low support threshold was required. However this produced a large number of frequent sub-graphs many of which were redundant. A feature selection operation (discussed in the following subsection) was thus applied to the identified frequent sub-graphs.

4.3 Feature Selection and Classifier Generation

Feature selection is a well understood process, in the context of Data Mining, for removing irrelevant data from a feature space so as to enhance computational efficiency. Feature selection has attracted a great deal of attention from the data research community, especially in the context of classification and prediction where the aim is to identify features that are "strong discriminators". For the corpus callosum application described here, a straightforward wrapper method was adopted whereby a decision tree generator was applied to the feature set (an approach also advocated by other practitioners, see for example [10]). The advantage of decision tree algorithms, with respect to feature selection, is that they inherently estimate the suitability of features for the separation of objects representing different classes. Features that are included as "choice points" in the decision tree were thus selected, while all remaining features were discarded. For the work described here, the well established C4.5 algorithm [16] was used.

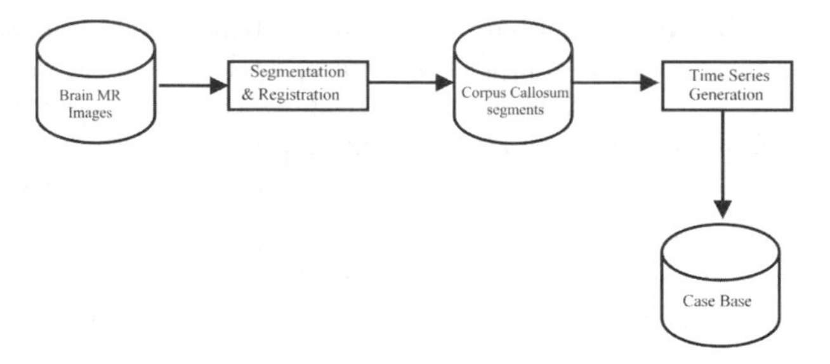

Fig. 3. Framework of time series based ROIBIC

5 Time Series Based Approach

In this section the time series based ROIBIC approach is described; a schematic is presented in Figure 3. As for the graph based approach the process commences with the segmentation and registration of the input images. The identified ROI are then encoded as time series. Each time series is conceptualised as a *prototype* or case contained in a Case Base (CB), to which a Case Based Reasoning (CBR) mechanism may be applied. Thus an unseen record is classified according to the "best match" discovered in the CB. The CBR community has proposed many techniques to identify the desired best match. In this paper the authors advocate a Dynamic Type Warping (DTW) time series comparison mechanism that operates regardless of the length of the individual time series [2].

5.1 The Time Series Representation

After the identification of the individual ROI, using segmentation, the registration process was undertaken by fitting each ROI into a Minimum Bounding

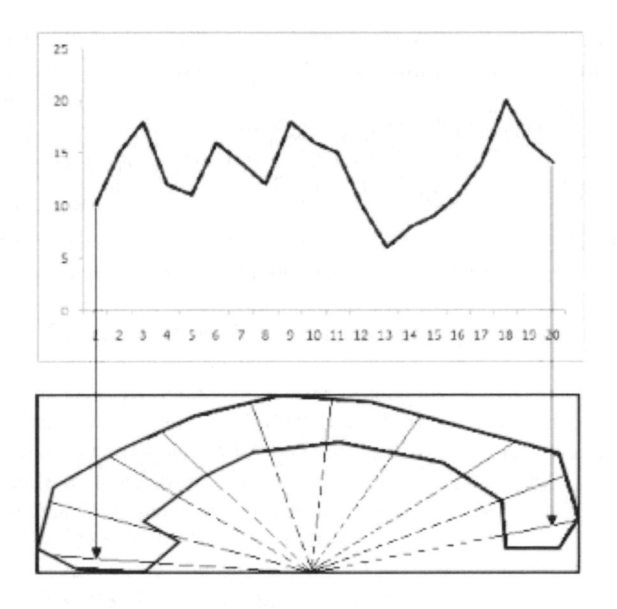

Fig. 4. Conversion of corpus callosum into time series

Rectangle (MBR). Individual time series is then derived, according to the boundary line circumscribing the ROI, using an ordered sequence of N vectors radiating out from a reference point. The time series was then expressed as a series of values (one for each of the N vectors) describing the size (length) of intersection of the vector with the ROI. It should be noted that the representation maintains the structural information (shape and size) of the ROI. It should also be noted that N is variable due to the differences of the shape and size of the individual ROI within the image data set.

With respect to the corpus callosum application the time series generation process is illustrated in Figure 4. The midpoint of the lower edge of the MBR was selected as the reference point. The vectors were derived by rotating an arc about the reference point pixel by pixel, thus the value of N will very across the image set. In this manner time series curves were generated of the form described in the top half of Figure 4 where the X-axis represents the vector (arc) number, and the Y-axis the "pixel-distance" where the vector intersects the corpus callosum.

5.2 The Dynamic Time Warping Algorithm

DTW [2] is a time series analysis technique for comparing curves. The advantage offered is that DTW is able to find the optimal alignment between two time series Q and C, of length n and m respectively. It is often used: to determine time series similarity for classification, or to find corresponding regions between two time series. The DTW-distance between the two time series Q and C is $D(M, N)$, which we calculate in a dynamic programming approach using:

$$D(i, j) = d(q_i, c_j) + \min\{D(i-1, j-1), D(i-1, j), D(i, j-1)\} \quad (1)$$

Backtracking along the minimum cost index pairs $w(i, j)_k$ starting from (M, N) yields the DTW *warping path.*

An example is given in Figure 5 where the warping path between two time series Q and C of different length is presented. Note that given two identical curves the warping path would be the straight line connecting the two opposite corners of the grid. The degree of similarity can thus be determined by comparing the calculated warping path with the "ideal path".

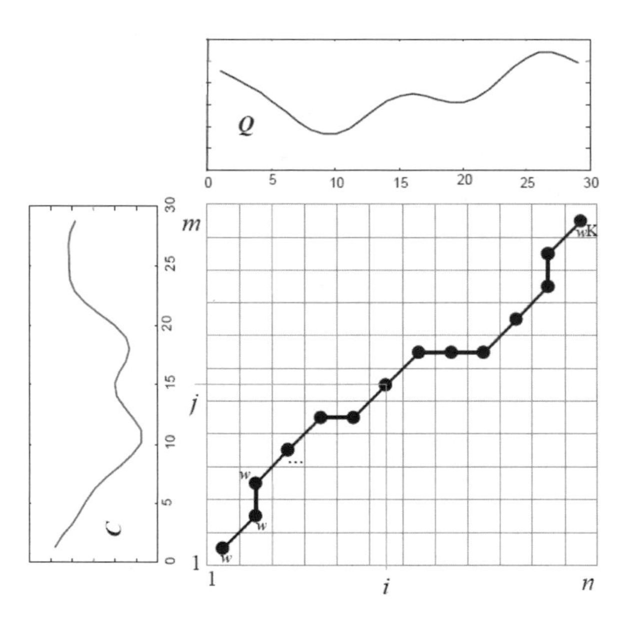

Fig. 5. An example warping path

6 Evaluation

The two advocated approaches to ROIBIC are evaluated and compared in this section with respect to the corpus callosum application. This section describes the evaluation of the proposed technique using an appropriate MRI image set. The evaluation was undertaken in terms of classification accuracy, sensitivity and specificity. Two studies have been used for the investigation: (i) a comparison between musician and non-musician MRI scans, and (ii) an epilepsy screening process. The studies are discussed in detail in Sub-sections 6.1 and 6.2 below.

6.1 Musicians v. Non-musicians

For the musicians study a data set comprising 106 MRI scans was used, 53 representing musicians and 53 non-musicians (i.e. two equal classes). The study was of interest because of the conjecture that the size and shape of the corpus callosum reflects human characteristics (such as a mathematical or musical ability). Table 1 shows the Ten Cross Validation (TCV) results obtained. The G-ROIBIC and T-ROIBIC columns indicate the results using the graph based and time series based approaches respectively. For the graph based approach a quad tree depth of six coupled with a 30% support threshold was used. For comparison purposes the results using other support threshold and depth settings are given in Table 2. The best result for each level is indicated in **bold** font. Inspection of Tables 1 and 2 demonstrate that the overall classification accuracy of the ROIBIC time series based approach significantly improves over that obtained using the graph based approach. In many TCV cases the time series based approach obtains 100% accuracy although visual inspection of the corpus callosums in the data set does not allow for the clear identification of any defining feature.

Table 1. TCV Classification Results for Musicians Study

Test	G-ROIBIC			T-ROIBIC		
set ID	Accuracy	Sensitivity	Specificity	Accuracy	Sensitivity	Specificity
1	92.45	94.12	90.91	91	100	85.71
2	96.23	94.55	98.04	100	100	100
3	95.28	96.15	94.44	91	100	85.71
4	93.40	94.23	92.59	100	100	100
5	97.17	96.3	98.08	100	100	100
6	94.34	96.08	92.73	100	100	100
7	97.17	96.3	98.08	100	100	100
8	95.28	96.15	94.44	100	100	100
9	96.23	94.55	98.04	100	100	100
10	95.28	96.15	94.44	100	100	100
Average	95.28	95.458	95.179	98.2	100	97.14
SD	1.54	0.95	2.7	3.8	0.0	6.03

Table 2. TCV Classification accuracy (%) using graph based ROIBIC

	Support Threshold (%)							
Levels	20	30	40	50	60	70	80	90
4	**70.75**	69.81	68.87	71.70	68.87	61.32	52.83	50.94
5	**90.57**	83.96	80.19	85.85	80.19	81.13	80.19	70.75
6	85.85	**95.28**	84.91	83.96	90.57	83.96	77.36	75.47
7	83.80	85.85	**89.62**	86.79	87.74	75.47	76.42	78.30

6.2 Epilepsy Screening

For the epilepsy study three data sets were used. The first comprised the control group from the musicians study together with 53 MRI scans from epilepsy patients. The Second data set used all 106 MRI scans from the musicians study and the 53 epilepsy scans. The third data set comprised the 106 MRI scans from the musicians study augmented with 106 epilepsy cases. The objective was to seek support for the conjecture that the shape and size of the corpus callosm is influence by conditions such as epilepsy ([4,17,22]). Tables 3 and 4 show the Ten Cross

Validation (TCV) classification results for the three epilepsy data sets. Again the G-ROIBIC and T-ROIBIC columns indicate the results obtained using the graph based and time series based ROIBIC approaches respectively. A quad-tree depth of six was again used, coupled with a 30% support threshold, as this had been found to give the best results in the case of the musicians study. Inspection of Tables 3 and 4 indicates that the graph based approach significantly out-performs the time series based approach. Best results are obtained using the large, 212 MRI scan data set, because this includes many more training examples.

Table 3. TCV Classification results for Epilepsy Study (Data Sets 1 and 2)

Test	106 MR scans						159 MR scans					
set	G-ROIBIC			T-ROIBIC			G-ROIBIC			T-ROIBIC		
ID	Acc.	Sens.	Spec.	Acc.	Sens.	Spec.	Acc.	Sens.	Spec.	Acc.	Sens.	Spec.
1	88.68	87.27	90.2	72.73	80.00	66.67	86.79	76.67	92.93	75.00	70.00	83.33
2	85.85	86.54	85.19	81.82	83.33	80.00	79.25	65.63	88.42	81.25	85.71	77.78
3	85.85	86.54	85.19	72.73	80.00	66.67	82.39	69.84	90.63	75.00	70.00	83.33
4	82.08	82.69	81.48	81.82	83.33	80.00	84.28	72.58	91.75	81.25	85.71	77.78
5	76.42	75.93	76.92	81,82	83.33	80.00	83.65	72.13	90.82	81.25	85.71	77.78
6	85.85	86.54	85.19	81.82	83.33	80.00	86.79	76.67	92.93	75.00	70.00	83.33
7	72.64	74	71.43	63.64	66.67	60.00	78.62	64.62	88.3	81.25	85.71	77.78
8	85.85	86.54	85.19	81.82	83.33	80.00	82.39	69.84	90.63	68.75	66.67	71.43
9	88.68	87.27	90.2	72.73	80.00	66.67	86.79	76.67	92.93	68.75	66.67	71.43
10	82.08	82.69	81.48	63.64	66.67	60.00	79.25	65.63	88.42	81.25	85.71	77.78
Average	83.40	83.60	83.25	75.46	79.0	72.0	83.02	71.03	90.78	76.88	77.19	78.18
SD	5.25	4.88	5.75	7.48	6.67	8.78	3.22	4.73	1.89	5.15	9.06	4.37

Table 4. TCV Classification results for Epilepsy Study (Data Set 3)

Test	212 MR scans					
Test	G-ROIBIC			T-ROIBIC		
set ID	Accuracy	Sensitivity	Specificity	Accuracy	Sensitivity	Specificity
1	85.38	85.71	85.05	81.82	88.89	76.92
2	84.91	86.27	83.64	77.27	80.00	75.00
3	87.74	89.22	86.36	81.82	88.89	76.92
4	89.15	89.52	88.79	77.27	80.00	75.00
5	86.32	87.38	85.32	68.18	70.00	66.67
6	85.85	86.54	85.19	72.73	77.78	69.23
7	84.91	86.27	83.64	77.27	80.00	75.00
8	84.43	84.11	84.76	81.82	88.89	76.92
9	83.96	83.33	84.62	72.73	77.78	69.23
10	87.74	89.22	86.36	81.82	88.89	76.92
Average	86.04	86.76	85.38	77.27	82.11	73.78
SD	1.68	2.12	1.52	4.79	6.51	3.89

6.3 Discussion

With respect to classification accuracy both algorithms performed well although the time series approach produced the best results for the musicians study while the graph based approach produced the best results for the epilepsy study. There is no obvious reason why this might be the case, visual inspection of the MRI scans does not indicate any obvious distinguishing attributes with respect to the size and shape of he corpus callosum. With respect to computational complexity image segmentation and the application of DTW for classification are both computationally expensive processes. The time complexity for the image segmentation was about 30 seconds per image. For the given data sets the application of DTW took 90 seconds on average to categorise the test set. The graph based approach was significantly faster.

7 Conclusion

Two approaches to Region of Interest Based Image Classification (ROIBIC) have been described. The first was founded on a graph representation to which graph mining techniques could be applied to obtain a feature space. The second used a time series based approach. The work was directed at the classification of MRI scans according to the nature of the corpus callosum featured within these images. Two studies were used for the evaluation: (i) distinguishing musicians from non-musicians, and (ii) epilepsy screening. However, the approach has more general applicability. The research team are also interested in alternative methods of pre-processing MRI data, and mechanism for the post-processing of results to provide explanations for specific classifications. The latter is seen as particularly significant in the context of medical research involving MRI scan data.

References

1. Allen, L., Richey, M., Chain, Y., Gorski, R.: Sex differences in the corpus callosum of the living human being. Journal of Neuroscience 11, 933–942 (1991)
2. Berndt, D., Clifford, J.: Using dynamic time warping to find patterns in time series. In: AAAI 1994 Workshop on Knowledge Discovery in Databases, Seattle, Washington, pp. 359–370 (1994)
3. Chapelle, O., Haffner, P., Vapnik, V.: Support vector machines for histogram-based image classification. IEEE Transactions on Neural Networks 10(5), 1055–1064 (1999)
4. Conlon, P., Trimble, M.: A study of the corpus callosum in epilepsy using magnetic resonance imaging. Epilepsy Res. 2, 122–126 (1988)
5. Cowell, P., Kertesz, A., Denenberg, V.: Multiple dimensions of handedness and the human corpus callosum. Neurology 43, 2353–2357 (1993)
6. Davatzikos, C., Vaillant, M., Resnick, S., Prince, J., Letovsky, S., Bryan, R.: A computerized approach for morphological analysis of the corpus callosum. Journal of Computer Assisted Tomography 20, 88–97 (1996)

7. Duara, R., Kushch, A., Gross-Glenn, K., Barker, W., Jallad, B., Pascal, S., Loewenstein, D., Sheldon, J., Rabin, M., Levin, B., Lubs, H.: Neuroanatomic differences between dyslexic and normal readers on magnetic resonance imaging scans. Archives of Neurology 48, 410–416 (1991)
8. Elsayed, A., Coenen, F., Jiang, C., García-Fiñana, M., Sluming, V.: Corpus Callosum MR Image Classification. In: Proc. AI 2009, pp. 333–346. Springer, Heidelberg (2009)
9. Gorkani, M., Picard, R.: Texture Orientation for Sorting Photos "at a glance". In: Proc. 12th Int'l Conf. on Pattern Recognition, pp. 459–464 (1994)
10. Grabczewski, K., Jankowski, N.: Feature selection with decision tree criterion. In: Proc. 5th Int. Conf. on Hybrid Intelligent Systems (HIS 2005), pp. 212–217 (2005)
11. Hampel, H., Teipel, S., Alexander, G., Horwitz, B., Teichberg, D., Schapiro, M., Rapoport, S.: Corpus callosum atrophy is a possible indicator of region and cell type-specific neuronal degeneration in Alzheimer disease. Archives of Neurology 55, 193–198 (1998)
12. Hijazi, M.H.Q.A., Coenen, F., Zheng, Y.: A Histogram Based Approach to Screening of Age-related Macular Degeneration. In: Proc. Medical Image Understanding and Analysis (MIUA 2009), pp. 154–158 (2009)
13. Hynd, G., Hall, J., Novey, E., Eliopulos, D., Black, K., Gonzalez, J., Edmonds, J., Riccio, C., Cohen, M.: Dyslexia and corpus callosum morphology. Archives of Neurology 52, 32–38 (1995)
14. Jiang, C., Coenen, F.: Graph-based Image Classification by Weighting Scheme. In: Proc. AI 2008, pp. 63–76. Springer, Heidelberg (2008)
15. Lyoo, I., Satlin, A., Lee, C.K., Renshaw, P.: Regional atrophy of the corpus callosum in subjects with Alzheimer's disease and multi-infarct dementia. Psychiatry Research 74, 63–72 (1997)
16. Quinlan, R.: C4.5: A program for machine learning. Morgan Kaufmann, San Francisco (1993)
17. Riley, J.D., Franklin, D.L., Choi, V., Kim, R.C., Binder, D.K., Cramer, S.C., Lin, J.J.: Altered white matter integrity in temporal lobe epilepsy: Association with Cognitive and Clinical Profiles (2010) (to appear in Epilepsia)
18. Salat, D., Ward, A., Kaye, J., Janowsky, J.: Sex differences in the corpus callosum with aging. Journal of Neurobiology of Aging 18, 191–197 (1997)
19. Smith, J., Li, C.: Image Classification and Querying Using Composite Region Templates. Int'l J. Computer Vision and Image Understanding 75(1/2), 165–174 (1999)
20. Szummer, M., Picard, R.: Indoor-Outdoor Image Classification. In: Proc. IEEE Int'l Workshop on Content-Based Access of Image and Video Databases, pp. 42–51 (1998)
21. Vailaya, A., Figueiredo, M., Jain, A., Zhang, H.: Image Classification for Content-Based Indexing. IEEE Transactions on Image Processing 10(1), 117–130 (2001)
22. Weber, B., Luders, E., Faber, J., Richter, S., Quesada, C.M., Urbach, H., Thompson, P.M., Toga, A.W., Elger, C.E., Helmstaedter, C.: Distinct regional atrophy in the corpus callosum of patients with temporal lobe epilepsy. Brain 130, 3149–3154 (2007)
23. Wang, J., Li, J., Wiederhold, G.: SIMPLIcity: Semantics-sensitive integrated matching for picture libraries. IEEE Transactions on Pattern Analysis and Machine Intelligence 23(9), 947–963 (2001b)
24. Weis, S., Kimbacher, M., Wenger, E., Neuhold, A.: Morphometric analysis of the corpus callosum using MRI: Correlation of measurements with aging in healthy individuals. American Journal of Neuroradiology 14, 637–645 (1993)
25. Yan, X., Han, J.: gspan: Graph-based substructure pattern mining. In: ICDM 2002: 2nd IEEE Conf. Data Mining, pp. 721–724 (2002)

Meta-learning for Post-processing of Association Rules

Petr Berka[1,2] and Jan Rauch[1]

[1] University of Economics, W. Churchill Sq. 4, 130 67 Prague
[2] Institute of Finance and Administration, Estonska 500, 101 00 Prague

Abstract. The paper presents a novel approach to post-processing of association rules based on the idea of meta-learning. A subsequent association rule mining step is applied to the results of "standard" association rule mining. We thus obtain "rules about rules" that help to better understand the association rules generated in the first step.

We define various types of such meta-rules and report some experiments on UCI data. When evaluating the proposed method, we use the *apriori* algorithm implemented in Weka.

1 Introduction

The term association rules was coined by R. Agrawal in the early 90th in relation to so called market basket analysis [2]. In this analysis, transaction data recorded by point-of-sale (POS) systems in supermarkets are analyzed in order to understand the purchase behavior of groups of customers, and use it to increase sales, and for cross-selling, store design, discount plans and promotions. This idea of association rules has been later generalized to any data in the tabular, attribute-value form. So data describing properties (values of attributes) of some examples can be analyzed in order to find associations between conjunctions of attribute-value pairs (categories). Let us denote these conjunctions as Ant and Suc and the association rule as

$$Ant \Longrightarrow Suc.$$

The two basic characteristics of an association rule are *support* and *confidence*. Support is the estimate of the probability $P(Ant \wedge Suc)$, (the frequency of $Ant \wedge Suc$ is the *absolute support*), confidence is the estimate of the probability $P(Suc|Ant)$. So an example of a rule based on Table 1 is

$$income(high) \wedge balance(high) \Longrightarrow loan(yes)$$

with the support 0.1667 and the confidence 1.

In association rule discovery the task is to find all syntactically correct rules $Ant \Longrightarrow Suc$ (i.e. rules, in which two different values of an attribute cannot occur) such that the support and confidence of the rules are above the userdefined thresholds *minconf* and *minsup*. There is a number of algorithms, that perform

T.B. Pedersen, M.K. Mohania, and A M. Tjoa (Eds.): DaWaK 2010, LNCS 6263, pp. 251–262, 2010.

Table 1. Running example data

client	income	balance	sex	unemployed	loan
c1	high	high	female	no	yes
c2	high	high	male	no	yes
c3	low	low	male	no	no
c4	low	high	female	yes	yes
c5	low	high	male	yes	yes
c6	low	low	female	yes	no
c7	high	low	male	no	yes
c8	high	low	female	yes	yes
c9	low	medium	male	yes	no
c10	high	medium	female	no	yes
c11	low	medium	female	yes	no
c12	low	medium	male	no	yes

this task. The main idea of these algorithm is to repeatedly generate a rule in a "top-down" way by rule specialization (i.e. by adding categories to an existing combination) and test, if this rule meets the thresholds *minconf* and *minsup*. The probably best-known algorithm called *apriori* proceeds in two steps. All frequent itemsets are found in the first step. A frequent itemset is a set of items that is included in at least *minsup* transactions. Then, association rules with a confidence of at least *minconf* are generated in the second step [2].

There is also an alternative approach to association rules mining, the so called GUHA method that originates from the research of a group of Czech researchers from mid. 60th [6]. The aim of the GUHA method is to offer all interesting facts hidden in the analyzed data and relevant to the given problem. The method is realized by GUHA-procedures. The input of the GUHA procedure consists of the analyzed data and of a simple definition of a set of relevant (i.e. potentially interesting) patterns. GUHA procedure automatically generates each particular pattern and tests if it is true in the analyzed data. The output of the procedure consists of all prime patterns. The pattern is prime if it is true in the analyzed data and if it does not immediately follow from the other more simple output patterns [6].

The most important GUHA procedure is the procedure ASSOC [6]. This procedure mines for patterns that can be understood as a generalization of patterns now called association rules [2]. The most used current implementation of procedure ASSOC is the procedure *4ft-Miner* [9]. This procedure mines for association rules of the form

$$\varphi \approx \psi \quad \text{or} \quad \varphi \approx \psi/\chi$$

where φ, ψ, and χ are Boolean attributes. The rule $\varphi \approx \psi$ means that φ and ψ are associated in the way given by the symbol $\approx$. The conditional rule $\varphi \approx \psi/\chi$ means that φ and ψ are associated in the way given by the symbol $\approx$ if the condition given by χ is satisfied. The symbol $\approx$ is called *4ft-quantifier*, φ is called *antecedent*, ψ is called *succedent* and χ is *condition*. The *4ft-quantifier* $\approx$

corresponds to a condition concerning a four-fold contingency table $4ft(\varphi, \psi, \mathcal{M})$ of φ and ψ in the analyzed data matrix $\mathcal{M}$.

The Boolean attributes φ, ψ, and χ are automatically derived from the columns of the analyzed data matrix. *Basic Boolean attributes* are created first. The basic Boolean attribute has a form of $A(\alpha)$ where A is an attribute i.e. a column of the analyzed data matrix and α is a set of its possible values. The basic Boolean attribute $A(\alpha)$ is true in the row o of the data matrix if the value $A(o)$ of attribute A in row o belongs to α, formally if $A(o) \in \alpha$. An example of basic Boolean attribute of the data matrix in Tab. 1 is balance(medium, high) which is true for clients c1, c2, c4, c5, c9, c10, c11, c12 and false for clients c3, c6, c7, c8. *Literal* is basic Boolean attribute $A(\alpha)$ or its negation $\neg A(\alpha)$. *Partial cedent* is a conjunction or a disjunction of literals. Antecedent, succedent and condition are conjunctions of partial cedents. There are very fine tools to define a set of association rules to be generated and verified.

The rule $\varphi \approx \psi$ is true in analyzed data matrix $\mathcal{M}$ if the condition related to 4ft-quantifier $\approx$ is satisfied in four-fold contingency table $4ft(\varphi, \psi, \mathcal{M})$ of φ and ψ in data matrix $\mathcal{M}$. It is a quadruple $\langle a, b, c, d \rangle$ where a is the number of rows of $\mathcal{M}$ satisfying both φ and ψ, b is the number of rows of $\mathcal{M}$ satisfying φ and not satisfying ψ etc., see Tab. 2. The conditional association rule $\varphi \approx \psi/\chi$ is true in data matrix $\mathcal{M}$ if the rule $\varphi \approx \psi$ is true in data matrix $\mathcal{M}/\chi$. Data matrix $\mathcal{M}/\chi$ consists of all rules of $\mathcal{M}$ satisfying χ.

Table 2. $4ft(\varphi, \psi, \mathcal{M})$

$\mathcal{M}$	ψ	$\neg\psi$
φ	a	b
$\neg\varphi$	c	d

There are 17 various 4ft-quantifiers implemented in *4ft-Miner*. An example is the 4ft-quantifier $\sim^+_{p,B}$ of *above average dependence* which is defined in [8] by the condition $\frac{a}{a+b} \geq (1+p)\frac{a+c}{a+b+c+d} \wedge a \geq B$ for $0 < p$ and $B > 0$. This means that the relative frequency of objects satisfying ψ among the objects satisfying φ is at least $100p$ per cent higher than the relative frequency of objects satisfying ψ among all the observed objects and that there are at least B objects satisfying both φ and ψ. Thus an example of a rule found in data shown in Table 1 can be

$$balance(medium, high) \wedge \neg(unemployed(yes)) \sim^+_{0.5,100} loan(yes) .$$

This rule says that among clients with medium or high balance on account which are not unemployed, there are 50 % higher frequency of clients which will get the loan that among all clients and that there are at least 100 such clients.

So

- GUHA method offers more types of relations (so called quantifiers) between *Ant* and *Suc*,
- GUHA method offers more expressive syntax of *Ant* and *Suc*.

But still the algorithms for mining this type of rules are based on generating and testing of huge set of potential rules.

The main drawback of the association rules mining is the fact, that the result of an analysis will consist of many (hundreds, thousands) rules which have to be visually interpreted and evaluated by the domain expert. So some kind of post-processing of the results would be very helpful for the user. And indeed, various approaches have been used to post-process the huge list of found associations: filtering, selection, visualization, grouping and clustering. In our paper, we present an alternative approach based on the idea of meta-learning.

The rest of the paper is organized as follows: section 2 defines the association meta-rules, section 3 shows experimental evaluation of the proposed method, section 4 reviews other approaches to association rules post-processing and section 5 gives directions for our future work.

2 Association Meta-rules

The inspiration of our method comes from the area of meta-learning. Meta learning is a subfield of machine learning where automatic learning algorithms are applied to meta-data about machine learning experiments. The mostly used approaches to meta-learning (or combining classifiers) are bagging, boosting and stacking [4]. In bagging each classifier in the ensemble votes with equal weight when classifying new example; in order to promote model variance, bagging trains each classifier in the ensemble using a randomly-drawn subset of the training set. In boosting the ensemble of classifiers is built incrementally by training each new classifier to emphasize the training instances that previous classifiers miss-classified. In stacking a meta-classifier is build on top of the results of so called base classifiers that are each separately trained to classify the data.

We propose to apply association rule mining algorithm to the set of original association rules obtained as a result of a particular data mining task. This idea thus follows the stacking concept that is used to combine classifiers, but that has not been presented yet for descriptive tasks. The input to the proposed meta-learning step will be association rules encoded in a way suitable for association rule mining algorithm; the result will be a set of association meta-rules uncovering relations between various characteristics of the original set of rules.

We will distinguish two types of association meta-rules: *qualitative* and *quantitative*. Qualitative rules will represent the meta-knowledge in the form "if original association rules contain a conjunction of categories Ant, then they also contain the conjunction of categories Suc", i.e qualitative rules have the form

$$Ant \Longrightarrow Suc.$$

Quantitative rules will represent the meta-knowledge in the form "if original association rules contain a conjunction of categories Ant, then they have quantitative characteristics Q", i.e

$$Ant \Longrightarrow Q.$$

or, "if original association rules have quantitative characteristics Q, then they contain a conjunction of categories Suc", i.e

$$Q \Longrightarrow Suc.$$

where Q can be e.g "confidence $\in [0.9, 1]$".

We can also search for conjunctions of categories, that frequently occur in the list of original association rules (let call them *frequent cedents*).

To find association meta-rules, standard association rule mining algorithms can be used. Encoding of the original rules is thus the key problem in our approach. Ant and Suc can be encoded either (1) using binary attributes, where each attribute represents one possible literal or (2) using the attributes from the original data set. In both cases we can (or need not) also consider whether the literal occurs in Ant or Suc. We can thus consider four different representation schemes. So to encode the rule

$$income(high) \wedge balance(high) \Longrightarrow loan(yes)$$

1. when using the encoding based on binary attributes without distinguishing between Ant and Suc, this rule will be represented using the categories `income_high(true)`, `balance_high(true)` and `loan_yes(true)`.
2. when using the encoding based on original attributes without distinguishing between Ant and Suc, this rule will be represented using the categories `income(high)`, `balance(high)` and `loan(yes)`.
3. when using the encoding based on binary attributes with distinguishing between Ant and Suc, this rule will be represented using the categories `Ant_income_high(true)`, `Ant_balance_high(true)` and `Suc_loan_yes(true)`.
4. when using the encoding based on original attributes with distinguishing between Ant and Suc, this rule will be represented using the categories `Ant_income(high)`, `Ant_balance(high)` and `Suc_loan(yes)`.

Another open question concerning the representation of a rule is whether categories not occurring in the rule should be treated as missing or as negative ones. In the first approach, attributes not used in the rule will be encoded using missing value code. In the second approach, when using the binary representation, categories not used in the rule will get the value false, and when using the original attributes, categories not used in the rule will get a new special value interpreted as not used. Our initial experiments show that using missing value code is more suitable as it will prevent the meta-learning step to generate a great number of meta-rules about non-occurrence of literals in the original rules, this option also corresponds to the original notion of association rules where only items that do occur in the market baskets are taken into consideration.

The selection of a proper representation formalism is closely related to the type of association rules to be analyzed. For apriori like rules, the formalism using the same attributes as for the original data is sufficient. On the contrary, to be able to represent the GUHA like rules (that can contain disjunctions of values of a single attribute or negations of literals) we have to encode each value of each attribute (i.e. each category) as a single binary attribute. This attribute takes value "true" if the encoded category occurs in positive literal, value "false" if the encoded category occurs in negative literal, or value "missing" if the encoded category does not occur in the rule. So the rule

$$balance(high \lor medium) \land \neg(unemployed(yes)) \Rightarrow loan(yes) \ / \ sex(male)$$

will be encoded using the categories `balance_high(true)`, `balance_medium (true)`, `unemployed_yes(false)`, `loan_yes(true)` and `sex_male(yes)`.

Quantitative characteristics can be encoded using numerical attributes that must be discretized in advance. There is no difference between the apriori-like and GUHA-like association rules in this encoding.

Anyway, all possible methods of rule encoding will result in building a data table (each rule represented by a single row) that can easily be analyzed using association rule mining algorithm to obtain the meta-rules. The obvious question of this approach is: does such post-processing make sense from the users point of view? We believe that it does, if we answer positively the following questions:

- Do the meta-rules give better insight into the list of "original" association rules?
- Is the list of meta-rules easier to evaluate?

We performed several experiments to find answers to these questions.

3 Experimental Evaluation

To evaluate our ideas, we performed several experiments on data. We carried out the experiments using Weka (a data mining system that is freely available from University of Waikato) [14]. In all of our experiments, we encoded the input rules using the original attributes without distinguishing if the category occurs in *Ant* or in *Suc* (in this case, the representation of the rules has the most similar structure to the original data) and encoding attributes not present in a rule as missing values.

3.1 Running Example

Let us start with a closer look on our running example. When applying the apriori algorithm (the Weka implementation) to the data shown in Table 1, we will obtain (for parameters $minsup = 0.2$, i.e. 2 instances and $minconf = 0.8$) 72 association rules, first 10 of them shown in Table 3. This set of rules has been post-processed in the first series of experiments. We choose the representation of *Ant* and *Suc* based on the original attributes and encode categories not occurring in the rule as "missing". Refer to Table 4 for the encoding of the first ten rules. Notice, that

- the numeric attributes support and confidence have been discretized; we used equifrequent discretization into 2 intervals for both support and confidence in this example,

– we added the attribute true (for technical reasons, to let Weka to find the frequent cedents).

At first we will look for *quantitative* meta-rules. Table 5 shows the listing of all quantitative meta-rules for the parameters $minsup = 0.1$ and $minconf = 0.8$; we intentionally used the same setting of parameters as for the analysis of the original data to compare the number of found rules and meta-rules.

Due to the way how the rules have been encoded for meta-learning, the meta-rules have the same syntax as the original rules. But their meaning is completely different. Recall that the rules are obtained from the original data but the meta-rules are obtained from rules. So the third rule from Table 3 says, that there are 4 clients in the analyzed data with high balance, all of them belonging to category `loan=yes`. But the "same" meta-rule (the rule 10 from Table 5) says that there are 22 rules having the category `balance=high` in *Ant* or in *Suc*, and 18 out of them have also the category `loan=yes` (in *Ant* or in *Suc*). We thus have found a (quite a large) subset of the original rules referring to the same characteristics of the clients.

Table 3. Association rules

```
1. income=high 5 ==> loan=yes 5 conf:(1)
2. loan=no 4 ==> income=low 4 conf:(1)
3. balance=high 4 ==> loan=yes 4 conf:(1)
4. income=high unemployed=no 4 ==> loan=yes 4 conf:(1)
5. income=high sex=female 3 ==> loan=yes 3 conf:(1)
6. income=low sex=female 3 ==> unemployed=yes 3 conf:(1)
7. unemployed=yes loan=no 3 ==> income=low 3 conf:(1)
8. balance=high unemployed=no 2 ==> income=high 2 conf:(1)
9. income=high balance=high 2 ==> unemployed=no 2 conf:(1)
10. income=high balance=high 2 ==> loan=yes 2 conf:(1)
. . .
72. income=high 5 ==> unemployed=no loan=yes 4 conf:(0.8)
```

Table 4. Encoded association rules

Id	true	income	balance	sex	unemp.	loan	abssup	conf
1	t	high	?	?	?	yes	(2.5-inf)	(0.915-inf)
2	t	low	?	?	?	no	(2.5-inf)	(0.915-inf)
3	t	?	high	?	?	yes	(2.5-inf)	(0.915-inf)
4	t	high	?	?	no	yes	(2.5-inf)	(0.915-inf)
5	t	high	?	female	?	yes	(2.5-inf)	(0.915-inf)
6	t	low	?	female	yes	?	(2.5-inf)	(0.915-inf)
7	t	low	?	?	yes	no	(2.5-inf)	(0.915-inf)
8	t	high	high	?	no	?	(-inf-2.5]	(0.915-inf)
9	t	high	high	?	no	?	(-inf-2.5]	(0.915-inf)
10	t	high	high	?	?	yes	(-inf-2.5]	(0.915-inf)

Table 5. Qualitative meta-rules

```
1. income=low loan=yes 7 ==> balance=high 7 conf:(1)
2. unemployed=yes loan=yes 7 ==> balance=high 7 conf:(1)
3. balance=high unemployed=yes 9 ==> income=low 8 conf:(0.89)
4. income=low balance=high 9 ==> unemployed=yes 8 conf:(0.89)
5. balance=high unemployed=no 8 ==> income=high 7 conf:(0.88)
6. income=high balance=high 8 ==> unemployed=no 7 conf:(0.88)
7. balance=medium loan=no 8 ==> unemployed=yes 7 conf:(0.88)
8. balance=medium unemployed=yes 8 ==> loan=no 7 conf:(0.88)
9. loan=no 18 ==> income=low 15 conf:(0.83)
10. balance=high 22 ==> loan=yes 18 conf:(0.82)
11. income=high 26 ==> loan=yes 21 conf:(0.81)
```

Table 6. Quantitative meta-rules

```
1. abssup=(-inf-2.5] 59 ==> conf=(0.915-inf) 59 conf:(1)
2. balance=high 22 ==> conf=(0.915-inf) 22 conf:(1)
3. loan=no 18 ==> conf=(0.915-inf) 18 conf:(1)
4. balance=high loan=yes 18 ==> conf=(0.915-inf) 18 conf:(1)
5. sex=female 15 ==> conf=(0.915-inf) 15 conf:(1)
6. income=low loan=no 15 ==> conf=(0.915-inf) 15 conf:(1)
7. balance=medium 12 ==> abssup=(-inf-2.5] 12 conf:(1)
8. balance=medium 12 ==> conf=(0.915-inf) 12 conf:(1)
9. sex=male 12 ==> abssup=(-inf-2.5] 12 conf:(1)
10. sex=male 12 ==> conf='(0.915-inf)' 12 conf:(1)
11. balance=medium 12 ==> abssup=(-inf-2.5] conf=(0.915-inf) 12 conf:(1)
```

The next step in our running example will be the mining for *quantitative* meta-rules. The input data (encoded rules) remain the same as in the previous step. We again used the parameters $minsup = 0.1$ and $minconf = 0.8$ and we obtained 11 meta-rules shown in Table 6. Like in the set of original rules and the set of qualitative meta-rules, we can again find in the listing a meta-rule dealing with the categories `balance=high` and `loan=yes`. The meta-rule no.4 says, that all original rules having `balance=high` and `loan=yes` in *Ant* or *Suc*, have the confidence greater than 0.915.

To be able to use the Weka system also for the last type of analysis, for looking for frequent cedents, we added a dummy category `true=T` to the data that encoded the original association rules. We are thus able to identify frequent cedents from the rules

$$true = T \Longrightarrow Suc,$$

that have sufficiently high confidence. Table 7 shows the 12 respective rules for $minsup = 0.25$ thus showing the cedents *Suc* that occur in at least 25 percents of the original association rules. We can e.g. see that the category `loan=yes` occurs in more than one half of the original rules.

Table 7. Frequent cedents

```
1. true=t 72 ==> loan=yes 37 conf:(0.51)
2. true=t 72 ==> income=low 30 conf:(0.42)
3. true=t 72 ==> unemployed=no 28 conf:(0.39)
4. true=t 72 ==> unemployed=yes 27 conf:(0.38)
5. true=t 72 ==> income=high 26 conf:(0.36)
6. true=t 72 ==> balance=high 22 conf:(0.31)
7. true=t 72 ==> income=high loan=yes 21 conf:(0.29)
8. true=t 72 ==> income=low unemployed=yes 21 conf:(0.29)
9. true=t 72 ==> income=high unemployed=no 20 conf:(0.28)
10. true=t 72 ==> unemployed=no loan=yes 20 conf:(0.28)
11. true=t 72 ==> loan=no 18 conf:(0.25)
12. true=t 72 ==> balance=high loan=yes 18 conf:(0.25)
```

3.2 Further Experiments

The next set of experiments was carried out on larger (and more realistic) data. We used several data sets from the UCI Machine Learning Repository [13]. The characteristics of the data (number of examples and number of attributes) are summarized in Table 8.

Table 9 summarizes the results of our analysis (both mining association rules and meta-rules) for different data sets. The numbers in the table show the number of found association rules, the number of qualitative meta-rules, the number of quantitative meta-rules, and the number of frequent cedents. To make the numbers comparable, we used the same settings of $minsup$ and $minconf$ during both learning and meta-learning for corresponding data ($minconf$ was in all experiments set to 0.2). The frequent cedents were obtained for $minconf = 0.1$. We used equifrequent discretization into 5 intervals for support and equidistant discretization into 4 intervals for confidence.

The results support our working hypothesis, that the number of meta-rules will be significantly smaller than the number of original rules. Thus the interpretation of meta-rules by domain expert will be significantly less time consuming and difficult compared to the interpretation of the original association rules.

Table 8. Description of used data

Data	no. examples	no. attributes
Brest cancer	286	10
Lenses	24	5
Monk1	123	7
Mushroom	8124	23
Tic-tac-toe	958	10
Tumor	339	18
Vote	435	17

Table 9. Summary of the results

Data	assoc rules	qualitative rules	quantiative rules	frequent cedents
Breast cancer	18742	167	341	80
Lenses	89	13	47	34
Monk1	124	29	30	33
Mushroom	100000	135	109	550
Tic-tac-toe	506	69	30	24
Tumor	100000	234	633	66
Vote	100000	6007	12	150

4 Related Work

The various approaches to post-processing of association rules can be divided into several groups. One group are methods for visualization, filtering or selection of the created rules. This are the standard options in most systems.

Second group contains methods that use some algorithms to further process the rules: clustering, grouping or using some inference methods fits into this group as well as our approach. An application of deduction rules to post-process the results of GUHA method is described in [8]; these rules define allow to remove association rules that are logical consequences of another association rules. Similar idea, but applied to "Agrawal-like" association rules can be found in [12]. This paper also describes clustering of association rules that have the same consequent; the distance between two rules is defined "semantically", i.e. as the number of examples covered only by one of the rules. Both semantical and syntactical (i.e. based on the lists of attribute-value pairs that occur in the rules) clustering of association rules can be found e.g. in [11].

The third possibility is to post-process the rules using some domain knowledge. So e.g. An et all use expert-supplied taxonomy of items for clustering the discovered association rules with respect to the taxonomic similarity ([1]), or Domingues and Rezende ([5]) iteratively scan the itemset rules and updates a taxonomy that is then used to generalize the association mining results.

An additional possibility is to filter out consequences of domain knowledge via application of logic of association rules [8]. This approach is introduced in [10].

5 Conclusions

We present a novel idea of using meta-learning approach to post-process the results of association rule mining. When looking at the two questions from the end of section 2, we can say, that the answer to the first question depends on the domain where association rule mining (and rule post-processing) is applied and that the answer must be given by the domain expert. The answer to the second question can be found in the table 9, where we can see that in all the experiments the number of meta-rules is significantly lower than the number of ordinary rules (and thus should take less time for the domain expert to go through it). Anyway, more experiments and the interpretation of the found meta-rules are necessary to validate the usefulness of the proposed method.

So far we focused on the classical apriori algorithm. Our future work will be oriented on following open issues:

- different types of association rules: the association rules analyzed so far are of the classical form as generated e.g. by the apriori algorithm. Another systems, e.g. LISp-Miner can produce different types of association rules; this brings us to the next issue.
- different types of meta-rules: also the meta-rules created so far are in the form of implications between two conjunctions of attribute-value pairs. When using LISp-Miner for building meta-rules, we can benefit from different types of associations implemented there.
- postprocessing of meta-rules: what will happen if we apply the proposed approach to the meta-rules, i.e. if we perform meta-meta learning?

Acknowledgement

The work is supported by the grant MSMT 1M06014 (from the Ministry of Education of the Czech Republic) and the grant GACR 201/08/0802 (from the Grant Agency of the Czech Republic).

References

1. An, A., Khan, S., Huang, X.: Objective and Subjective Algorithms for Grouping Association Rules. In: Third IEEE Conference on Data Mining (ICDM 2003), pp. 477–480 (2003)
2. Agrawal, R., Imielinski, T., Swami, A.: Mining Association Rules Between Sets of Items in Large Databases. In: SIGMOD Conference, pp. 207–216 (1993)
3. Baesens, B., Viaene, S., Vanthienen, J.: Post-processing of association rules. In: The Sixth ACM SIGKDD International Conference on Knowledge Discovery and Data Mining (KDD 2000), Boston, Massachusetts, August 20-23 (2000)
4. Bauer, E., Kohavi, R.: An Empirical Comparison of Voting Classification Algorithms: Bagging, Boosting, and Variants. Machine Learning 36(1/2), 105–139 (1999)
5. Domingues, M.A., Rezende, S.O.: Using Taxonomies to Faciliate the Analysis of the Association Rules. In: Second International Workshop on Knowledge Discovery and Ontologies (KDO 2005), ECML/PKDD, Porto (2005)
6. Hájek, P., Havránek, T.: Mechanising Hypothesis Formation - Mathematical Foundations for a General Theory. Springer, Heidelberg (1978)
7. Jorge, A., Poas, J., Azevedo, P.J.: Post-processing Operators for Browsing Large Sets of Association Rules. Discovery Science 2002, 414–421
8. Rauch, J.: Logic of association rules. Applied Intelligence 22, 9–28 (2005)
9. Rauch, J., Šimůnek, M.: An Alternative Approach to Mining Association Rules. In: Lin, T.Y., Ohsuga, S., Liau, C.J., Tsumoto, S. (eds.) Proc. Foundations of Data Mining and Knowledge Discovery. Springer, Heidelberg (2005)

10. Rauch, J.: Considerations on Logical Calculi for Dealing with Knowledge in Data Mining. In: Ras, Z.W., Dardzinska, A. (eds.) Advances in Data Management, pp. 177–202. Springer, Heidelberg (2009)
11. Sigal, S.: Exploring interestingness through clustering. In: Proc. of the IEEE Int. Conf. on Data Mining (ICDM 2002), Maebashi City (2002)
12. Toivonen, H., Klementinen, M., Roikainen, P., Hatonen, K., Mannila, H.: Pruning and grouping discovered association rules. In: Workshop notes of the ECML 1995 Workshop on Statistics, Machine Learning and Knowledge Discovery in Databases, Heraklion, pp. 47–52 (1995)
13. UCI Machine Learning Repository, `http://www.ics.uci.edu/~mlearn/MLRepository.html`
14. Weka - Data Mining with Open Source Machine Learning Software, `http://www.cs.waikato.ac.nz/ml/weka/`

A Relational Approach for Discovering Frequent Patterns with Disjunctions

Corrado Loglisci, Michelangelo Ceci, and Donato Malerba

Department of Computer Science, University of Bari
Via E.Orabona 4, 70126, Bari-Italy
{loglisci,ceci,malerba}@di.uniba.it

Abstract. Traditional pattern discovery approaches permit to identify frequent patterns expressed in form of conjunctions of items and represent their frequent co-occurrences. Although such approaches have been proved to be effective in descriptive knowledge discovery tasks, they can miss interesting combinations of items which do not necessarily occur together. To avoid this limitation, we propose a method for discovering interesting patterns that consider disjunctions of items that, otherwise, would be pruned in the search. The method works in the relational data mining setting and conserves anti-monotonicity properties that permit to prune the search. Disjunctions are obtained by joining relations which can simultaneously or alternatively occur, namely relations deemed similar in the applicative domain. Experiments and comparisons prove the viability of the proposed approach.

1 Introduction

Discovery of frequent patterns in large collections of transactions or tuples has become one of the broadly investigated topics in data mining[1,4]. Patterns represent statistical regularities of co-occurrences (expressed as conjunctions) of the items present in the transactions. The most interesting patterns are those that express conjunctions which occur in at least a user-defined number of transactions. Typically, such conjunctions are obtained by the intersection of the transactions in which the items occur under the assumption that the items occur independently of each other. This poses some limitations to the mining process and, in particular, leaves unexplored two potentialities of the pattern discovery: i) discovering interesting patterns when items are not present in a sufficient number of transactions, and ii) discovering forms of relationships among items different from the classical conjunctions. Indeed, the two potentialities are not independent each other since the discovery of patterns including other relationships between items may lead to discover patterns that otherwise would be discarded.

Although traditional frequent patterns discovery approaches are based on the items which co-occur, other forms of relationships among items have been actually investigated in the literature. In particular, some works propose to accommodate a domain-dependent taxonomy over the items in the mining process in

T.B. Pedersen, M.K. Mohania, and A M. Tjoa (Eds.): DaWaK 2010, LNCS 6263, pp. 263–274, 2010.

addition to the classical conjunction. This allows to consider the generalization relationship (*is-a relationship*) among the items[13]. The presence of items at higher levels of the taxonomy in the patterns implies the presence of the items at lower levels related with the former through is-a relationships. For instance, given a complete taxonomy for which, milk *is-a* food, coffee *is-a* food, fruit juice *is-a* food, the pattern $\langle food, soap\rangle$ can be interpreted as the pattern $\langle (milk \vee fruitjuice \vee coffee), soap\rangle$ where the occurrence of $food$ implies the occurrence of at least one among $milk$, $bread$ or $fruitjuice$, that is $food = (milk \vee fruitjuice \vee coffee)$. The accommodation of a taxonomy thus allows to represent relationships among items in the form of fixed disjunctions.

Independently from the accommodation of a domain taxonomy in the mining process, the discovery of patterns with items in disjunction has been already investigated in the literature with approaches that permit to mine disjunctive association rules from transactions or tuples [10][12]. In [10], the authors provide a statistical framework based on a set operations (union and intersection) among transaction sets to identify itemsets (called contexts) that can potentially contain disjunctions. Then, these contexts are combined and explored to generate a preliminary set of disjunctive rules. Finally, the application of propositional logic techniques on this set allows to infer rules with items related by inclusive logical disjunction and exclusive logical disjunction. Differently, in [12], the authors extend traditional algorithms to mine associations among item groups formed by items in disjunction. Each group is generated by aggregating items on the basis of their conceptual distance. The items are accommodated in a weighted directed graph provided as background information, whereas the conceptual distance between two items is expressed as the weight of their relative edge. The conceptual distance is thus exploited to aggregate two rules, which present conceptually close items, into only one rule: the final rule will incorporate a group of close items in relationship of disjunction and thus it will be more frequent.

Although the first approach allows to discover disjunctive patterns without requiring background information about the items, it could join unrelated items (e.g., $\langle milk \vee jackets\rangle$) and then produce rules that are difficult to understand and which do not exploit the potentialities of the disjunction of representing the occurrence of at least one between two related events (e.g., $\langle milk \vee coffee\rangle$). An important common aspect of both approaches is that they work on tuples, namely on items represented in form of attribute-value pairs which lead to consider the disjunction only among the discrete, categorical or taxonomic values. Although simple and reasonably more effective, this representation, also known as *propositional*, can turn out to be too restrictive in applications where data are naturally complex, and moreover, trasforming such data in tuples could lead to information loss. Several studies in the literature have proved that in those cases resorting to the *relational* representation [6] permits to directly deal with the complex structure of data, to conduct a realistic investigation which distinguishes the main subjects of analysis from other subjects as well as to represent their interaction. Examples of such subjects can be found in spatial analysis where the location and the extension of spatial objects define spatial *relations*, such as those topological (e.g., the region A is contained in the region B - $contained\_in(A, B)$), and spatial *properties*, such as those geometrical (e.g., shape of a region - $rectangle\_shape(A)$) [7]. Existing approaches to disjunctive patterns discovery do not consider complex data, and, in particular, they analyze neither possible interactions among them nor the sets of possible descriptive properties.

In this paper we propose a relational data mining approach for discovering frequent patterns that include disjunctions. Patterns are represented in terms of atoms [3]. The approach allows to mine frequent patterns with disjunctions among atoms that can express relations (e.g., $contained\_in(A, B) \vee overlaps(A, B)$) or properties ($rectangle\_shape(A) \vee square\_shape(A)$) of the analyzed data. It extends an existing logic-based method for conjunctive pattern mining [8] to the discovery of disjunctive patterns, where disjunctions are generated among *similar* relations or properties. Similarity between relations or properties is defined in the user defined background knowledge in form of conceptual distance. The approach takes advantage of the representation and reasoning techniques developed in the field of inductive logic programming (ILP). In particular, the expressive power of logic formalism is profitably used to represent relations, properties and background knowledge in the natural form of *n-ary* logic predicates. This way of using the disjunction permits to combine the occurrences of the involved relations in order to produce patterns with higher frequency, that, potentially, can be more interesting.

The paper is organized as follows. In the next section, motivation and overview of the proposed approach are presented. In Section 3, the approach is presented in detail. In Section 4 experimental results on real world data are reported. Finally, conclusions are drawn and future works are presented.

2 Motivation and Overview of the Approach

The motivation behind the usage of disjunctive forms is that the set of patterns discovered with traditional approaches strongly depends on frequency-based thresholds (e.g., support, confidence, lift) so, when these assume high values, many interesting patterns are missed: conjunctions of atoms, for which the considered statistical measure does not exceed the minimum threshold, are ignored. The introduction of the disjunctive forms would permit to include the atoms which occur simultaneously with or alternatively to other atoms with the effect of increasing the values of the considered measures associated to the patterns. For instance, by supposing that the atom $overlaps(A, B)$ may occur also when $contained\_in(A, B)$ does not occur, the pattern $\langle district(A), (contained\_in(A, B) \vee overlaps(A, B)), marketplace(B)\rangle$ might be frequent while both $\langle district(A), contained\_in(A, B), marketplace(B)\rangle$ and $\langle district(A), overlaps(A, B), marketplace(B)\rangle$ might not be frequent.

This advocates the starting point of our approach, which is that of considering infrequent conjunctive patterns. These patterns are re-evaluated and extended to the disjunctive form by inserting disjunctions which involve atoms already present in the patterns. Disjunctions are created among atoms which are semantically related in the application domain. The semantic relatedness is intended as background knowledge on the atoms and permits us to numerically quantify the dissimilarity or conceptual distance between atoms. It guarantees that meaningful disjunctions are created. In this work we exploit the ILP system SPADA [8] to identify infrequent conjunctive patterns, but this does not exclude the possibility of using other methods for mining infrequent relational patterns in the initial processing step.

The proposed approach follows a three-stepped procedure. First, it extracts the infrequent conjunctive patterns which can be considered in disjunctive patterns. In particular, the patterns whose frequency is lower than the classical minimum threshold but exceeds a new ad-hoc threshold are selected. These thresholds determine therefore the set of patterns to be extended to the disjunctive form. Second, by following the main intuition proposed in [12], background knowledge is accommodated to exploit the information on the dissimilarity among the atoms in the process of generation of disjunctive patterns. Third, disjunctive patterns are produced by iteratively integrating disjunctions into the patterns by means of a pair-wise joining. The final result consists of patterns, in form of conjunctions of disjunctions of atoms, whose frequency is greater than the traditional minimum threshold. For instance, given the patterns P_1 : $\langle district(A), contained\_in(A,B), marketplace(B)\rangle$, P_2 : $\langle district(A), overlaps(A,B)$, $marketplace(B)\rangle$ and let $contained\_in(\cdot,\cdot)$ and $overlaps(\cdot,\cdot)$ be two "similar" atoms according to the background knowledge, P_1 and P_2 can be joined in $\langle district(A), (contained\_in(A,B) \vee overlaps(A,B)), marketplace(B)\rangle$.

Working in the relational setting adds additional sources of complexity to the problem of joining patterns due to the *linkedness* property [9]. In fact, in the relational representation atoms in a pattern are dependent each other due to the presence of variables (differently from the items in the propositional representation [12]). In this work, patterns to be joined should differ in only one atom (if the atoms are similar) and share the remaining atoms up to a redenomination of variables. For instance, consider the patterns P_1 : $\langle district(A), contained\_in(A,B)$, $crossed\_by(A,C), marketplace(B)\rangle, P_2 : \langle district(A), crossed\_by(A,B), overlaps(A,C)$, $marketplace(C)\rangle$. The pattern $\langle district(A)$, $(contained\_in(A,B) \vee overlaps(A,B))$, $crossed\_by(A,C), marketplace(B)\rangle$ can be extracted since B in $contained\_in(A,B)$ is involved in $marketplace(B)$ of the first pattern, as well as C in $overlaps(A,C)$ is involved in $marketplace(C)$ of the second pattern.

3 Mining Disjunctive Relational Patterns

Before formally defining the problem we face in this work, some notions are necessary. In the relational setting, when handling complex data, different roles can be played by different *sorts* of data. In our formulation complex data are distinguished into target objects of analysis (TO) and non-target objects of analysis (NTO). The former are data on which patterns are enumerated and contribute to compute the frequency of a pattern, while the latter contribute to define the former and they can be involved in a pattern. We denote the set of TO

as S and the sets of NTO by means of the sets R_k ($1 \leq k \leq M$), where M is the number of sorts of data that are not considered to be TO. NTOs, belonging to a set R_k, can be organized hierarchically according to a user defined taxonomy. Target objects and non-target objects are represented in Datalog language [3] as ground atoms and populate the extensional part D_E of a deductive database D. A ground atom is an n-ary logic predicate symbol applied to n constants.

Some predicate symbols are introduced in order to express both properties and relationships of TO and NTO. They can be categorized into four classes: 1) *key predicate* identifies the TO in D_E (e.g., in the examples above, *district(·)*); 2) property predicates are binary predicates which define the values taken by an attribute of a TO or of an NTO; 3) structural predicates are binary predicates which relate NTO as well as TO with others NTO (e.g., in the examples above, *contained_in*(·,·)); 4) *is_a* predicate is a binary taxonomic predicate which associates NTO with a symbol contained in the user defined taxonomy.

The intensional part D_I of the deductive database D includes the definition of the domain knowledge that permits us to express the dissimilarity among atoms in the form of *Datalog* weighted edges of a graph. An example of the Datalog weighted edge is the following:

external_touch_to - (crosses - 0.88)

It states that the dissimilarity between the relationships *external_touch_to(·,·)* and *crosses(·,·)* is 0.88. More generally, it represents an undirected edge e between two vertices v_i, v_j (e.g., *external touch to*, *crosses*) with weight w_{ij} (e.g., 0.88) and it is denoted as $e(v_i, v_j, w)$. A finite sequence of undirected edges $e_1, e_2, \ldots, e_m$ which links two vertices v_i, v_j is called *path* and denoted as $\rho(v_i, v_j)$. The complete list of such undirected edges represents the background information on the dissimilarity among atoms and allows to join patterns by introducing disjunctions *(externa_touch_to(A,B)* $\vee$ *crosses(A,B))*.

Discovered patterns are conjunctions of Datalog non-ground atoms and disjunctions of non-ground atoms, which can be expressed by means of a set notation. A Datalog non-ground atom is an n-ary predicate symbol applied to n terms (either constants or variables), at least one of which is a variable. A formal definition of pattern of our interest is reported in the following:

Definition 1. *A disjunctive pattern P is a set of atoms and disjunctions of atoms* $p_0(t_0^1)$, $(p_1(t_1^1, t_1^2)|p_2(t_2^1, t_2^2)|\ldots), \ldots, (p_k(t_k^1, t_k^2)|\ldots|p_{k+h}(t_{k+h}^1, t_{k+h}^2))$ *where p_0 is the key predicate, while p_i, $i = 1, \ldots, k+h$, is either a structural predicate or a property predicate or an* is_a *predicate. Symbol "|" indicates disjunctions.*

Terms t_i^j are either constants, which correspond to values of property predicates, or variables, which identify target objects or non-target objects. Each p_i is a predicate occurring in D_E (extensionally defined predicate).

Some examples of disjunctive patterns are the following:

$P_1 \equiv district(A), (comes\_from(A, B)|external\_ends\_at(A, B)), shape(A, rectangle)$

$P_2 \equiv district(A), (external\_ends\_at(A, B)|runs\_along\_boundary\_and\_goes\_in(A, B)),$
$transport\_net(A, roads)$

where the variables A denote target objects, and variables B denote some non-target objects, while the predicates $district(A)$ identify the key predicate in P_1 and P_2, $shape(A, rectangle)$ and $transport\_net(A, roads)$ are property predicates and the others are structural predicates. All variables are implicitly existentially quantified.

We now can give a formal statement of the problem of discovering relational frequent patterns with disjunctions:
1. *Given:* the extensional part D_E of a deductive database D, and two thresholds $minSup \in [0; 1]$, $nSup \in [0; 1]$, the former represents a minimum frequency value while the latter represents maximum frequency value ($nSup < minSup$), *Find:* the collection I_R of the relational infrequent patterns whose support is included in $[nSup; minSup)$.
2. *Given:* the collection I_R, the intensional part D_I of a deductive database D, and two thresholds $minSup$ and $\gamma \in [0; 1]$ (γ defines the maximum dissimilarity value of atoms involved in the disjunctions), *Find:* relational disjunctive patterns whose frequency exceeds $minSup$ and whose dissimilarity of atoms involved in the disjunctions does not exceed γ.

3.1 Mining Infrequent Conjunctive Patterns

The intuition underlying the discovery of pattern with disjunctions is that of extending infrequent conjunctive patterns with disjunctive forms until the threshold $minSup$ is exceeded. Each conjunctive pattern P is associated with a statistical parameter $sup(P, D)$ (support of P on D), which is the percentage of *units of analysis* in D *covered by* P. More precisely, a unit of analysis of a target object $s \in S$ is a subset of ground atoms in D_E defined as follows:

$$D[s] = is\_a(R(s)) \cup D[s|R(s)] \cup \bigcup_{r_i \in R(s)} D[r_i|R(s)], \tag{1}$$

where $R(s)$ is the set of NTO directly or indirectly related to s, $is\_a(R(s))$ is the set of is_a atoms which define the sorts of $r_i \in R(s)$, $D[s|R(s)]$ contains properties of s and relations between s and some $r_i \in R(s)$, $D[r_i|R(s)]$ contains properties of r_i and relations between r_i and some $r_j \in R(s)$. By assigning a pattern P with an existentially quantified conjunctive formula $eqc(P)$ obtained by transforming P into a Datalog query, the units of analysis $D[s]$ are *covered by* a pattern P if $D[s] \models eqc(P)$, namely $D[s]$ logically entails $eqc(P)$).

Conjunctive patterns are mined with SPADA which however enables the discovery of relational patterns whose support exceeds $minSup$ (frequent patterns). SPADA performs a breadth-first search of the space of patterns, from the most general to the more specific ones, and prunes portions of the space which contain only infrequent patterns, which are the conjunctive patterns of our interest. The pruning strategy guarantees that all infrequent patterns are removed and, at this aim, uses a generality ordering based on the notion of θ-subsumption [11]:

Definition 2. *P_1 is more general than P_2 under θ-subsumption ($P_1 \succeq_\theta P_2$) if and only if P_1 θ-subsumes P_2, i.e. a substitution θ exists, such that $P_1\theta \subseteq P_2$.*

For instance, given $P1 \equiv district(A), crosses(A,B)$, $P2 \equiv district(A), crosses(A,B), is\_a(B, transport\_net)$, $P3 \equiv district(A), crosses(A,B), is\_a(B, transport\_net), along(A,C)$ we observe that P_1 θ-subsumes P_2 ($P_1 \succeq_\theta P_2$) and P_2 θ-subsumes P_3 ($P_2 \succeq_\theta P_3$) with substitutions $\theta_1 = \theta_2 = \oslash$. The generality order is monotonic with respect to the pattern support, so whenever $P1$ will be infrequent the patterns more specific of it (e.g., $P2, P3$) will be infrequent too.

The search is based on the level-wise method and implements a two-stepped procedure: i) generation of candidate patterns with k atoms (k-th level) by considering the frequent patterns with $k-1$ atoms ((k-*1*)-th level); ii) evaluation of the frequency with k atoms. So, the patterns whose support does not exceeds $minSup$ will be not considered for the next level: the patterns discarded (infrequent) at each level are rather considered for the generation of disjunctions. The collection I_R is thus composed of a subset of infrequent patterns, more precisely those with support greater than or equal to $nSup$ (and less than $minSup$). A detailed description on SPADA can be found in [8].

3.2 Extending Relational Patterns with Disjunctions

The generation of disjunctive patterns is performed by creating disjunctions among similar atoms in accordance to the background knowledge: two patterns which present similar atoms are joined to form only one. The implemented algorithm (see Algorithm 1) is composed of two sub-procedures: the first one (lines 2-12) creates a graph $\mathcal{G}_\mathcal{D}$ with the patterns of I_R by exploiting the knowledge defined in D_I, while the second one (lines 13-32) joins two patterns (vertices) on the basis of the information (weight) associated to their edge.

In particular, for each pair of patterns which have the same length (namely, at the same level of the level-wise search method) it checks whether they differ in only one atom and share the remaining atoms up to a redenomination of variables (line 3). Let α and β be the two atoms differentiating P from Q (α in P, β in Q), a path ρ which links α to β (or viceversa) is searched among the weighted edges according to D_I: in the case the sum ω of the weights found in the path is lower than the maximum dissimilarity γ the vertices P and Q are inserted into $\mathcal{G}_\mathcal{D}$ and linked through an edge with weight ω (lines 4-9). Note that when there is more than one path between α and β, then the path with lowest weight is considered. Intuitively, at the end of the first sub-procedure, $\mathcal{G}_\mathcal{D}$ will contain, as vertices, the patterns which meet the condition at the line 3, and it will contain, as edges, the weights associated to the path linking the atoms differentiating the patterns.

Once we have $\mathcal{G}_\mathcal{D}$, a list $\mathcal{L}_\mathcal{D}$ is populated with the vertices and edges of $\mathcal{G}_\mathcal{D}$: an element of $\mathcal{L}_\mathcal{D}$ is a triple $\langle P, Q, \omega \rangle$ composed of a pair of vertices-patterns (P,Q) with their relative weight. Elements in $\mathcal{L}_\mathcal{D}$ are ranked in ascending order with respect to the values of ω so that the pairs of patterns with lower dissimilarity will be joined for first. This guarantees that disjunctions with very similar atoms will be preferred to the others (line 13). For each element of $\mathcal{L}_\mathcal{D}$ whose weight ω is lower than γ the two patterns P, Q are joined to generate a pattern J composed by the conjunction of the same atoms in common to the two patterns P, Q and of

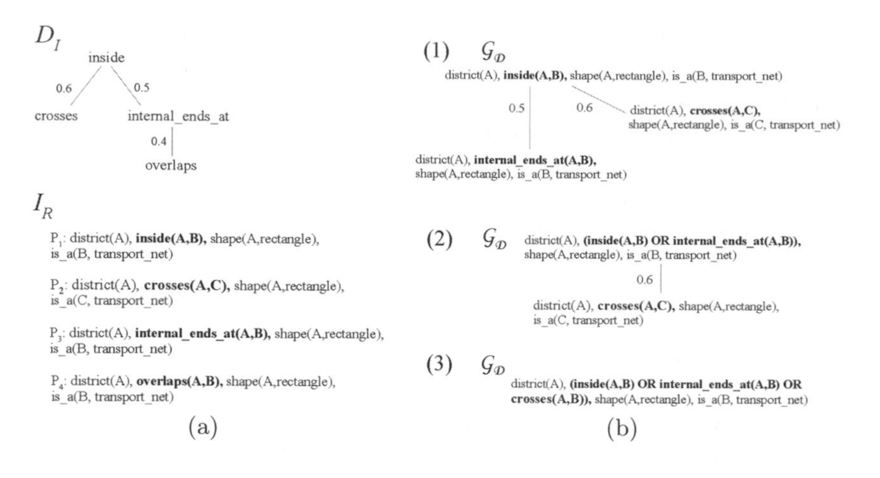

Fig. 1. Extending relational pattern with disjunctions: an example (γ=0.7)

the disjunction formed by the two different (but similar) atoms (lines 14-15). This joining procedure permits to have patterns with the same length of the original ones and which occur when at least one of original patterns occurs. Therefore, if a pattern J is obtained by joining P and Q, it covers a set of units of analysis equal to the union of those of P and Q: the support of J is determined as in line 16 and, generally, it is higher than the support of P and Q. In the case the support of J exceeds *minSup* then it can be considered statistically interesting and no further processing is necessary (lines 16-17). Otherwise, J is again considered and inserted into $\mathcal{G}_\mathcal{D}$ as follows. The edges which linked another pattern R of $\mathcal{G}_\mathcal{D}$ to P and Q are modified in order to keep the links from R to J: the weight of the edges between one pattern R and J will be set to the average value of the weights of all the edges which linked R to P and Q (lines 19-27). The modified graph $\mathcal{G}_\mathcal{D}$ contains conjunctive patterns (those of I_R) and pattern with disjunctions (those produced by joining). Thus, $\mathcal{G}_\mathcal{D}$ is re-evaluated for further joins and the algorithm proceeds iteratively (line 29-30) until no additional disjunctions can be done (namely, when $\mathcal{L}_\mathcal{D}$ is empty or the weights ω are higher than γ). At each iteration, the patterns P and Q are removed from $\mathcal{G}_\mathcal{D}$ (line 32).

An explanatory example is illustrated in Figure 1. Consider the background knowledge D_I on the dissimilarity among four atoms and the set I_R containing four infrequent conjunctive patterns as illustrated in Figure 1a and γ equal to 0.7. The first sub-procedure of the algorithm 1 analyzes P_1, P_2, P_3, P_4 and discovers that they differ in only one atom, while the other atoms are in common. Then, it creates the graph $\mathcal{G}_\mathcal{D}$ by collocating P_1, P_2, P_3 in three different vertices and linking them through edges whose weights are taken from the paths ρ in D_I. P_4 is not considered because the vertex *overlaps* has dissimilarity with *internal_ends_at* higher than γ (row (1) in Figure 1b). The second sub-procedure starts by ordering the weights of the edges: the first disjunction is created by joining P_1 and P_3 given that the dissimilarity value is lower than γ and the lowest (row (2) in Figure 1b). Next, the pattern so created and P_2 are checked

Algorithm 1. Extending Relational Pattern with Disjunctions

1: **input:** $I_R, D_I, \gamma, minSup$ **output:** $\mathcal{J}$ // $\mathcal{J}$ set of disjunctive patterns
2: **for all** $(P, Q) \in I_R \times I_R$, $Q \neq P$ **do**
3: **if** $P.length = Q.length$ and $check\_atoms(P, Q)$ **then**
4: $(\alpha, \beta) := atoms\_diff(P, Q)$ //α, β atoms differentiating P,Q
5: **if** $\rho(\alpha, \beta) \neq \oslash$ **then**
6: $\omega := \sum_{e(v_i, v_j, w_{ij}) \; in \; \rho(\alpha,\beta)} w_{ij}$
7: **if** $\omega \leq \gamma$ **then**
8: $addNode(P, \mathcal{G}_\mathcal{D})$; $addNode(Q, \mathcal{G}_\mathcal{D})$; $addEdge(P, Q, \omega, \mathcal{G}_\mathcal{D})$
9: **end if**
10: **end if**
11: **end if**
12: **end for**
13: $\mathcal{L}_\mathcal{D} \leftarrow$ edges of $\mathcal{G}_\mathcal{D}$ // list of edges of $\mathcal{G}_\mathcal{D}$ ordered in ascending mode w.r.t. ω
14: **while** $\mathcal{L}_\mathcal{D} \neq \oslash$ and $\forall e(P, Q, \omega) \in \mathcal{G}_\mathcal{D}$ $\omega \leq \gamma$ **do**
15: $J \leftarrow join(P, Q)$; $J.support := P.support + Q.support - (P \cap Q).support$;
16: **if** $J.support \geq minSup$ **then**
17: $\mathcal{J} := \mathcal{J} \cup \{J\}$
18: **else**
19: **for all** R *such that* $\exists\ e(P, R, \omega_1) \in \mathcal{G}_\mathcal{D}$ and $\exists\ e(Q, R, \omega_2) \in \mathcal{G}_\mathcal{D}$ **do**
20: $addEdge(R, J, (\omega_1 + \omega_2)/2, \mathcal{G}_\mathcal{D})$
21: **end for**
22: **for all** R *such that* $\exists\ e(P, R, \omega_1) \in \mathcal{G}_\mathcal{D}$ and $\nexists\ e(Q, R, \omega_2) \in \mathcal{G}_\mathcal{D}$ **do**
23: $addEdge(R, J, \omega_1, \mathcal{G}_\mathcal{D})$
24: **end for**
25: **for all** R *such that* $\exists\ e(Q, R, \omega_2) \in \mathcal{G}_\mathcal{D}$ and $\nexists\ e(P, R, \omega_1) \in \mathcal{G}_\mathcal{D}$ **do**
26: $addEdge(R, J, \omega_2, \mathcal{G}_\mathcal{D})$
27: **end for**
28: $\mathcal{L}_\mathcal{D} \leftarrow$ edges of $\mathcal{G}_\mathcal{D}$
29: *update* $\mathcal{L}_\mathcal{D}$
30: **end if**
31: $removeNode(P, \mathcal{G}_\mathcal{D})$; $removeNode(Q, \mathcal{G}_\mathcal{D})$
32: **end while**

for joining. Both have the same length and differ in only one atom. Although the first presents a disjunction and the second presents a "simple" atom, dissimilarity is lower than γ and a new disjunctive pattern is created (row (3) in Figure 1b).

4 Experiments

The described approach has been implemented as the upgrading of the system SPADA to discover relational patterns with disjunctions: the system (afterwards *jSPADA*) is now able to mine relational conjunctive patterns and disjunctive patterns as well. The experiments were performed in order to evaluate the viability of jSPADA and to compare it with SPADA from a quantitative and qualitative standpoint[1]. In this section we present the application of both systems in spatial data mining [2] in order to discover statistical regularities in the spatial objects which can be exploited in decision making for transportation planning.

More precisely, frequent relational patterns are mined from a dataset concerning census and digital maps of Stockport, one of the ten districts in Greater Manchester, to investigate the accessibility *to* the Stepping Hill Hospital *from* the actual residence of people living within in the area served by the hospital. To define the accessibility we used the Ordnance Survey data on transport network, namely the layers of roads, railways and bus priority lines. Frequent patterns can relate five areal spatial objects or *districts* (non-target objects) which are close to the Stepping Hill Hospital with one-hundred and fifty-two districts distant from the hospital (target objects) through the transport network lines (non-target objects). D_E contains 1147 ground atoms for 152 target objects.

Property predicates represent discretized numerical census data in TO and describe the households (people) with car, more precisely these are: *no_car()*, *one_car()*, *two_cars()*, *three_more_cars()*. Structural predicates represent binary topological relations between districts and roads, railways or bus lines, and correspond to the twelve feasible relations between a region and a line according to the 9-intersection model [7]. Here, background knowledge D_I has been defined on the structural predicates and the dissimilarity values have been manually determined by applying the Sokal-Michener dissimilarity measure on the matrix representation of the twelve relations[5]. In this sense, the goal of jSPADA is of discovering disjunctive patterns defined among the twelve relations which can express information otherwise discarded by SPADA.

Experiments were performed by tuning the thresholds $minSup$, $nSup$, γ and the results are reported in Figure 2. A comparison between SPADA and jSPADA has been conducted by varying $minSup$, while, for jSPADA, the values of $nSup$ and γ are set to 0.005 and 0.6 respectively. As we see the histogram values in Figure 2a, jSPADA discovers an higher number of patterns than that of SPADA. Indeed, jSPADA returns a set which includes those frequent conjunctive (generated by SPADA) and those disjunctive generated by re-evaluating the infrequent conjunctive ones. Thus, as $minSup$ increases, the range $[nSup; minSup)$ becomes wider and, generally, more disjunctive patterns are extracted while the number of conjunctive frequent patterns decreases.

As expected, also the threshold $nSup$ has influence on the patterns discovered by jSPADA. Indeed, from the figures 2c, 2d ($minSup = 0.025$ and $\gamma = 0.6$) we note that jSPADA is highly sensitive to $nSup$ since the number of disjunctive patterns is reduced of one order of magnitude (from 20 to 0) as $nSup$ is increased by factor of two (from 0.01 to 0.02). By comparing the plots a), c) and d) we note that, by varying $minSup$, have a limited capacity in unearthing infrequent patterns (but potentially interesting) than when varying $nSup$. This confirms the viability of the approach to discover new forms of interesting patterns. Another quantitative analysis can be done with respect to the dissimilarity of the disjunctions (Figure 2b). At high values of γ disjunctions can be created also between atoms whose

[1] Data and results are accessible at http://www.di.uniba.it/~loglisci/jSPADA/

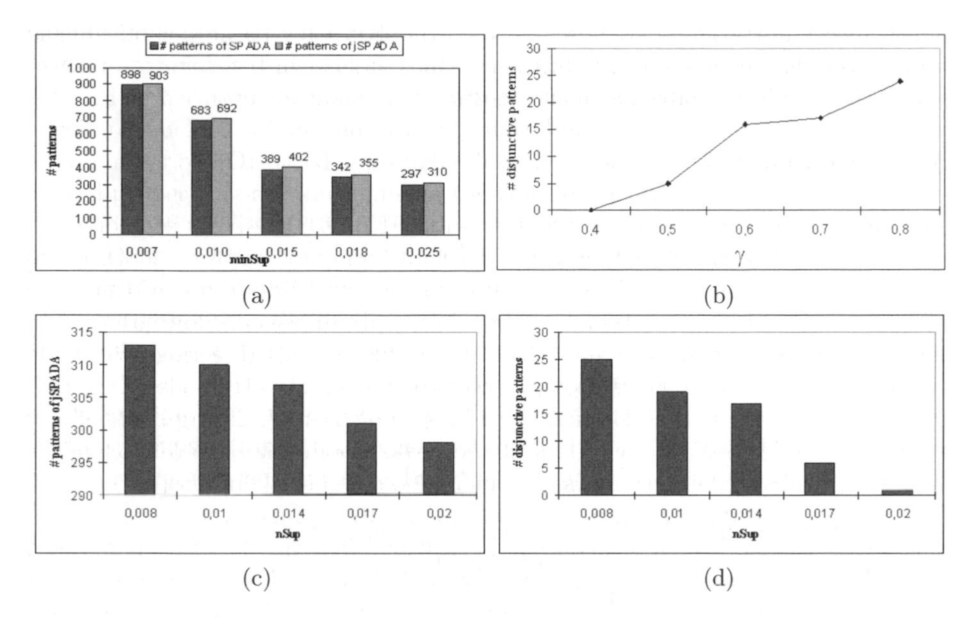

Fig. 2. Number of patterns discovered by tuning *minSup*, *nSup*, γ

similarity is small, so the patterns present disjunctions with several atoms and the final set is larger. On the contrary, lower values of γ permit of identifying disjunctions only between very similar atoms, so the disjunctions present less atoms and the final set is smaller: when γ is set to 0.4 no disjunction is created since the minimum value of similarity between atoms amounted to 0.44.

A comparison between jSPADA and SPADA can also be done from a qualitative viewpoint. jSPADA enables the discovery of patterns which enrich the information conveyed by the patterns of SPADA. For instance, the pattern discovered by SPADA

$P_1 : district(A), comes\_from(A, B), is\_a(B, road), comes\_from(A, C), is\_a(C, road)$ $[support : 12\%]$ is enriched by

P_2 discovered by jSPADA:

$P_2 : district(A), [comes\_from(A, C)|external\_ends\_at(A, C)], is\_a(C, road),$
$comes\_from(A, B), is\_a(B, rail)$ $[support : 16\%]$

which introduces the disjunctions $comes\_from(A, C)|external\_ends\_at(A, C)$ between two structural predicates. P_2 expresses the information that the road named as C can be connected to the district named as A through two possible simultaneous or alternative ways, $comes\_from(A, C)$ (C starts in A and terminates outside A) and $external\_ends\_at(A, C)$ (C starts outside A and terminates inside A). Remarkably the support of P_2 is higher than that of P_1. jSPADA permits also the discovery of completely novel patterns that SPADA neglects. One of these is the following:

$P_3 : district(A), [external\_ends\_at(A, B)|along(A, B)|comes\_from(A, B)],$
$three\_more\_cars(A, [0.033; 0.114])$ $[support : 11.1\%]$

which introduces a property predicate (i.e., the percentage of households owing more three cars included in [0.033;0.114]) and expresses in the disjunction three possible forms of accessibility to the district A by the transport line B.

5 Conclusion

In this paper we present a relational data mining approach that discovers frequent patterns that consider disjunctive forms. We advocate to the relational approach to properly deal with the complexity of real-world data. The approach enables the discovery of disjunctive patterns by re-evaluating the infrequent conjunctive patterns and extending them with disjunctions created through the exploitation of a background knowledge. We applied the algorithm to the domain of the spatial analysis and the experimental results prove the advantages of the proposed algorithm with respect to traditional algorithms of frequent pattern mining. As future work, we intend to apply jSPADA to other domains.

Acknowledgement. This work is partial fulfillment of the research objectives of the projects "DM19410 - The Molecular Biodiversity LABoratory Initiative" and "ATENEO 2008 - Scoperta di conoscenza in domini relazionali".

References

1. Agrawal, R., Mannila, H., Srikant, R., Toivonen, H., Verkamo, A.I.: Fast discovery of association rules. In: Advances in Knowledge Discovery and Data Mining, pp. 307–328. AAAI/MIT Press (1996)
2. Appice, A., Ceci, M., Lanza, A., Lisi, F.A., Malerba, D.: Discovery of spatial association rules in geo-referenced census data: A relational mining approach. Intell. Data Anal. 7(6), 541–566 (2003)
3. Ceri, S., Gottlob, G., Tanca, L.: Logic Programming and Databases. Springer, Heidelberg (1990)
4. Dehaspe, L., Toivonen, H.: Discovery of frequent datalog patterns. Data Min. Knowl. Discov. 3(1), 7–36 (1999)
5. Diday, E., Esposito, F.: An introduction to symbolic data analysis and the sodas software. Intell. Data Anal. 7(6), 583–601 (2003)
6. Dzeroski, S., Lavrac, N.: Relational Data Mining. Springer, Heidelberg (2001)
7. Egenhofer, M.J., Franzosa, R.D.: Point set topological relations. International Journal of Geographical Information Systems 5, 161–174 (1991)
8. Lisi, F.A., Malerba, D.: Inducing multi-level association rules from multiple relations. Machine Learning 55(2), 175–210 (2004)
9. Lloyd, J.W.: Foundations of Logic Programming, 2nd edn. Springer, Heidelberg (1987)
10. Nanavati, A.A., Chitrapura, K.P., Joshi, S., Krishnapuram, R.: Mining generalised disjunctive association rules. In: CIKM, pp. 482–489. ACM Press, New York (2001)
11. Plotkin, G.D.: A note on inductive generalization. Machine Intelligence 5, 153–163 (1970)
12. Roddick, J.F., Fule, P.: Semgram - integrating semantic graphs into association rule mining. In: Proc. of AusDM, vol. 70, pp. 129–137 (2007)
13. Srikant, R., Agrawal, R.: Mining generalized association rules. In: VLDB, pp. 407–419. Morgan Kaufmann, San Francisco (1995)

An Occurrence Based Approach to Mine Emerging Sequences

Kang Deng and Osmar R. Zaïane

Department of Computing Science, University of Alberta
Edmonton, Alberta, T6G 2E8
{kdeng2,zaiane}@ualberta.ca

Abstract. An important purpose of sequence analysis is to find the distinguishing characteristics of sequence classes. Emerging Sequences (ESs), subsequences that are frequent in sequences of one group and less frequent in the sequences of another, can contrast sequences of different classes and thus facilitating sequence classification. Different approaches have been developed to extract ESs, in which various mining criterions are applied. In our work we compare Emerging Sequences fulfilling different constraints. By measuring ESs with their occurrences, introducing gap constraint and keeping the uniqueness of items, our ESs demonstrate desirable discriminative power. Evaluating against two mining algorithms based on support and no gap constraint subsequences, the experiments on two types of datasets show that the ESs fulfilling our selection criterions achieve a satisfactory classification accuracy: an average F-measure of 93.2% is attained when the experiments are performed on 11 datasets.

Keywords: Emerging Sequences, Classification, Occurrence Count.

1 Introduction

Sequence comparison is an significant Data Mining Task [4], where the distinguishing subsequences play an important role in the contrast. Given two sequence groups, Emerging Sequences (ESs) is defined as subsequences that are frequent in sequences of one group and less frequent in the sequences of another, and thus distinguishing or contrasting sequences of different classes [3]. With the discriminative power of emerging sequences, prediction models trained by using ESs perform well and achieve satisfactory classification accuracies on labeling sequence instances.

Different approaches have been developed to extract ESs, in which various mining criterions are applied. For instance, in bioinformatics, researchers align genome by using substrings [7], in which, items have to appear immediately next to each other in the original sequence. However, Lo et al. [8] removed the restriction that related events must occur close together in a sequence, i.e. the distance between two events could be arbitrarily large. Furthermore, most sequential mining algorithms regard support of features (the number of transactions that contain the feature) as the selection standard, while the total occurrence of the feature is also crucial.

T.B. Pedersen, M.K. Mohania, and A M. Tjoa (Eds.): DaWaK 2010, LNCS 6263, pp. 275–284, 2010.

In our research, we compare Emerging Sequences fulfilling different constraints and try to find the important factors for ESs. Besides the frequency distinction, we discover that the following criterions are also significant in the sequence classification:

- Occurrences of subsequences are more informative than supports.
- Items are unique (i.e. not repeated) in a given subsequence.
- Any two adjacent items in the subsequence should be close in the original sequence.

To mine subsequences fulfilling the above conditions, we provide an algorithm as well as a pruning strategy based on a previous work [6]. After ES candidates are extracted, we perform feature selection by F-ratio [10], by which the distinguishing subsequences are selected to represent the original sequence groups. To evaluate the discriminative power of emerging sequences, a SVM classifier [2] is trained by using ESs to classify sequence instances. In this learning framework, higher prediction accuracy indicates better emerging sequences.

For comparison, we perform controlled experiments on two recent and well-known sequence mining algorithms. One algorithm, ConsGapMiner [6] can control the gap between related events, while another algorithm [8] removes the gap constraint and extracts iterative patterns. We choose two types of datasets, one is the UNIX user command sequences [1], the other is a software behavior history [9].

The experiments demonstrate the effectiveness of our emerging sequences: the classifier based on ESs outperforms the other two baseline algorithms. The prediction accuracy measured by the average F-measure is 93.2% when the experiments are performed on 11 datasets. On the datasets CVS-Omission and MySQL [9], our model perfectly labels the sequences with a prediction accuracy of 100%.

In the next section, we introduce some terminology. In Section 3, we describe the sequence mining algorithm and the feature selection strategy. We present the prediction performance of our proposed approach in Section 4. Finally, Section 5 presents our conclusions.

2 Preliminaries

Let $I = \{i_1, i_2, \ldots, i_k\}$ be a set of all items, or the alphabet, a sequence is an ordered list of items from I. Given a sequence $S = \langle s_1, s_2, \ldots, s_m \rangle$ and a sequence $S' = \langle s'_1, s'_2, \ldots, s'_n \rangle$, we say that S' is a subsequence of S or S contains S', denoted as $S' \sqsubseteq S$, if there exist integers $1 \leq j_1 < j_2 < \ldots < j_n \leq m$ such that $s'_1 = s_{j_1}$, $s'_2 = s_{j_2}$, ..., $s'_n = s_{j_n}$.

Definition 1 (Subsequence Occurrence). *Given a sequence $S = \langle s_1, \ldots, s_n \rangle$ and a subsequence $S' = \langle s'_1, s'_2, \ldots, s'_m \rangle$ of S, an occurrence of S' is a sequence of indices $\{i_1, i_2, \ldots, i_m\}$, whose items represent the positions of elements in S.*

For instance, if sequence $S = \langle B, C, B, C, A, C \rangle$, and its subsequence $S' = \langle B, C \rangle$. There are 5 occurrences of S' in S: $\{1, 2\}$, $\{1, 4\}$, $\{1, 6\}$, $\{3, 4\}$ and $\{3, 6\}$.

Definition 2 (Gap Constraint). *The gap constraint is specified by a positive integer g. In a subsequence occurrence $o_s = \{i_1, i_2, i_3, \ldots, i_m\}$, the difference of any two adjacent indices is $i_{k+1} - i_k$. If $i_{k+1} - i_k \leq g + 1$, we say the occurrence o_s fulfills the g-gap constraint.*

For example, if $g = 1$, the occurrences of S' $\{1, 2\}$ and $\{3, 4\}$ fulfill the 1-gap constraint (also 0-gap) but $\{1, 4\}$, $\{1, 6\}$ and $\{3, 6\}$ do not.

Definition 3 (Support and Occurrence Count). *Given a sequence dataset $\mathcal{D}_c$, where c is a class label, $\mathcal{D}_c$ consists of a set of sequences. The support of a subsequence α is the number of sequences in $\mathcal{D}_c$ that contain α, while the occurrence count is the number of non-overlapping occurrences of α in $\mathcal{D}_c$.*

For example, in Table 1, if the gap constraint is 1, the support of the sequence $\alpha = \langle a, b \rangle$ in $\mathcal{D}_{pos}$ is 3, meaning all sequences contain α while fulfilling 1-gap constraint. The occurrence count of α is 4, because α appears twice in Sequence 1. One thing we need to notice is that the total occurrences of α fulfilling 1-gap constraint in Sequence 1 is 5. However, some of them are overlapped, so the non-overlapping count is 2.

In this paper, related support and count, denoted as $support(\alpha, \mathcal{D}_c)$ and $count(\alpha, \mathcal{D}_c)$ respectively, are used to measure the frequency of subsequences. As for the example above, $support(\alpha, \mathcal{D}_c) = \frac{3}{3}$ and $count(\alpha, \mathcal{D}_c) = \frac{4}{3}$.

Table 1. A sequence dataset example

sequence ID	sequences	labels
1	*aabbcab*	*pos*
2	*cadb*	*pos*
3	*bcab*	*pos*
4	*acabd*	*neg*
5	*bda*	*neg*

The notion of Emerging Sequences (ESs) was introduced by Zaïane et al. [11], here we generalize this notion and define:

Definition 4 (Emerging Sequences). *Given two contrasting sequence classes, Emerging Sequences (ESs) are subsequences that are frequent in sequences of one group and less frequent in the sequences of another, and thus distinguishing or contrasting sequences of different classes.*

3 Sequence Mining and Feature Selection

To distinguish one group of sequence data from another, representative subsequences must be extracted. In this section, we explain how we first extract the ES candidates; then implement a dynamic feature selection to mine the most discriminative subsequences.

3.1 Mining Criterion

To mine the representative subsequences, one fundamental question is: "what kind of sequences should we choose?" An essential selection criterion is that features should be discriminative. Let $\mathcal{D}_{pos}$ and $\mathcal{D}_{neg}$ be two classes of sequences; the occurrence counts of a ES candidate α in both classes, denoted as $count(\alpha, \mathcal{D}_{pos})$ and $count(\alpha, \mathcal{D}_{neg})$, need to meet the following conditions:

$$count(\alpha, \mathcal{D}_{pos}) > \theta \quad (1)$$

$$count(\alpha, \mathcal{D}_{neg}) \leq \theta \quad (2)$$

where θ is the minimum count threshold.

Instead of supports, we use the occurrence counts of subsequences to measure their discriminative power, because repetitive features within a sequence is important. For instance, a UNIX user may repeatedly type the same command pattern within one session.

Another mining principle we apply is that items are unique in one sequence pattern. Since the multiple occurrences of patterns in each original sequence are counted, it is not necessary to consider subsequences with repetitive items.

The last standard is that items have to appear closely with each other in the original sequence, as items far apart are less relevant in the decision making. An example is the relationships between words in a long sentence. A verb probably serves as the predicate of a subject if they are close to each other. Therefore, gap constraints need to be considered in subsequence mining.

3.2 ES Candidates Extraction

To control the gap constraint when mining emerging sequences, our mining model is based on a previous work, ConsGapMiner [6]. In their approach, however, they choose support as the selection criterion, and items are not unique in the sequence pattern.

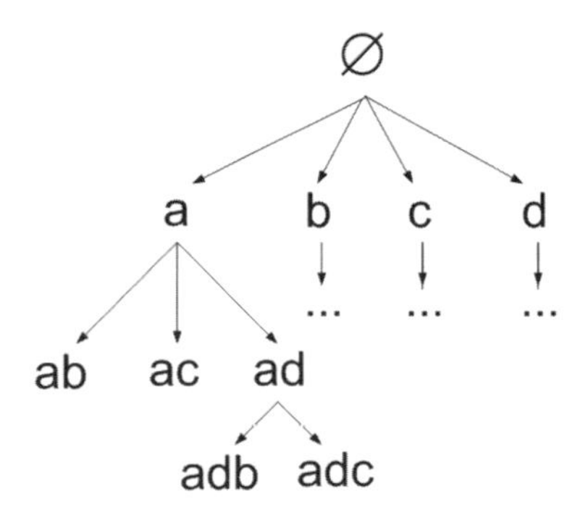

Fig. 1. Candidate Generation Tree

We enumerate ES candidates by a Depth First Search. Given an ES candidate, an item from the vocabulary is appended at the ending of the current subsequence, so a new candidate is generated. Figure 1 shows the candidate generation tree. Given a vocabulary $I = \{a, b, c, d\}$, the root of the tree is an empty set. For a subsequence $\langle a \rangle$, if it fulfills the discriminative conditions 1, 2 and the gap constraint, three new candidates are generated by appending b, c and d respectively. Sequence pattern $\langle a, a \rangle$ is not taken into consideration, because items ought to be unique in our emerging sequences. This pruning strategy can greatly reduce the searching space and improve the scalability of the tree generation algorithm.

3.3 Support Calculation

Given an ES candidate, the next problem is to validate if the candidate fulfills the discriminative conditions and the gap constraint. In [6], a bitset operation-based algorithm is proposed because the bit is the basic operation unit in computers.

A bitset is a sequence of bits which each takes the value 0 or 1, indicating the subsequence occurrences in the original sequence. The bitset of the length-1 subsequence is easy to generate, i.e. 1 indicates the appearance of the item, while other digits are 0s. Given the sequence $\langle aabbcab \rangle$, the bitset of the subsequence $\langle a \rangle$ is simply 1100010. And the occurrence count of a length-1 subsequence is also straightforward: it is the number of 1s in the bitset and the gap constraint is irrelevant. Therefore, the occurrence count of $\langle a \rangle$ is 3.

For a subsequence whose length is larger than 1, the calculation of its occurrence count is more complicated, because the gap constraint is taken into account. It has three steps:

1. Perform right shift operation on the bitset of its parent to generate the mask bitset.
2. Attain the subsequence bitset by **AND** operation.
3. Calculate the occurrence count based on the bitset.

The first step is to generate the mask bitset by the parent of the target subsequence. From the last example, given a subsequence $\alpha = \langle a, b \rangle$, the bitset of its parent $\alpha_p = \langle a \rangle$ is known. To calculate the mask bitset when the gap constraint is g, we right shift the bitset of α_p for $g + 1$ times, then perform **OR** operation on the results. As the bitset of α_p is 1100010, the process is as follows:

$$\begin{array}{r} 1100010 >> 0110001 \\ 0110001 >> 0011000 \\ \hline \textbf{OR} \quad >> 0111001 \end{array}$$

So the mask bitset is 0111001.

Based on the mask bitset and the bitset of the last item of α, the bitset of α is generated by **AND**ing them. Taking the last example, the mask bitset is 0111001 and the bitset of $\langle b \rangle$ is 0011001, by **AND**ing them:

$$\begin{array}{r} 0111001 \\ 0011001 \\ \hline \textbf{AND}\ 0011001 \end{array}$$

the bitset of $\alpha = \langle a, b\rangle$ is 0011001.

Finally, it is the calculation of the occurrence count. Since there is more than one item in the subsequence, we cannot simply count the number of 1s in the bitset. Given a bitset and a gap constraint g, for each 1 in the bitset, the following g digit(s) must be set to 0. Then the occurrence count is the number of 1s in the bitset. For the last example, the bitset 0011001 is converted to 0010001, so the occurrence count of the pattern $\alpha = \langle a, b\rangle$ in the sequence $\langle aabbcab\rangle$ is 2.

3.4 Feature Selection

By combining the candidate generation tree and the bitset operation, numerous ES candidates are extracted. In this subsection, we refine the result and select the most discriminative subsequences as ESs.

Given several sequence groups $\{\mathcal{G}_1, \mathcal{G}_2, \ldots, \mathcal{G}_m\}$ and a set of subsequences $\{s_1, s_2, \ldots, s_k\}$, the objective is to find the most discriminative subsequences. For an ideal emerging sequence, its occurrence counts in several groups should differ greatly, i.e. the variance between groups should be much larger than that within each group. To solve this classic analysis of variance (ANOVA) problem, we apply F-ratio to measure the discriminative power:

$$F - ratio = \frac{MS_{between}}{MS_{within}} \tag{3}$$

where $MS_{between}$ is the mean square (variance estimate) explained by the different groups, and MS_{within} is mean square (variance estimate) that is due to chance (unexplained). Given the number of groups m and the total number of sequences N, $MS_{between}$ and MS_{within} are defined as:

$$MS_{between} = \frac{\sum_i n_i(\overline{c_i} - \overline{c})^2}{m - 1} \tag{4}$$

$$MS_{within} = \frac{\sum_{ij}(c_{ij} - \overline{c_i})^2}{N - m} \tag{5}$$

where n_i is the number of sequences in group i, c_{ij} is the occurrence count in the jth sequence of the ith group, $\overline{c_i}$ is the mean of the occurrence counts in Group i, and $\overline{c}$ is the mean of those for all samples. As m and N are fixed for all subsequences, the F-ratio can be simplified as:

$$F - ratio = \frac{\sum_i n_i(\overline{c_i} - \overline{c})^2}{\sum_{ij}(c_{ij} - \overline{c_i})^2} \tag{6}$$

From Equation 6, we can see that, for an ES candidate, when the variance of the occurrence count between groups is large and that within groups is small, its F-ratio become large. Based on the F-ratio, we rank the ES candidates, and the highly-ranked ones are more discriminative.

To avoid numerous ESs, we then perform a dynamic feature selection strategy [5], i.e. only the top-m subsequences, based on F-ratio, are kept. It guarantees that each sequence can be represented by at least m ESs (the high-ranked ones) and the database does not become too large due to the possible sheer number of candidate subsequences.

4 Experimental Results

In the last section, emerging sequences are selected by our occurrence count based mining framework. To verify the discriminative power of ESs, we then perform the controlled experiments on ESs.

4.1 Evaluation Methodology

To perform the experiments, the sequence datasets are transformed to transactional datasets in order to be in a suitable form for learning algorithms, i.e. each sequence is represented by a set of attribute-value pairs, where the attribute represents an emerging sequence, and the value is its occurrence count in this sequence. Then, a classifier is trained by using the transactional datasets. In this paper, we choose a well-developed classification package LIBSVM [2] as the prediction model. Finally, we perform a 6-folder cross validation on the classification framework. The average prediction accuracy, represented by f-measures, indicates the performance of the selected features.

For comparison, we chose two other recent and well-known mining algorithms to extract different features from the original datasets:

- Minimal Distinguishing Subsequences (MDSs): this kind of sequences are mined by ConSGapMiner [6]. There are two main differences between MDSs and our ESs: 1. they use support as the selection criterion, while occurrence count is applied in our mining model. 2. Items are unique in ESs but can be repetitive in MDSs.
- Iterative Patterns (IPs) [8]: IPs achieve satisfactory performance on classifying software behaviour sequences. As opposed to our ESs, they remove the restriction that related events must occur close together in a sequence, i.e. the distance between two events could be arbitrarily large.

These two kinds of features are selected and used in the validation framework above. By comparing the prediction accuracies of those features, we can verify the effectiveness of our selection criterions:

- Occurrences of subsequences are more informative than supports.
- Items are unique in the subsequences (i.e. not repeated).
- Any two adjacent items in the subsequence should be close in the original sequence.

4.2 UNIX User Command Dataset

The first type of datasets we use is the UNIX user commands dataset from the UCI Machine Learning Repository [1]. It contains 9 sets of sanitized user data drawn from the command histories of 8 UNIX computer users at Purdue University. This dataset only keeps command names, flags, and shell meta characters, while removing filenames, user names, directory structures etc. For each user, we select 100 sequences. In each experiment, two users' commands are chosen, and the F-measures and standard deviations are presented in Table 2.

Table 2. Classification performances on the UNIX dataset

Datasets	Length	Size	MDSs	IPs	ESs
user 0 and 1	28	176	0.770 ± 0.053	0.776 ± 0.065	$\mathbf{0.806} \pm 0.048$
user 1 and 8	33	268	0.938 ± 0.045	$\mathbf{0.959} \pm 0.024$	0.958 ± 0.035
user 2 and 3	39	231	$\mathbf{0.970} \pm 0.018$	0.948 ± 0.078	0.965 ± 0.032
user 3 and 6	36	231	0.918 ± 0.058	0.913 ± 0.044	$\mathbf{0.929} \pm 0.044$
user 5 and 6	49	278	0.865 ± 0.092	0.865 ± 0.060	$\mathbf{0.903} \pm 0.062$
user 5 and 7	44	284	0.908 ± 0.028	0.905 ± 0.031	$\mathbf{0.920} \pm 0.029$

In Table 2, column Length represents the average sequence length of this user pair, while column Size means the vocabulary size. We observe that the prediction accuracy of our ESs-based approach is comparable or better than the other two features. This demonstrates that our selection criterions are effective with various average sequence length and vocabulary size.

4.3 Software Behaviour Dataset

The second type of datasets is the set of software behaviour sequences. Software behavior is the way a program executes. From the start of the program until its termination, the execution events are recorded. A software behaviour, composed by a sequence of normal individual execution events, could be broken down by their interaction in an undesirable order. Therefore, the objective of the analysis is to distinguish deviant software behaviours from regular ones by sequence mining.

Lo et al. [8] focus on this type of data and proposed Iterative Patterns (IPs) mining algorithm, which achieves satisfactory performance and improves the prediction accuracy greatly. So in this subsection, we focus on the comparison between IPs and ESs. The same datasets as in [8] are chosen to perform the experiment. For more information about the dataset, please refer to [8].

Table 3 presents the comparison between the IPs-based and the ESs-based SVM classifiers. Compared with IPs which was designed specifically for software behaviour datasets, our emerging sequences also achieve satisfactory classification accuracy.

Table 3. Classification performances on the software behaviour dataset

Datasets	Length	Size	IPs	ESs
CVS-Mix	9	16	0.935 ± 0.060	$\mathbf{0.945} \pm 0.058$
CVS-Omission	10	16	1 ± 0	1 ± 0
CVS-Ordering	9	16	0.857 ± 0.031	$\mathbf{0.951} \pm 0.032$
MySQL	24	16	1 ± 0	1 ± 0
X11	4	8	$\mathbf{0.979} \pm 0.015$	0.888 ± 0.024

5 Conclusion

In this paper, we focus on Emerging Sequences (ESs), which are frequent in sequences of one group and less frequent in the sequences of another, and thus distinguishing or contrasting sequences of different classes. After comparing ESs of various characteristics, we find that: 1. the occurrence count can measure the discriminative power of ESs more precisely; 2. the uniqueness of items in a subsequence is important; 3. the gap constraint is relevant in the decision making. A mining model is proposed to extract ESs fulfilling our selection criterions.

The experiments demonstrate the effectiveness of our emerging sequences: an average F-measure of 93.2% is achieved when the experiments are performed on 11 datasets. In the datasets CVS-Omission and MySQL [9], our model perfectly labels the sequences with a prediction accuracy of 100%.

However, since our candidate generation algorithm is based on the Depth First Search Tree, the scalability of our current approach is not desirable. As a future work, we are looking into the possibility of a more efficient mining algorithm or pruning strategies, while preserving and potentially improving the prediction accuracy of the classification model.

Acknowledgement

The execution program and datasets for the iterative patterns were provided by the original author, Dr. Lo et al. We would like to acknowledge their help in this regard.

References

1. Asuncion, A., Newman, D.: UCI machine learning repository (2007)
2. Chang, C.-C., Lin, C.-J.: LIBSVM: a library for support vector machines (2001), `http://www.csie.ntu.edu.tw/~cjlin/libsvm`
3. Deng, K., Zaïane, O.R.: Contrasting sequence groups by emerging sequences. Discovery Science, 377–384 (2009)
4. Han, J., Kamber, M.: Data Mining, Concepts and Techniques. Morgan Kaufmann, San Francisco (2001)
5. Jazayeri, S.V., Zaïane, O.R.: Plant protein localization using discriminative and frequent partition-based subsequences. In: ICDM Workshops, pp. 228–237 (2008)

6. Ji, X., Bailey, J., Dong, G.: Mining minimal distinguishing subsequence patterns with gap constraints. Knowl. Inf. Syst. 11(3), 259–286 (2007)
7. Kurtz, S.: Reputer: fast computation of maximal repeats in complete genomes (1995)
8. Lo, D., Cheng, H., Han, J., Khoo, S.-C.: Classification of software behaviors for failure detection: A discriminative pattern mining approach. In: KDD (2009)
9. Lo, D., Cheng, H., Han, J., Khoo, S.-C.: Technical report, School of Information Systems, Singapore Management University (2009), `http://www.mysmu.edu/faculty/davidlo/kdd09.htm`
10. Lomax, R.G., Hahs-Vaughn, D.L., Lomax, R.G.: Statistical Concepts: A Second Course, 3rd edn. Routledge, New York (2007)
11. Zaïane, O.R., Yacef, K., Kay, J.: Finding top-n emerging sequences to contrast sequence sets. Technical Report TR07-03, Department of Computing Science, University of Alberta (February 2007)

Mining Closed Itemsets in Data Stream Using Formal Concept Analysis

Anamika Gupta, Vasudha Bhatnagar, and Naveen Kumar

Department of Computer Science, University of Delhi, India
{agupta,vbhatnagar,nk}@cs.du.ac.in

Abstract. Mining of frequent closed itemsets has been shown to be more efficient than mining frequent itemsets for generating non-redundant association rules. The task is challenging in data stream environment because of the unbounded nature and no-second-look characteristics.

In this paper, we propose an algorithm, CLICI, for mining all recent closed itemsets in landmark window model of online data stream. The algorithm consists of an online component, which processes the transactions arriving in the stream without candidate generation and updates the synopsis appropriately. The offline component is invoked on demand to mine all frequent closed itemsets. User can explore and experiment by specifying the support threshold dynamically.

The synopsis, CILattice, stores all recent closed itemsets in the stream. It is based on Concept Lattice - a core structure of Formal Concept Analysis (FCA). Closed itemsets stored in the form of lattice facilitate generation of non-redundant association rules and is the main motivation behind using lattice based synopsis.

Experimental evaluation using synthetic and real life datasets demonstrates the scalablility of the algorithm.

Keywords: Closed Itemsets, Data Stream, Landmark Window Model Formal Concept Analysis.

1 Introduction

A data stream is an unbounded sequence of data often coming at a high speed The problem of mining data stream is more challenging than mining static datasets in view of following aspects[2,3,10]. First, stream is a continuous flow of data and hence data must be processed at a rate faster than its arrival. Second, each element of stream must be examined only once. Third, memory usage should be bounded even though the stream is continuously growing. Further, the results should be instantly available in real time and error in the results, if any, should be bounded. Since stream evolves with time, capturing recent information is another vital issue in data stream mining.

Mining of frequent itemsets (FI) for association rules has been studied extensively in both static datasets [6] and data stream [3]. Researchers have recently explored the idea of mining frequent closed itemsets (FCI) instead of frequent itemsets for discovering non-redundant association rules [11,12,16]. Set of FCI

T.B. Pedersen, M.K. Mohania, and A M. Tjoa (Eds.): DaWaK 2010, LNCS 6263, pp. 285–296, 2010.

has been shown to be a complete, loss-less and reduced representation of set of FI [11]. Mining FCI instead of FI saves computation efforts and memory usage. Several algorithms like Closet, CHARM, Closet+, CHARM-L, FP-Close, DCI-Closed [6] have been proposed to generate FCIs in static datasets.

Mining frequent closed itemsets in data stream throws newer challenges. Since every transaction in stream is a closed itemset [13], it leads to addition of at least one entity in the synopsis. The computational challenge is an immediate consequence of the large size of the synopsis and may result into loss of data because of higher per-transaction-processing time.

Traditionally, algorithms for mining frequent closed itemsets (FCI) in data stream use either sliding window model [4,7,9,14] or landmark window model [10] to meet the dual objective of 'maintaining recency' and 'constrained memory usage'. Sliding window model constrains the size of the synopsis by fixing the number of transactions in the window, leading to nearly constant per transaction processing time. However, the model falls short of monitoring the continuous variation of data stream.

Landmark window model considers entire data starting from a particular landmark to the curent time [2]. Although this model enables monitoring of gradual changes in the data stream, capturing recent data and keeping the size of synopsis under control is a challenging task.

State of the Art. Moment [14], CFI-Stream [7], NewMoment [9] and GC-Tree [4] are some of the known algorithms for mining FCI in sliding window model of data stream. Moment, CFI-Stream, and GC-Tree algorithms store current transactions in a memory based window. On arrival of a new transaction, these algorithms make multiple scans of the window for finding support of discovered itemsets. NewMoment stores bitwise representation of all the transactions along with 1-itemsets to find support of itemsets. Moment and CFI-Stream generate all candidates while processing new transaction, leading to increased per-transaction-processing time.

To the best of our knowledge, FP-CDS [10] algorithm has been recently proposed for discovery of frequent closed itemsets in landmark window model. The algorithm works in a batch mode, dividing the landmark window into several 'basic' windows, using them as updating units. However, in this process it ignores the recency of discovered itemsets. Recency in landmark window model has been handled by estDec [2], an algorithm for mining frequent itemsets. The effect of older transactions is diminished by decaying their old occurences and later pruning the decayed ones periodically, thus maintaining only the recent data.

Formal Concept Analysis Technique. Recently, data mining researchers have exploited Formal Concept Analysis(FCA), a field of applied mathematics [5], for discovery of closed itemsets. Zaki [16], Pasquier et al. [11] and Stumme et al. [12] proved that intent of concept (section 2) represents closed itemset and all concept generating algorithms generate closed itemsets as well. FCA based algorithms generate all the concepts (closed itemsets) and have been shown to work well on small datasets [8,15].

Substantial reduction in the set of generated rules is the main motivation for exploring concept lattice in association rule mining. FCA based mining algorithms store the closed itemsets in the form of a concept lattice. Concept lattice facilitates the efficient generation of non-redundant association rules [16]. One look characteristic of the updation algorithms for concept lattice is another advantage of using lattice in discovery of closed itemsets.

High memory and computational requirement of FCA based algorithms prohibits their use in data stream environment. Each node of the concept lattice stores the extent (set of transactions) alongwith the intent (closed itemset) which contribute to high memory usage. Further, processing of a new transaction involves computation of intersection of its extent with extent of different nodes in the lattice making it computationally expensive.

Further, decaying and pruning away older data to capture recent changes in the dataset has not been addressed in these algorithms and hence makes them unsuitable for data stream.

Our Contribution. In this paper, we present an algorithm for mining recent closed itemsets in landmark window model of on-line data stream. The proposed algorithm has an online component that processes the transactions without candidate generation and stores the results in a synopsis data structure. Offline component is invoked on demand and mines the closed itemsets from the synopsis based on dynamically specified support threshold. The salient features of the proposed algorithm CLICI (Concept Lattice based Incremental Closed Itemset) are listed below:

1. The algorithm mines *all* recent closed itemsets in landmark window model of online data stream. It fades out the obsolete information of old transactions using a decay function and later prunes the decayed information, thereby ensuring the recency of closed itemsets and in turn keeping size of the synopsis under control.
2. The algorithm processes the transactions without any candidate generation.
3. The algorithm is based on sound mathematical foundation of Formal Concept Analysis and stores the closed itemsets in a lattice based synopsis, CILattice. Following advantages accrue due to use of lattice based synopsis.
   (a) Generation of non-redundant association rules is naturally facilitated.
   (b) Data is scanned only once for maintaining the synopsis.
   (c) Only closed itemsets are stored in CILattice, unlike concept lattice where set of transactions are stored additionally.
4. CILattice maintains *all* closed itemsets, thereby providing a valuable facility for experimentation with varying support threshold without any overhead.
5. Experimental evaluation using synthetic and real life datasets demonstrates the scalablility of the algorithm.

The remainder of this paper is organized as follows. Section 2 defines closed itemsets and presents the background of Formal Concept Analysis (FCA). Section 3 describes the algorithm in detail for mining closed itemsets in landmark window model. Section 4 presents the experimental results and section 5 concludes the paper.

2 Background

Given a database D of N transactions and a set I of n items in D. A transaction $t \in D$ is a set of items and is associated with a unique identifier TID. A set of one or more items belonging to I is termed as an itemset. A $k - itemset$ is an itemset of cardinality k. An itemset X is **frequent closed itemset (FCI)** if it is frequent and there exists no proper superset Y of X such that support of Y is same as that of X.

Formal Concept Analysis. Following Ganter and Wille [5], we give some definitions and a theorem used in the paper.

Definition 1. *A formal context $K = (G, M, I)$ consists of two sets G (objects) and M (attributes) and a relation I between G and M. For a set $X \subseteq G$ of objects, the set of all attributes common to the objects in X is defined as $X' = \{m \in M | gIm \text{ for all } g \in X\}$. Correspondingly, for a set Y of attributes, the set of objects common to the attributes in Y is defined as $Y' = \{g \in G | gIm \text{ for all } m \in Y\}$. A formal concept of the context (G, M, I) is a pair (X, Y) with $X \subseteq G$, $Y \subseteq M$, $X' = Y$ and $Y' = X$. X is called the extent and Y is the intent of the concept (X, Y).*

Definition 2. *If (X_1, Y_1) and (X_2, Y_2) are concepts of a context, (X_1, Y_1) is called a subconcept of (X_2, Y_2), provided that $X_1 \subseteq X_2$ (which is equivalent to $Y_2 \subseteq Y_1$). In this case, (X_2, Y_2) is a superconcept of (X_1, Y_1) and we write $(X_1, Y_1) \leq (X_2, Y_2)$. The relation $\leq$ is called the hierarchical order of the concepts. The set of all concepts of (G, M, I) ordered in this way is called the concept lattice of the context (G, M, I).*

Definition 3. *For an object $g \in G$, $g' = \{m \in M | gIm\}$ is object intent of g. Correspondingly, $m' = \{g \in G | gIm\}$ is the attribute extent of the m.*

Theorem 1. *Each Concept of a context (G, M, I) has the form (X'', X') for some subset $X \subseteq G$ and the form (Y', Y'') for some subset $Y \subseteq M$. Conversely all such pairs are Concepts. This implies every extent is the intersection of attribute extents and every intent is the intersection of object intents.*

For proof of the theorem, please refer to [5].

3 CLICI Algorithm

CLICI algorithm mines all recent closed itemsets without candidate generation in landmark window model of data stream. Use of landmark window model facilitates continuous monitoring of changes in the stream. Maintaining all recent closed itemsets allow user to experiment with varying support thresholds. The algorithm has an online component that processes the transactions in stream and stores the results in a synopsis called CILattice. The algorithm inserts the new transaction in the lattice, if not already there. Offline component is invoked on demand and mines the closed itemsets from the synopsis based on dynamically specified support threshold.

3.1 Terminology and Data Structure

Let D_N denote the current data stream with N transactions seen so far and I denote the set of n items in D_N. The incoming transactions are inserted in a data structure, CILattice, which has two components i) a lattice $\mathcal{L}$ ii) header table *Itable*. $\mathcal{L}$ is a complete lattice, with topnode $\top$ and bottom node $\bot$. A node X of $\mathcal{L}$ represents a closed itemset I_X and stores its frequency f_X along with links to its parents and children nodes. *Itable* is an array storing items and pointers to the nodes corresponding to first occurrence of that item in $\mathcal{L}$, which aids efficient traversal during search and insert procedure. The definitions and observations used later in the algorithm are given below:

Let $A(X)$, $D(X)$, $P(X)$ and $C(X)$ denote the set of ancestors, descendants, parents and children respectively of the node X in $\mathcal{L}$.

Definition 4. *A node X is ancestor of node Y iff $I_X \subset I_Y (I_X \neq I_Y)$.*

Definition 5. *A node X is descendant of node Y iff $I_X \supset I_Y (I_X \neq I_Y)$.*

Definition 6. *A node X is parent of a node Y if $X \in A(Y)$ and $\nexists$ any $Z \in A(Y) : X \in A(Z)$ and $Z \neq X$.*

Definition 7. *A node X is a child of node Y if $X \in D(Y)$ and $\nexists$ any $Z \in D(Y) : X \in D(Z)$ and $Z \neq X$.*

It is obvious from the above definitions that ancestor nodes are generalizations of the descendant nodes. Further, parents of a node are its immediate ancestors (generalizations) and children of a node are its immediate descendants (specializations).

Observation 1. *If node X has a child node Y in $\mathcal{L}$ then closed itemset of Y is minimal superset of all descendants of X. Similarly closed itemset of X is maximal subset of all ancestors of Y.*

Definition 8. *First Node F_i of an item i is a node X in the $\mathcal{L}$ where $i \in I_X$ and $\nexists$ a node Y such that $i \in I_Y$ and $Y \in A(X)$.*

Naturally there is exactly one F_i for each item $i \in I$ and Itable stores the pair (i, F_i). Fig. 1 shows a toy database and the corresponding data structure $< \mathcal{L}, ITable >$.

Example 1. Fig. 1 shows a toy database and the corresponding $< \mathcal{L}, Itable >$.

3.2 Capturing Recent Closed Itemsets

CLICI algorithm maintains all closed itemsets starting from the set landmark point. The effect of older transactions is diminished by decaying their old occurences and later pruning the decayed ones, thus maintaining only the recent closed itemsets. This feature controls the size of lattice also.

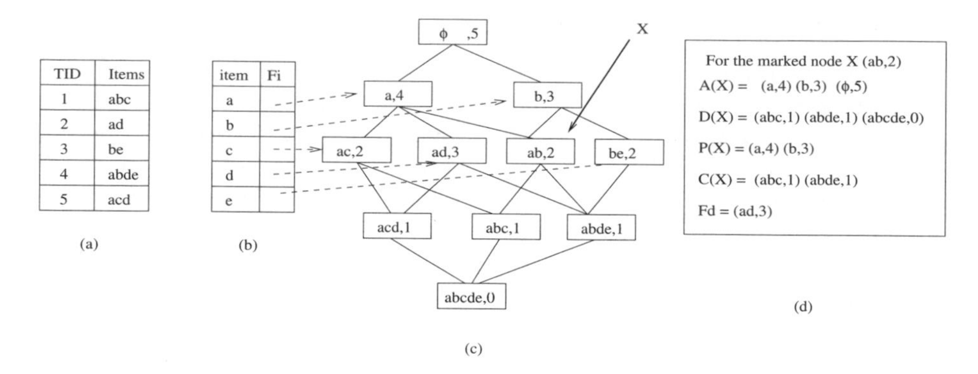

Fig. 1. (a) Toy Database (b) Itable (c) $\mathcal{L}$ (d) Set of ancestors, descendants, parents and children of node X

Chang and Lee [2] define a decay factor d which is associated with each transaction in the stream and is defined as $d = b^{-1/h}$ $(b > 1, h \geq 1, b^{-1} \leq d < 1)$, where decay-base b determines the amount of weight reduction per a decay-unit and decay-base-life h is defined by the number of decay-units that makes the current weight be b^{-1}. When a new transaction t_k arrives at time k, number of transactions in the current stream $|D|_k$ is updated as $|D|_k = |D|_{k-1} * d + 1$. Decayed frequency of all nodes corresponding to itemsets of t_k is updated as $(f_k = f * d^{k-(MRtid_{pre})} + 1)$, where f denote the decayed frequency of node and MRtid is the transaction identifier of most recent transaction that contain t_k. $MRtid_k$ is set to k. Decayed support count of nodes is calculated as $f/(|D|_k)$.

Periodically, the lattice is traversed and all nodes having support count less than threshold for pruning (S_{prn}) are removed from $\mathcal{L}$. Note here that S_{prn}, which is also user specified parameter, is distinctly different from support threshold. While support threshold deals with the frequency of the itemset, S_{prn} takes into account the effect of age of the itemset. Higher values of S_{prn} leads to considerable reduction in the size of the synopsis. Since, data characteristics of the stream change with time, guessing the correct value of S_{prn} is a difficult task. An alternative approach, though somewhat crude, is to fix the size of synopsis based on total available memory and remove decayed nodes from the lattice so as to keep the recent and repetitively occurring itemsets intact.

3.3 Processing of Transaction

Each incoming transaction t is a closed itemset [13]. In case transaction t exist already as a node in $\mathcal{L}$, then the support of the relevant nodes is updated as mentioned in section 3.2. Otherwise processing of t results into insertion of one or more nodes in $\mathcal{L}$.

If t exists as a node in $\mathcal{L}$ then we write $t \in \mathcal{L}$ for simplicity purposes.

We present below the procedures for search, insert and delete a node.

Search Procedure. This procedure checks whether incoming transaction t exists as a node in $\mathcal{L}$ or not. The bruteforce approach of searching $\mathcal{L}$ for node containing t would be to start search either from $\top$ or $\bot$. This approach is computationally expensive as the lattice has a tendency to grow in size as the stream progresses. Proposition 1 permits an optimization by searching *First Node* corresponding to any one of the items in the transaction and its descendents.

Proposition 1. *Let t be the transaction to be searched in $\mathcal{L}$. It is necessary and sufficient to search among F_i and its descendants, for an arbitrarily chosen i, $i \in t$.*

Proof. If $t = I_{F_i}$ for an arbitrarily chosen i, $i \in t$, then $t \in \mathcal{L}$.

By definition 5 and 8, $i \in I_{F_i}$ $\forall i \in t$. $D(F_i)$ is a set containing all supersets of I_{F_i}. If $t \in \mathcal{L}$ then t is a superset of I_{F_i} and hence t exists as a descendant of F_i. Thus t is necessarily a descendant of each of the items contained in t.

By definition 5 and 8, no other node except nodes belonging to $D(F_i)$ contain i. Since $i \in t$ so if $t \in \mathcal{L}$ then t is among the descendants of F_i. Hence it is sufficient to search among descendants of F_i.

Thus, if t contains an item i for which F_i does not exist, then $t \notin \mathcal{L}$. Otherwise it needs to be searched among F_i and its descendants corresponding to any one of the arbitrarily chosen item i in t.

Insert Procedure. If transaction t does not already exists as a node in $\mathcal{L}$ then it is added as a new node N in $\mathcal{L}$, with $I_N = t$. Subsequently N is linked to its child nodes as well as parents node in $\mathcal{L}$. Insert Procedure determines the minimal superset of I_N to find the child nodes and the maximal subsets of I_N to find the parents of I_N (Observation 1). Proposition 2 states that Node N can have only one child in $\mathcal{L}$.

Proposition 2. *If N is the node corresponding to transaction t that has been just added to lattice $\mathcal{L}$, then $|C(N)| = 1$.*

Proof. We prove this assertion by contradiction. Let, if possible, $|C(N)| = m > 1$. Let $C(N) = C_1, C_2, \ldots, C_m$ be the set of children of N in $\mathcal{L}$. By definition, $I_N = I_{C_1} \cap I_{C_2} \ldots \cap I_{C_m}$. According to theorem 1, intersection of extents of nodes in lattice is always an extent of a node in lattice i.e. there must exist a node X in $\mathcal{L}$ such that $I_X = I_{C_1} \cap I_{C_2} \ldots \cap I_{C_m}$. But then X and N are the same nodes. This contradicts the fact that N is a new node. Hence the assumption that $|C(N)| = m > 1$ is not correct.

Further m cannot be zero because bottom node $\bot$ of $\mathcal{L}$ contains all the items and is always a superset of transaction t. Hence $m = 1$. ∎

The insert procedure traverses $\mathcal{L}$ using *ITable* for finding the child node and parent nodes of N. The use of Proposition 2 reduces the effort involved in searching children nodes of N as it can have only one child in $\mathcal{L}$. However, potential number of ancestors can be $2^{|I_N|}$. The exact number of immediate ancestors i.e. parents is unpredictable and may involve rigorous searching in the lattice $\mathcal{L}$. This process of searching immediate ancestors i.e. parents of N is speeded up using F_i of

Itable, $i \in t$. F_i provides the entry point for search in $\mathcal{L}$. However the involved task depends on the relationship of F_i and N. Following three cases arise:

Case 1: F_i is ancestor of N i.e. $I_{F_i} \subset I_N$. Child of N is among the descendants of F_i. Parent(s) of N is either the node F_i itself or some of F_i's descendant(s). If closed itemset of any of the child node X of F_i is superset of I_N then F_i is the parent of N and X is the child of N. Otherwise we search amongst the descendants of F_i till we find a node X such that I_X is superset of I_N and closed itemset of one of the parent Y of X is subset of I_N. In that case, X is the child of N and Y is the parent of N.

Case 2: F_i is descendant of N i.e. $I_{F_i} \supset I_N$. F_i is the child of N and *First Node* F_i corresponding to $i \in I_N$ is set to N. We search among the parents of F_i for finding parents of N. If closed itemset of those parents is subset of I_N, then those parents are parents of N also. If no such parent exist then top node $\top$ of $\mathcal{L}$ is the parent of N.

Case 3: F_i is neither ancestor nor descendent of N i.e. $I_{F_i} \not\subset I_N$, $I_{F_i} \not\supset I_N$ and $I_{F_i} \cap I_N \neq \phi$. In this case, common parent of F_i and N with closed itemset as $I_{F_i} \cap I_N$ may exist in $\mathcal{L}$ or it may not exist.

Case 3.1: $I_{F_i} \cap I_N$ exists as closed itemset of a node X in $\mathcal{L}$. If such a node X exists in $\mathcal{L}$ then X is the parent of N. Child of N is among the descendants of F_i. We check all the descendants of F_i till we get a node whose closed itemset is minimal superset of closed itemset of N, that node becomes the child of N.

Case 3.2: $I_{F_i} \cap I_N$ does not exist as closed itemset of a node in $\mathcal{L}$. A new node X is created with $I_X = I_{F_i} \cap I_N$ and added to $\mathcal{L}$. X becomes the parent of N and child of N is among the descendants of F_i. We check all the descendants of F_i till we get a node whose closed itemset is minimal superset of closed itemset of N, that node becomes the child of N. Next step is to find the child and parents of X. Parents of X are among ancestors of F_i. We check all the ancestors of F_i. If we get a node Y such that closed itemset of Y is superset of X and closed itemset of one of the parent Z of Y is subset of X, then Z becomes the parent of X. If we find an ancestor Y of F_i such that Y is neither ancestor nor descendent of X, then a new node U is created with $I_U = I_Y \cap I_X$. Then we repeat the process to find child and parents of node U.

The algorithm for inserting a new transaction is given in Algorithm 27.

Delete procedure. $\mathcal{L}$ is traversed from bottom and all nodes having decayed support less than S_{prn} are removed by invoking delete procedure. Bottom node is never removed from the $\mathcal{L}$ so as to maintain the whole structure of $\mathcal{L}$. We may note here that if decayed support count of a node is less than S_{prn} then decayed support count of all its subsets is also less than S_{prn} i.e. if a node has been removed from $\mathcal{L}$ then all its descendents (except bottom) have already been removed from $\mathcal{L}$.

Let N denotes the node to be deleted. Node N will have one child node i.e. bottom and can have more than one parent. Delete procedure removes link of

N from its parents and child nodes. Two cases may arise: i) child node of N (i.e. bottom node) has no parent. In that case, parents of N become parents for bottom. ii) parent nodes of N has no child. In that case, bottom become child for parents of N.

```
Input: ℒ - lattice of closed itemsets, t - transaction to insert
Output: ℒ - updated lattice
Process:
create a new node N with I_N = t
for all item i ∈ t in ℒ do
   if F_i is ancestor of N then
      find child and parents of N among the descendants of F_i
   else
      if F_i is descendant of N then
         child of N is F_i and parents of N are among the ancestors of F_i
      else
         if common parent X of F_i and N exist in ℒ then
            parent of N is X and child of N is among the descendants of F_i
         else
            create a new node X where I_X = I_{F_i} ∩ I_N
            parent of N is X and child of N is among descendants of F_i
            children of X are F_i and some ancestor of N.
            parents of X are among the ancestors of F_i.
            if X is neither ancestor nor descendent of node Y ∈ ℒ and I_X ∩ I_Y ≠ ϕ
            then
               create a new node U such that I_U = I_Y ∩ I_X
               find child and parents of node U
            end if
         end if
      end if
   end if
   Update frequency of N.
end for
```

Algorithm 1. Insert Procedure

Update ITable. If N is added as an ancestor node of F_i in $\mathcal{L}$, First Node F_i corresponding to $i \in I_N$ is set to N. If a node N is deleted from $\mathcal{L}$ then First Node F_i corresponding to $i \in I_N$ is set to bottom as all the descendents of N (except bottom) have already been removed from $\mathcal{L}$.

4 Experimental Analysis

Since there is no algorithm known for mining all recent closed itemsets in landmark window model, it is not possible to perform a comparative analysis. We evaluate our algorithm for scalability using several synthetic and real life datasets. All experiments were done on a 2GHz AMD Dual-Core PC with 3 GB main memory, running redhat linux operating system. All algorithms are implemented in C++ and compiled using g++ compiler without optimizations. In all experiments, the transactions of each dataset are examined one by one in sequence to simulate the environment of an online data stream.

Experiments on Synthetic data

We generated four different datasets T3I4D100K, T5I4D100K, T8I4D100K, T10I4D100K using IBM data generator [1]. Three numbers of each dataset denote the average transaction length (T), average maximum potential frequent itemset size (I) and the total number of transactions (D) respectively.

Fig. 2 (a) shows the effect of pruning threshold S_{prn} on per-transaction-processing time in the dataset T10I4D100K, as stream progresses. Higher values of S_{prn} lead to substantial reduction in the size of synopsis. Hence per-transaction-processing time is nearly constant. As value of S_{prn} decreases, size of the synopsis increases, leading to increase in rate of growth of per-transaction-processing time. However if size of synopsis is fixed, per transaction processing time is nearly

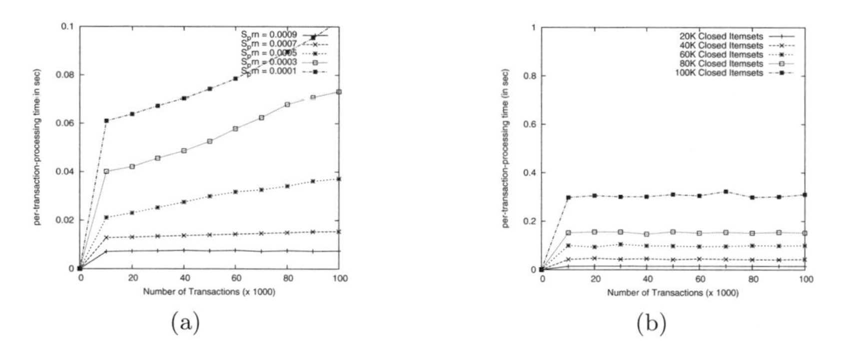

Fig. 2. Per-transaction-processing time corresponding to (a) different values of S_{prn} (b) fixed size of synopsis in T10I4D100K dataset

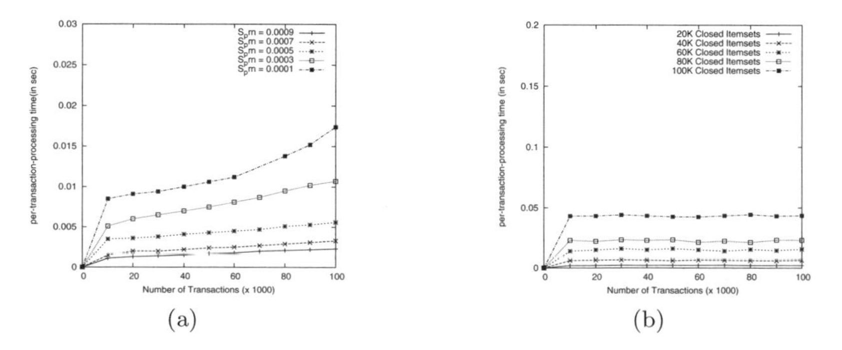

Fig. 3. Per-transaction-processing time corresponding to (a) different values of S_{prn} (b) fixed size of synopsis in BMS-Web-View-1 dataset

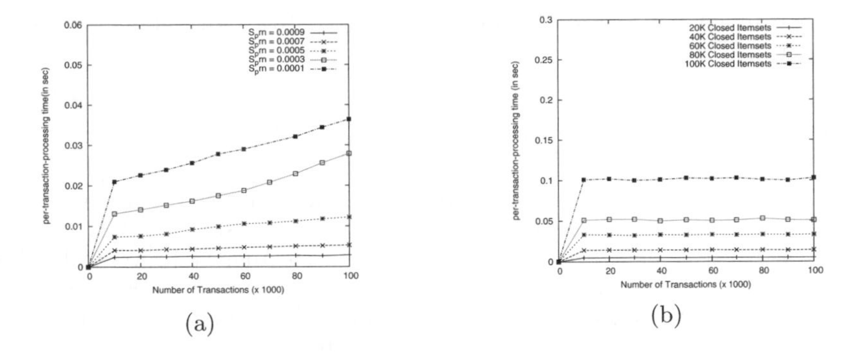

Fig. 4. Per-transaction-processing time corresponding to (a) different values of S_{prn} (b) fixed size of synopsis in BMS-POS dataset

constant. Fig 2 (b) shows the results for different sizes of synopsis. This establishes the scalability of the algorithm.

Experiments on Real Life datasets
We experimented on two real life datasets, BMS-Web-View-1 and BMS-POS [17]. BMS-Web-View-1 contains a few months of clickstream data from an e-commerce web site [17]. There are 59602 transactions with 497 items; average transaction length is 2.5 and maximum transaction length is 267. BMS-POS contains several years of point of sale data from a large electronics retailer [17]. There are 515,597 transactions with 1657 items; average transaction length is 6.5 and maximum transaction length is 164.

Fig. 3 (a) and Fig. 4 (a) shows graph depicting per-transaction-processing time when CLICI runs on BMS-Web-View-1 and BMS-POS datasets respectively. Different values of pruning threshold S_{prn} are tested and it is observed that higher values of S_{prn} leads to reduced size of the synopsis and hence reduced per-transaction-processing time. It is also observed that rate of growth of per-transaction-processing time increases with decrease in pruning threshold. Fig. 3 (b) and Fig. 4 (b) present results on different sizes of the synopsis, 20K, 40K, 60K, 80K, 100K closed itemsets. As depicted in the graphs, per-transaction-processing time remains nearly constant on a particular size of the synopsis and increases with increase in size of synopsis.

5 Conclusion and Future Work

We have proposed CLICI algorithm to mine all recent closed itemsets in landmark window model of data stream. The algorithm is based on Formal Concept Analysis, a well established discipline in applied mathematics. The proposed algorithm maintains a lattice of recent closed itemsets in the stream and delivers frequent closed itemsets to the user on demand, based on the dynamically specified support threshold. Use of lattice as a synopsis facilitates efficient formulation of non-redundant association rules as shown in earlier works. Since the synopsis is independent of the support threshold, user is encouraged to explore and experiment.

References

1. Agarwal, R., Srikant, R.: Fast Algorithms for Mining Association Rules. In: 20th International Conference on Very Large Databases, pp. 487–499 (1994)
2. Chang, J., Lee, W.: Finding Recent Frequent Itemsets Adaptively over Online Data stream. In: 9th ACM SIGKDD, pp. 487–492. ACM Press, New York (2003)
3. Cheng, J., Ke, Y., Ng, W.: A Survey on Algorithms for Mining Frequent Itemsets over Data stream. KAIS Journal 16(1), 1–27 (2008)
4. Chen, J., Li, S.: GC-Tree: A Fast Online Algorithm for Mining Frequent Closed Itemsets. In: Proceeding of PAKDD Workshop of HPDMA, pp. 457–468 (2007)
5. Ganter, B., Wille, R.: Formal Concept Analysis: Mathematical Foundations. Springer, Heidelberg (1999)
6. Han, J., Cheng, H., Xin, D., Yan, X.: Frequent Pattern Mining: Current Status and Future Directions. Journal of Data Mining and Knowledge Discovery 15, 55–86 (2007)
7. Jiang, N., Gruenwald, L.: CFI-Stream: Mining Closed Frequent Itemsets in Data stream. In: 12th ACM SIGKDD, Poster Paper, pp. 592–597. ACM Press, New York (2006)
8. Kuznetsov, S.O., Obiedkov, S.A.: Comparing Performance of Algorithms for Generating Concept Lattices. JETAI 14, 189–216 (2002)
9. Li, H., Ho, C., Lee, S.: Incremental Updates of Closed Frequent Itemsets Over Continuous Data stream. Expert Systems with Applications 36, 2451–2458 (2009)
10. Liu, X., Guan, J., Hu, P.: Mining Frequent Closed Itemsets from a landmark window over online data stream. Journal of Computers and Mathematics with Applications 57(6), 927–936 (2009)
11. Pasquier, N., et al.: Efficient Mining of Association Rules using Closed Itemset Lattices. Journal of Information Systems 24(1), 25–46 (1999)
12. Stumme, G., et al.: Computing Iceberg Concept Lattices with Titanic. Journal on Knowledge and Data Engineering 42(2), 189–222 (2002)
13. Valtchev, P., Missaoui, R., Godin, R.: A framework for incremental generation of closed itemsets. Discrete Applied Mathematics 156(6), 924–949 (2008)
14. Chi, Y., Wang, H., Yu, P.S., Muntz, R.R.: Catch the Moment: Maintaining Closed Frequent Itemsets over a Stream Sliding Window. Journal of Knowledge and Information Systems 10, 265–294 (2006)
15. Yahia, S.B., Hamrouni, T., Nguifo, E.M.: Frequent Closed Itemset Based Algorithms: A thorough structural and analytical survey. ACM SIGKDD Explorations Newsletter 8, 93–104 (2006)
16. Zaki, M.J.: Generating Non-Redundant Association Rules. In: 6th ACM SIGKDD International Conference on Knowledge Discovery and Data Mining, pp. 34–43. ACM Press, New York (2000)
17. Zheng, Z., Kohavi, R., Mason, L.: Real World Performance of Association Rule Algorithms. In: Proceedings of the 2001 International Conference Knowledge Discovery and Data Mining, SIGKDD 2001 (2001)

XML Data Fusion

Frantchesco Cecchin[1], Cristina Dutra de Aguiar Ciferri[2], and Carmem Satie Hara[1]

[1] Federal University of Paraná – Curitiba, PR – Brazil
{frantchesco,carmem}@inf.ufpr.br
[2] University of São Paulo – São Carlos, SP – Brazil
cdac@icmc.usp.br

Abstract. Ensuring high quality data when collecting and integrating information from heterogeneous sources into a data warehouse is a challenging problem. In this paper, we propose a model for XML data fusion, which allows the integrator to define data cleaning rules for solving value conflicts that may have been detected during the integration process. These rules resemble decisions that are made by users when data are manually curated and, once defined, conflicts detected in subsequent integration processes that are within the context of existing rules can be automatically solved without user intervention. We also introduce a notion of fusion policy validation that prevents conflicting resolution rules to be defined. To validate our proposal, we developed XFusion, a rule-based cleaning tool that stores curated data in a integrated repository.

1 Introduction

Nowadays companies of all sizes and from different segments maintain a repository of data imported from a number of sources. The integration of imported data in a single repository provides a unified view of the available information, and also constitutes the basis for applying data analysis techniques, such as data mining and multidimensional analysis. In fact, data warehousing has emerged as an area in recognition of the value and role of information, providing integrated and high quality information targeted at decision support.

Imported data are often inconsistent. Thus, for achieving full integration, data usually goes through an iteration of integration and cleaning processes. Here, integration refers to the problem of identifying overlapping data in different sources. This problem has been the subject of extensive research on relational [8], entity-relationship [10], and XML [15] data models. Cleaning refers to the process of solving attribute value conflicts. The problem arises when two or more sources contain information on the same entity or attribute, but disagree on their values. A number of approaches have been proposed in the literature for addressing this problem, including data profiling, data mining, constraint-based, and ontology-based techniques [12,17,8]. Recently, the process of combining multiple records representing the same real-world object into a single, consistent, and clean representation has been denoted as data fusion [3].

T.B. Pedersen, M.K. Mohania, and A M. Tjoa (Eds.): DaWaK 2010, LNCS 6263, pp. 297–308, 2010.

The majority of existing systems for data fusion considers data structured on relational format. Nevertheless, given that XML has become the standard for data exchange on the Web, it is natural to also consider this format for the integration process. In addition to the fact that currently most data sources provide their data in XML, features that make this format suitable for data exchange are also desirable for data cleaning. One of these features is its hierarchical structure, which naturally represents relationships between entities.

Data cleaning usually requires some manual user intervention, even though this is an error-prone and time-consuming process. In this paper we propose a data fusion model for minimizing the amount of user mediation for data cleaning in integration processes. The model is based on establishing a policy, composed of a set of rules, which resemble decisions that are usually made by users when data are manually curated. Once the policy is defined, conflicts detected in subsequent integration processes that are within its context can be automatically solved without user intervention.

The data integration scenario we consider is depicted in Figure 1. The input for an `integration` tool is a set a data sources $S_1, \ldots, S_n$. This tool is responsible for identifying corresponding entities among sources, and also for detecting value conflicts. The user defines a set of rules for solving these conflicts, which are stored in a `policy` base. Rules are then applied by a `cleaning` tool. "Discarded" data values are stored in a `resolution log`, and a clean, consistent view of the data is provided to the user by the `data repository`.

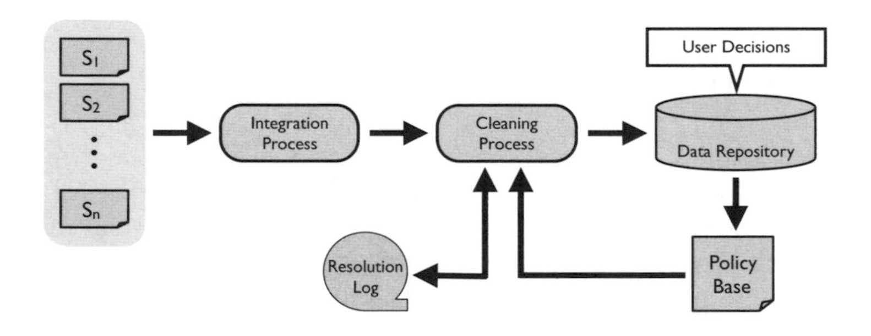

Fig. 1. Integration process with policy-based conflict resolutions for data cleaning

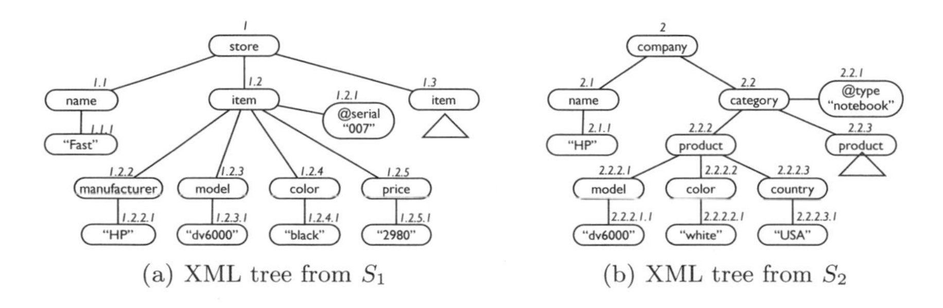

(a) XML tree from S_1 (b) XML tree from S_2

Fig. 2. Samples of XML tree representation

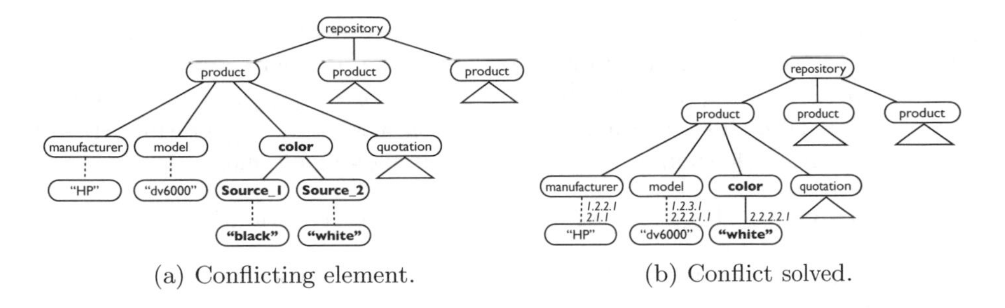

(a) Conflicting element. (b) Conflict solved.

Fig. 3. Data repository tree with temporary inconsistency [14] and after cleaning

Example 1. Consider two data sources providing data on `products` as depicted in Figure 2. Data from source S_1 are extracted to populate a data repository, depicted in Figure 3(a), as follows. Each `item` element is mapped to a `product`, along with its subelements `manufacturer`, `model`, and `color`. The `item`'s `price` is mapped to a child of the `product`'s `quotation`, which stores price values from different stores. For S_2, the value for `manufacturer` is extracted from `company`'s `name`, and values for the remaining subelements are given by subelements of `category/product`. We assume that `product` elements in the repository are identified by their `manufacturer` and `model`. Since both S_1 and S_2 provide data on products that coincide on the values of these elements, they are merged in the repository. Nevertheless, they disagree on the value of the product's `color`. Following the integration model proposed in [14], value conflicts are explicitly represented as depicted in Figure 3(a), along with their provenance, i.e. the sources that provided the conflicting data.

A cleaning strategy for solving the conflict may determine that whenever a data item provided from S_2 disagrees with any other source, we should choose S_2's value over the others, since S_2 contains data provided by the product manufacturer while other sources are resellers. As a result of applying this strategy, the data repository keeps a single consistent value for all product's subelements, as shown in Figure 3(b). In our model, a rule expressing a cleaning strategy is stored in the `policy base`, while the discarded value of product's `color` (`‘‘black’’`) is kept in the `resolution log`, which helps us consider the value in future conflict resolution processes. As an example, suppose that data from a new source S_3 is imported into the repository, and that S_3 also provides the value `‘‘black’’` for the same product. If the strategy for solving the conflict is modified for choosing the value provided by the majority of the sources, we would be unable to determine that the value for `color` in the repository should be changed to `‘‘black’’`. This action is only possible because our model stores the discarded value of product's color in the resolution log.

Rules for solving value conflicts can be defined on different contexts of an XML tree. They may involve a single element or a set of subtrees. The ability to define contexts on subtrees may generate inconsistencies among conflict resolution rules. As an example, suppose that the rule for choosing S_2 over other data sources is to be applied on all subtrees for which the product's `manufacturer` is `HP`, and a second rule for choosing the value provided by the majority of the sources is defined on all subtrees for which the product's `model` is `dv6000`. Since both rules apply to the product in our running example, we need a notion of policy validation for avoiding such inconsistencies and for deterministically determine which rule should be applied for solving a given one.

Contributions. In this paper we make the following contributions.

- We propose a model for XML data fusion based on a set of rules for solving value conflicts in data integration processes. Cleaning rules resemble decisions commonly made by users for handling value conflicts, and minimize manual intervention for data curation.
- We define a fusion policy validation on a set of conflict resolution rules. Our fusion policy validation prevents inconsistent rules to be defined on the same element of the repository.
- We present a tool, called XFusion, that has been developed based on our model. It supports both XML data integration and cleaning processes.

Organization. The rest of the paper is organized as follows. Section 2 presents preliminary definitions used as the basis for our proposal, while Section 3 introduces our model for XML data fusion. Fusion policy validation is the subject of Section 4, and Section 5 presents the XFusion tool. Section 6 discusses related work and Section 7 concludes the paper.

2 Preliminary Definitions

Before describing our fusion model, we present a definition of XML trees, the integration model considered in this paper, and strategies for data fusion previously proposed in the literature.

2.1 XML Trees and Integration Model

An XML document is typically modeled as a node-labeled tree T, in which the set of nodes V can be of one of three types: element, attribute and text nodes. For each node n in T we define the following functions: (1) $lab(n)$ assigns a label to n if n is an element or attribute node, and a distinct label L if n is a text node; (2) $val(n)$ assigns a string to attribute and text nodes, and is undefined for element nodes; (3) $ele(n)$ and $att(n)$ define the edge relation of T: if n is an element then $ele(n)$ is a *list* of elements and text nodes in V and $att(n)$ is a *set* of attributes in V; if n is an attribute or a text node then $ele(n)$ and $att(n)$ are undefined; (4) $id(n)$ assigns a unique identifier to n, which represents the path from the root r of the tree to n. We assume that each XML tree has a distinct identifier S which denotes its source, and that $id(r) = S$. That is, the root's node identifier coincides with the XML tree source identity.

Examples of XML trees are given in Figures 2(a) and 2(b), where each node n is represented with its identifier ($id(n)$), a label ($lab(n)$) if it is an element or attribute node, and a value ($val(n)$) if it is an attribute or text node. The encoding adopted by the identifier function id is called *Dewey Order* [5], which provides a global node ordering. Since in our model each data source has a distinct source identifier, which coincides with the root identifier, in Figure 2(a), $id(r) = 1$ and in Figure 2(b), $id(r) = 2$.

In this paper, we adopt the model proposed in [14] for identifying corresponding entities among data sources, and for explicitly representing value conflicts. This model assumes that the data repository has a fixed schema, and a set of XML keys [4] for identifying elements in a document. The repository is populated with data imported from data sources through a transformation language that maps source data to the repository schema. Whenever two source elements are mapped to an entity in the repository that coincide on their key values, they are merged. The repository stores provenance information on every imported data item, and explicitly represents value conflicts detected after the merging.

Example 2. Consider again XML data sources depicted in Figures 2(a) and 2(b), and the XML tree resulting from merging them given in Figure 3(a). Following the syntax proposed in [4], the XML keys that determine how elements are merged in the repository can be defined as follows.

- $k_1 : (\epsilon, (product, \{manufacturer, model\}))$: in the context of the entire document (ϵ denotes the root), a `product` is identified by its `manufacturer` and `model` number;
- $k_2 : (product, (color, \{\}))$: within the context of any subtree rooted at a `product` node, there exists at most one `color` element; that is, it is identified by an empty set of values.

Observe that in the repository, `manufacturer` is populated with nodes reached by path $/item/manufacturer$ in S_1, and nodes reached by path $/name$ in S_2. Similarly, element `model` is populated with nodes reached by path $/item/model$ in S_1 and path $/category/product/model$ in S_2. Given that values of nodes reached by both paths in sources S_1 and S_2 coincide, they are be merged in the repository according to k_1. The resulting tree is depicted in Figure 3(a).

In the repository, leaf nodes are annotated with provenance information. These annotations are important not only to determine the origin of data, but they also allow the portion of source XML tree used to populate the data repository to be reconstructed [14].

XML keys involve path expressions. An algorithm for deciding path containment for the fragment of XPath involved in defining keys is presented in [4], while [11] and [6] investigate the problem for larger fragments of XPath. It has been shown that for some of these fragments containment can be checked in PTIME. The problem of determining intersection of XPath expressions has also been investigated in the context of query optimization [7].

2.2 Strategies for Data Fusion

There are a number of strategies proposed in the literature for solving value conflicts. We adopt a subset of strategies proposed in [2], described as follows.

Trust Your Friends. This strategy is based on a reliability criterion. The user assigns a confidence rate for each source, and a value conflict is solved by choosing the one provided by the source with the highest confidence rate.

Meet In The Middle. This is a strategy to mediate the conflict by generating a new value that is a compromise among all conflicting values, e.g., an average of all conflicting numeric values.

Cry With The Wolves. This strategy is defined for choosing the value reported by the majority of data sources.

Roll The Dice. This strategy randomly chooses one value among the conflicting ones.

Pass It On. This is a non-resolving strategy. Although in most cases the user wants a single value for each data item, for some items she may want to keep all the conflicting values in the repository. When this is the case, this choice can explicitly be made applying the *Pass It On* strategy.

In [2] these strategies are integrated to the relational model by developing functions that can be used within SQL sentences to solve inconsistencies from the resulting data set. Next section presents our model, which extends this strategy-based conflict resolution approach for XML.

3 XML Fusion Model

Our model for solving conflicts detected during the integration process is based on the definition of a fusion policy, which consists of a set of data conflict resolution rules, defined as follows.

Definition 1. *A* **conflict resolution rule** *is a pair* $\langle \sigma, \Sigma \rangle$, *where*

(1) σ *is a path expression representing the context covered by the rule;*

(2) Σ *is a non empty list of strategies for handling instance-level conflicts on nodes reached by following the context path* σ.

The context of a rule is defined by a path expression σ and therefore it may cover not only a single element or attribute node, but a *set* of nodes reached by following σ. Furthermore, a rule may define a *list* of strategies for solving a conflict. Thus, if the first strategy is not able to single out a value for a given data item, the following strategies are considered one by one until either the end of the list is reached or the conflict is solved. If the conflict is solved, we say that the rule *effectively solves the value conflict.*

Example 3. Consider the value conflict between a product's `color` depicted in Figure 3(a). Suppose the following rule has already been defined in the fusion policy: $\langle /product[manufacturer = \text{``}HP\text{''}]/color, [\textit{Trust Your Friends, Cry With the Wolves}] \rangle$. It determines that whenever there is a value conflict on element `color` of `product`, and the `manufacturer` of `product` is `''HP''`, then the strategy *Trust Your Friends* should be applied, followed by *Cry With the Wolves.* Assuming that the confidence rate of S_2 is higher than S_1, strategy *Trust Your Friends* is applied and the value `''white''` from S_2 is chosen to be stored in the repository, as shown in Figure 3(b).

Observe that some strategies for solving conflicts may depend on the provenance of the data, as exemplified in the previous example by strategy *Trust Your Friends*. Data provenance should also be kept for discarded values. In our model this information is kept in a *resolution log*, which is defined as follows.

Definition 2. *A* **resolution log** *is a set of records, where each record refers to a data value v that has been discarded during the cleaning process. Given that v has been populated from an element e of a data source S, the record that refers to v contains the following attributes: (1) key values of the element or attribute with value v in the repository; (2) the discarded value v; (3) id(e) in the original source S; (4) the path from the root to e in the source S; (5) the strategy $s \in \Sigma$ applied for solving the conflict.*

We need to keep the keys for the element for which a value has been discarded in order to retrieve all the discarded values for the same data item in the repository. This may be necessary for automatically reapplying a conflict resolution rule in future cleaning processes. Both the identity and the original path of the element which provides v are stored in the log for keeping provenance information. By storing the `strategy` executed to solve the conflict, we can trace back why `value` v has been discarded.

Example 4. Consider again the value conflict depicted in Figure 3(a) and the conflict resolution rule in Example 3. The record for S_1's discarded value `''black''` stored in the resolution log contains the following data: `(key: /product[manufacturer=''HP'' and model=''dv6000'']/color, value: ''black'', id: 1.2.4.1, path: /item/color, strategy: Trust Your Friends)`. Recall that from the value of *id* it is also possible to get the source identification, given that the first number of the sequence corresponds to the root element, which coincides with the source identifier.

To illustrate how the log is used in future fusion processes, consider that source S_3, as defined in Example 1, has been integrated into the repository, and that it has the same confidence rate as S_2. Given that both sources S_2 and S_3 have the same confidence rate, and that S_3 provides the value `''black''` for `color`, strategy *Trust Your Friends* is not able to solve the conflict. Then, the next strategy, *Cry With The Wolves* is applied. In this case, `''black''` is chosen to be stored in the repository, since this strategy chooses the value reported by the majority of the sources. Our model is only able to make such a decision because the log maintains S_1's discarded value for `color`.

Given the definitions of conflict resolution rules and repository log, we are now ready to define our fusion model.

Definition 3. *A* **data fusion model** $\mathcal{D}$ *is a 5-tuple* $\langle \mathcal{R}, \mathcal{T}, \mathcal{K}, \mathcal{P}, \mathcal{L} \rangle$*, where:*

(1) $\mathcal{R}$ *is a set of pairs* $(S, rank)$*, where S is an XML data source, and rank its confidence rate; the value of rank is greater for sources with higher reliability;*

(2) $\mathcal{T}$ *is the data repository tree with a set of nodes V such that each leaf node* $v \in V$ *is annotated with a set of pairs* (id_n, p)*, where* id_n *is the identifier of a*

node n in a source XML tree T_S *used to populate v, and p is the path in* T_S *from its root to n;*

(3) $\mathcal{K}$ *is the set of XML keys defined on* $\mathcal{T}$*. Every element node in* $\mathcal{T}$ *can be uniquely identified according to keys in* $\mathcal{K}$*;*

(4) $\mathcal{P}$ *is a set of conflict resolution rules that define strategies for solving data conflicts;*

(5) $\mathcal{L}$ *is the resolution log for storing data discarded during a fusion process.*

An example of a data repository tree is given by the XML tree depicted in Figure 3(b). Observe that leaf nodes have been annotated with node identifiers from *Source 1* and *Source 2*. Paths traversed from the root have been omitted for simplicity.

4 Fusion Policy Validation

Recall that conflict resolution rules are defined on contexts described as path expressions. Since a path expression σ denotes a *set* of nodes in a XML tree reached by following σ, there may exist nodes that are covered by more than one rule. In order to deterministically single out a rule for solving a value conflict, we introduce a notion of *policy validity.*

In the following definition, we denote as $Nodes(r)$ the set of nodes covered by a rule r. That is, given a rule $r = (\sigma, \Sigma)$ and an XML tree T, $Nodes(r)$ is the set of nodes in T reached by following σ in T.

Definition 4. *Given two rules* r_1 *and* r_2*, we say that* r_1 **is valid with respect to** r_2 *if they satisfy one of the following conditions:*
(1) $Nodes(r_1) \subset Nodes(r_2)$ *or*
(2) $Nodes(r_1) \supset Nodes(r_2)$ *or*
(3) $Nodes(r_1) \cap Nodes(r_2) = \emptyset$*.*

Intuitively, rules can be related either by *specialization* (Case 1) or *generalization* (Case 2), or not related at all (Case 3). Following the traditional definition of class hierarchy, Cases 1 and 2 allow rules to be defined on different levels of a tree hierarchy. That is, there may exist a general rule for solving value conflicts, but it may be overridden by rules defined for treating specific cases, that are restrict to subsets of nodes covered by the general rule, and are used in the cleaning process instead of the general rule.

Rules with intersecting coverage that are not related by specialization / generalization are not allowed. This is because there is no deterministic way of deciding which rule should be applied for solving conflicts on nodes covered by both rules. For determining the validity of a fusion policy, each conflict resolution rule should be valid with respect to all others. Checking validity involves checking both path expressions containment and intersection. Some previous work on these subjects are presented in Section 2. The model we propose in this paper is orthogonal to the path language adopted. The following definition establishes an order for applying conflicting resolution rules in a fusion policy.

Definition 5. *Let v be an element or attribute with conflicting values, and $\mathcal{P}$ a fusion policy, consisted of a set of rules that are valid with respect to each other. Rule $r_v \in \mathcal{P}$ is applied for solving the value conflict on v if:*
(1) $v \in Nodes(r_v)$ and r_v effectively solves the value conflict;
(2) there exists no rule $r_i \in \mathcal{P}$ that satisfies condition (1), such that $Nodes(r_i) \subset Nodes(r_v)$.

Example 5. Consider the data repository depicted in Figure 4, and four conflict resolution rules, defined as follows:
$r_a = (/product[manufacturer = \text{“}HP\text{”} \ and \ model = \text{“}tx1220\text{”}]/color, \Sigma_a)$
$r_b = (/product[manufacturer = \text{“}HP\text{”}]/color, \Sigma_b)$
$r_c = (/product/quotation[store = \text{“}Fast\text{”}]/price, \Sigma_c)$
$r_d = (/product[manufacturer = \text{“}Sony\text{”}]/quotation/price, \Sigma_d)$

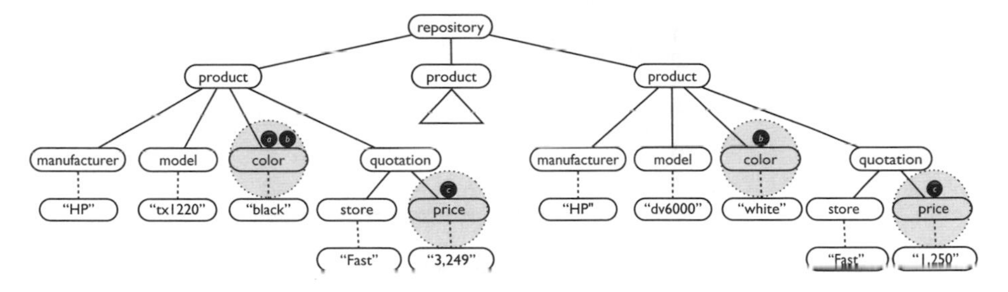

Fig. 4. Example of conflict between resolution rules

Rules r_a and r_b are defined on `color` elements, while r_c and r_d are defined on `price`. Observe that r_a is a specialization of r_b, since $Nodes(r_a) \subset Nodes(r_b)$, and they are valid with respect to each other. On the other hand, r_c is *not* valid with respect to r_d since it is not true that for all possible XML trees $Nodes(r_c) \cap Nodes(r_d) = \emptyset$, although in the tree depicted in Figure 4 this condition holds.

5 XFusion

In order to validate the proposed XML fusion model and fusion policy validation, we have developed the XFusion tool. The tool integrates several data sources into a data repository, and presents to the user the detected value conflicts along with a set of available strategies for solving them.

Two screenshots of XFusion's interface are presented in Figure 5. Screenshot A is the main screen of the tool. It shows in the *Data Repository* panel the data repository in a tree format, and in the *Integrated Sources* panel the source's names that have been considered in the integration process. In this panel, the DBA represents the source of data items that have not been imported from external sources, but locally produced. The tool assigns a distinct color for each of the integrated sources, so that value conflicts are shown along with "colored" representation of sources that provided them (little squares that precede each value). Furthermore, screenshot A contains buttons to perform actions over the

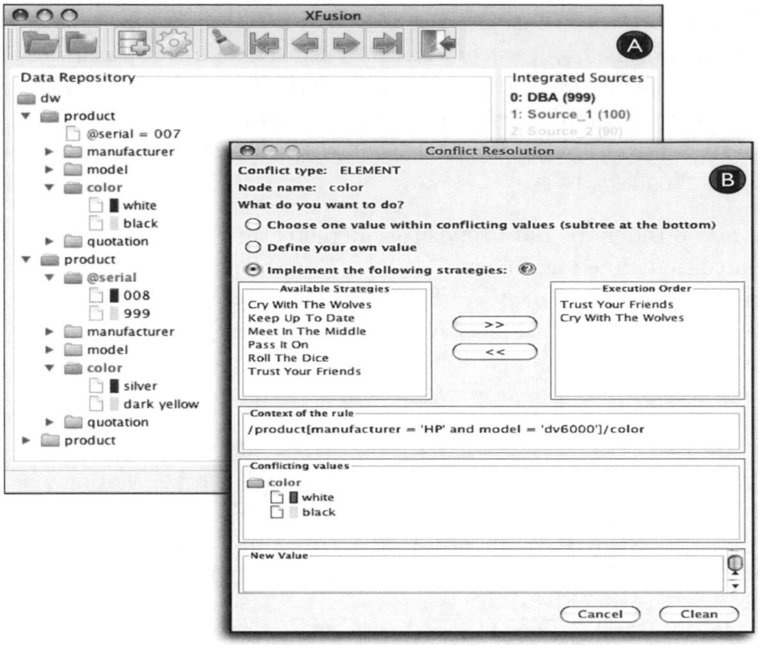

Fig. 5. XFusion screenshots: **A** – Main screen; **B** – Resolution screen

data repository, such as add new source ( ), resolve conflicting values ( ), and navigate through conflicts ( ). To add a new data source, the user must provide a mapping that determines how source data are extracted and then inserted into the repository. She also defines the confidence rate to be assigned to the new source.

Data fusion is the main functionality of the tool based on the model proposed in this paper. When the user selects an existing conflict and clicks on the resolve button, the screen depicted in screenshot B of Figure 5 is shown. Observe that the user has three main options for solving a conflict: choose one among the conflicting values, manually insert a new value, or apply some of the available strategies, which are described in Section 2.2. Strategies are chosen by clicking on direction buttons in the middle of the screen, determining the order in which they should be considered.

Below the strategies boxes, the *Context of the rule* is presented. This path is originally set to uniquely identify the conflicting element or attribute, according to the XML keys defined on the repository. Nevertheless, the user can edit the path for applying the list of strategies on different contexts. In the current implementation, we only consider simple XPath expressions (without wildcards) with simple predicates involving elements, attributes and string values. When the user edits the context of the rule, our tool validates the rule with respect to all existing ones according to the policy validation described in Section 4. Finally, when she clicks on *Clean* button, the new rule is inserted into the policy base and its execution propagates the chosen value to the repository, and the discarded ones to the resolution log. XFusion allows the user to define rules incrementally. That is, in the first iteration a single strategy may be defined on a context path and applied. The user can then check whether the strategy has been effective for solving all conflicts within the rule's context. If not, she may decide to extend the rule by defining additional strategies to be applied.

XFusion has been implemented in Java, using the Swing graphical package. For manipulating XML documents we used JDOM. The data repository is stored on eXist-db [9], a native XML database system. XFusion implements the XML integration model proposed in [14] combined with our fusion model, showing the feasibility of our approach.

6 Related Work

Data integration and cleaning have been studied extensively by the database community [1,3]. Most of previous works consider data on relational format, but recently it has been stressed the need for investigating the problem of solving conflicts on semi-structured data. XClean [17] is a system that allows declarative and modular specification of a cleaning process. As oppose to our approach, which adopts strategies, XClean is based on operators. Systems like Potter's Wheel [16] and Fusionplex [13] are also strategy-based systems, but they allow the definition of a single cleaning strategy. To the best of our knowledge, our fusion model is the first to define a general framework for applying strategy-based techniques for solving value conflicts that maintains discarded data values in a log repository. This approach allows strategies to be applied in subsequent integration processes, and also keeps provenance information for tracing back cleaning processes. Our model also builds on previous works for determining XPath expressions containment [11,6] and intersection [7] in order to determine fusion policy validity.

7 Conclusion

In the relational model, fusion strategies are usually defined on the context of an attribute value. The model proposed in this paper naturally extends this notion by allowing strategies to be defined on subtrees of an XML document. Our notion of policy validation also extends relational fusion policies by allowing strategies to be defined on subsets (supersets) of nodes reached by previously defined rules, by specialization (generalization). Furthermore, we guarantee that there are no two rules that can be applied on a data item, except for those related by specialization/generalization. The ability to define value conflict resolution rules on subtrees can drastically reduce the amount of user mediation on data cleaning processes, given that conflicts detected in subsequent integration processes that are within the context of existing rules can be automatically solved. The repository log plays an important role in keeping the necessary data for supporting this functionality, while also storing provenance data for tracing purposes. The model has been validated by developing XFusion, a tool based on the proposed model, which shows that it can be incorporated in a data integration and cleaning application.

Some issues that need to be further investigated include: (1) integration of the model to a technique for incrementally updating the data repository when new versions of data sources become available; (2) extensions to the fusion policy, by supporting new strategies and a declarative definition of rules application, allowing for instance, conditional execution of strategies; (3) experimental study for solving conflicts in real-world applications and for determining the cost of applying our rule-based cleaning policy.

Acknowledgments. This work has been supported by the following Brazilian research agencies: FAPESP, CNPq, CAPES, INEP and FINEP.

References

1. Bhattacharya, I., Getoor, L.: Collective entity resolution in relational data. IEEE Data Eng. Bull. 29(2), 4–12 (2006)
2. Bleiholder, J., Naumann, F.: Conflict handling strategies in an integrated information system. In: Proceedings of IIWeb (2006)
3. Bleiholder, J., Naumann, F.: Data fusion. ACM Comp. Surveys 41(1), 1–41 (2008)
4. Buneman, P., Davidson, S., Fan, W., Hara, C., Tan, W.C.: Reasoning about keys for XML. Information Systems 28(8), 1037–1063 (2003)
5. Chan, L.M., Mitchell, J.S.: Introduction to the Dewey Decimal Classification (2003), `http://www.oclc.org/dewey/versions/ddc22print/intro.pdf`
6. Genevès, P., Layaïda, N.: Deciding XPath containment with MSO. Data & Knowledge Eng. 63(1), 108–136 (2007)
7. Hammerschmidt, B.C., Ad Volker Linnemann, M.K.: On the intersection of XPath expressions. In: Proc of IDEAS, pp. 49–57 (2005)
8. Lim, E.P., Srivastava, J., Prabhakar, S., Richardson, J.: Entity identification in database integration. Information Sciences 89(1) (1996)
9. Meier, W.: eXist-db open source native XML database (2000), `http://exist.sourceforge.net`
10. Menestrina, D., Benjelloun, O., Garcia-Molina, H.: Generic entity resolution with data confidences. In: Proc. of VLDB Work. on Clean Databases (2006)
11. Miklau, G., Suciu, D.: Containment and equivalence for a fragment of XPath. J. of the ACM 51(1), 2–45 (2004)
12. Milano, D., Scannapieco, M., Catarci, T.: Using ontologies for XML data cleaning. In: OTM Workshops, pp. 562–571 (2005)
13. Motro, A., Anokhin, P.: Fusionplex: resolution of data inconsistencies in the integration of heterogeneous information sources. Info. Fusion 7(2), 176–196 (2006)
14. do Nascimento, A.M., Hara, C.S.: A model for XML instance level integration. In: Proc. of SBBD, pp. 46–60 (2008)
15. Poggi, A., Abiteboul, S.: XML data integration with identification. In: Proc. of DBPL (2005)
16. Raman, V., Hellerstein, J.M.: Potter's wheel: An interactive data cleaning system. In: Proc. of VLDB, pp. 381–390 (2001)
17. Weis, M., Manolescu, I.: Declarative XML data cleaning with XClean. In: Krogstie, J., Opdahl, A.L., Sindre, G. (eds.) CAiSE 2007 and WES 2007. LNCS, vol. 4495, pp. 96–110. Springer, Heidelberg (2007)

An Efficient Duplicate Record Detection Using q-Grams Array Inverted Index

Alfredo Ferro, Rosalba Giugno, Piera Laura Puglisi, and Alfredo Pulvirenti

Dept. of Mathematics and Computer Sciences
University of Catania
{ferro,giugno,lpuglisi,apulvirenti}@dmi.unict.it

Abstract. Duplicate record detection is a crucial task for data cleaning process in data warehouse systems. Many approaches have been presented to address this problem: some of these rely on the accuracy of the resulted records, others focus on the efficiency of the comparison process. Following the first direction, we introduce two similarity functions based on the concept of q-grams that contribute to improve accuracy of duplicate detection process with respect to other well known measures. We also reduce the number and the running time of record comparisons by building an inverted index on a sorted list of q-grams, named *q-grams array*. Then, we extend this approach to perform a clustering process based on the proposed q-grams array. Finally, an experimental analysis on synthetic and real data shows the efficiency of the novel indexing method for both record comparison process and clustering.

Keywords: Duplicate record detection, q-grams, inverted index, bitmaps, clustering.

1 Introduction

The quality of data can significantly affect the reliability of decision support systems. Real-world data are neither carefully controlled for quality nor defined in a consistent way due to the fact that they have huge size and come from multiple sources [10]. Thus, data mining techniques have to take into account incomplete or missing values, constraints violations, noisy and inconsistent data. Misspellings and different conventions result in a multiple and not unique representation of objects. The process of resolving such identification problems refers to the *data heterogeneity* term [3].

There are two different types of data heterogeneity: *structural* and *lexical*. Structural heterogeneity occurs when the fields of the tuples in the database have different structures in different databases [8]. For example, in one database the customer address might be recorded in one field named `addr`, while in another database the same information might be stored in multiple fields (`street`, `city`, `state` and `zipcode`). Lexical heterogeneity occurs when the tuples have identically structured fields across databases, but the data use different representations to refer to the same real-world object. For example, in one database

T.B. Pedersen, M.K. Mohania, and A M. Tjoa (Eds.): DaWaK 2010, LNCS 6263, pp. 309–323, 2010.

the first and last name of a person can be stored using this format: *Ilary Patricia Doe*, while another database can use different convention: *I. P. Doe*. Based on this consideration, two records can be considered *equivalent* if they are semantically equal. The similarity between records is computed by metrics which measure the semantic equivalence through a score. Record pairs with high similarity scores (above a specified threshold) are treated as duplicates. In this paper, we focus on the problem of lexical heterogeneity.

In addition to the accuracy of classifying records pairs into matches and mismatches, the central issue consists of improving the speed of comparisons. For this reason, many techniques have been proposed to reduce the quadratic complexity of comparing two tables and the running time of a single comparison.

After a brief review of existing techniques, we propose a new method, called DDEBIT, that efficiently solves the Duplicate record DEtection problem using BITmaps. By introducing two similarity functions based on q-grams [9] and Monge-Elkan [17] distances, DDEBIT improves the accuracy of comparison process with respect to other well known approaches. Next, we show that DDEBIT can also be used to cluster records in a table. Finally, in the experimental section we demonstrate the efficacy of our method on synthetic and real data.

2 Related Work

Duplicate record detection typically relies on string comparison techniques such as, edit distance [13], Smith-Waterman [19], Jaro [11], Monge-Elkan [17] and others [18,23]. Although effective, all these measures are computationally expensive (quadratic on the length of the strings) and therefore, they are not suitable for record detection cause of the size of databases. Therefore, approximate similarity metrics have been designed [21,20,9]. They are based on the notions of q-grams and positional q-grams [22].

Related to efficiency, different methods have been proposed to support duplicate detection process. Common techniques, such as *Standard Blocking* [12], divide the database into blocks and compare only the records that fall into the same block. The blocking is performed by grouping records having the same *blocking key*, that could be an attribute value or a concatenation of more attribute values. The risk is to miss some matches due to errors in the blocking step (records assigned to the wrong block) or having false mismatches caused by the failure of comparing records that do not agree on the blocking field.

The *Sorted Neighborhood Approach* [14] is based on the idea that the probability that similar records will be closer after sorting data, according to the value of the blocking key, is high.

Suffix-array indexing [1] uses a suffix array [15] as an inverted index to dynamically generate blocks of associated records using blocking keys.

In [2], blocking key values are converted into lists of bigrams (q-grams with q=2) alphabetically sorted. Then, sub-lists of all possible combinations are built using a threshold and inserted in an inverted index. The number of sub-lists depends on both the length of the key value and the threshold. In [5], Christen and Gayler present *similarity-aware* and *materialized similarity-aware* inverted indexes for large real-world data sets. The basic idea is to store the similarity in the inverted index and use it to reduce the computation at query time.

Monge and Elkan [17] proposed a technique called *Canopy clustering* to improve the performance of record comparisons by assuming that duplicate detection is transitive. They use a union-find structure in which duplicate records are merged into a cluster and only a representative of the cluster is kept for subsequent comparisons. Several methods define metric similarity as a canopy distance [7,4]. In particular McCallum et al. [16] propose the use of *canopies* to cluster large high-dimensional datasets. The key idea is to use a cheap and approximate distance measure to efficiently group records into overlapping clusters. Then, a more expensive function can be used to achieve a better accuracy when similar record pairs are compared.

3 DDEBIT: A Duplicate Record Detection Algorithm Based on Bitmaps and q-Grams

Here, we introduce a new approach to efficiently solve the duplicate record detection problem by extending the strategy adopted in [16]. The method is based on the use of two similarity functions together with an efficient indexing technique.

First, a lightweight function filters out record pairs with a lower 'global' similarity. This produces a set of candidate pairs whose global similarity is above a minimum *loose threshold*. Next, a more accurate function performs comparisons between candidate pairs that passed the above filter. This function detects potentially duplicated records whose 'local' similarity is above a *tight threshold*.

The intuition behind this method is that record pairs with a higher global similarity are likely to be duplicated and can be successively compared to determine a better accuracy using a finer and local similarity function. Moreover, the increase of accuracy due to the combination of global and local similarities reduces the number of potential false positives and negatives.

To support an efficient duplicate detection process in large datasets, we propose a two-phase approach:

1. **Create index.** We implement an indexing technique based on a sorted list of q-grams named *q-grams array*. For each q-gram, we generate an inverted index by using bitmaps.
2. **Record detection.** After indexing q-grams in the previous phase, we use bitmaps to perform a fast comparison process or an efficient clustering.

Next Sections describe the method in more details.

3.1 DDEBIT Similarity Functions

Before introducing the new similarity functions, we report some basic properties of q-grams.

Basic Definitions. Let Σ be a finite alphabet of size $|\Sigma|$. Let $s_1 \in \Sigma^*$ be a string of length n. The **q-grams** are short characters substrings of length q of the database strings. Given a string s_1, its **positional q-grams** are obtained by sliding a window of length q over the characters of s_1. Q-grams at the beginning and the end of the string can have fewer than q characters. Therefore, new characters "#" and "$" not in Σ are used to extend the string (1) by prefixing it with q-1 occurrences of # and (2) suffixing it with q-1 occurrences of $.

For example, the positional q-grams of length q=2 for string *tom_smith* are: { (1,#*t*), (2,to), (3,om), (4,m_), (5,_s), (6,sm), (7,mi), (8,it), (9,th), (10,*h*$) }. The set of all positional q-grams of a string s_1 is the set of all the $|s_1|$ + q - 1 pairs constructed from all q-grams of s_1. The intuition behind the use of q-grams is that when two strings s_1 and s_2 have a small edit distance, they have many q-grams in common. The use of positional q-grams will involve comparing positions of matching q-grams within a certain distance. With the appropriate use of hash-based indexes, the average time required for computing the q-gram overlap between two strings s_1 and s_2 is $O(\max(|s_1|, |s_2|))$.

Follows the formal definitions of the proposed similarity functions. These are inspired by Monge-Elkan [17] and q-grams [9] distances. The first one better detects typographical errors and block movements whereas, the q-gram distance allows a quick computation of the edit distance to filter non-similar strings.

Definition 3.11. (Lightweight similarity). *Let R be a relation and X, Y be strings representing two tuples in R. Let q_X and q_Y be the sets of distinct q-grams of X and Y, respectively. The function $d_{light,q}$ is defined as follows:*

$$d_{light,q}(X,Y) = \frac{|q_X \bigcap q_Y|}{\max(|q_X|, |q_Y|)} \tag{1}$$

This function computes the ratio of the number of distinct q-grams that two strings have in common over the number of q-grams of the longest string. This score represents a 'global' similarity on the compared strings. It does not take into account the position of common q-grams within the strings. In Figure 1 (a), for each string, the distinct bigrams (q=2) are listed in a lexicographical order. The computation of $d_{light,2}$, for all possible string pairs, is reported in Figure 1 (b). In some cases, d_{light} returns a low value also for pairs that could represent the same real world entity. Here, the strings S_2 and S_3 differ on an high number of q-grams, caused by the word 'professor' and a typographical error (Stewen in place of Steven). To refine the score of this type of string pairs, a more accurate function combines d_{light} with Monge-Elkan distance [17].

Definition 3.12. (Accurate similarity). *Let R be a relation and X, Y be strings representing two tuples in R. Let X_i and Y_j be the tokens or atomic strings (i.e. a sequence of alphanumeric characters delimited by punctuation characters) of strings X and Y, respectively. Let $|X|$ and $|Y|$ be the number of tokens of X and Y respectively, and $|X| \geq |Y|$. The function $d_{acc,q}$ is defined as follows:*

ID	String	List of distinct and sorted q-grams	Distinct q-grams
S_1	Prof. Brown Steven	#b #p #s br en ev of ow pr ro st te ve wn f$ n$	16
S_2	Stewen Brown	#b #s br en ew ow ro st te we wn n$	12
S_3	Professor Brown Steven	#b #p #s br en ev es fe of or ow pr ro so ss st te ve wn n$ r$	21
S_4	Steven Brown, executive	#b #e #s br cu ec en ev ex iv ow ro st te ti ut ve xe wn e$ n$	21
S_5	Dr. Brown, executive	#b #d #e br cu dr ec ex iv ow ro ti ut ve xe wn e$ n$ r$	19

(a)

$d_{light,2}$ ($d_{\_acc,2}$)	S_1	S_2	S_3	S_4	S_5
S_1	1 (1)	0,62 (0,62)	0,71(0,96)	0,5 (0,7)	0,36 (0,53)
S_2		1 (1)	0,47 (0,60)	0,47 (0,57)	0,31 (0,50)
S_3			1 (1)	0,57 (0,70)	0,38 (0,40)
S_4				1 (1)	0,76 (0,83)
S_5					1 (1)

(b)

Fig. 1. (a) List of sorted q-grams. (b) $d_{light,2}$ and $d_{acc,2}$ scores for all string pairs.

$$d_{acc,q}(X,Y) = \frac{1}{|X|} \sum_{i=1}^{|X|} \max_{j=1}^{|Y|} d_{light,q}(X_i, Y_j) \tag{2}$$

Before computing $d_{acc,q}$, X_i and Y_j are divided into q-grams and $d_{light,q}$ is computed for all token pairs in order to find the best matches. The sum of these maximal scores is then normalized by the maximum number of tokens given by $|X|$. Intuitively, this function computes the number of common q-grams locally, giving more importance to tokens similarity than global strings similarity. Notice that, to increase the accuracy of similarity of tokens X_i which are prefixes of Y_j, $d_{light,q}$ within $d_{acc,q}$ is replaced with a function $d_{prefix}(X_i, Y_j) = 1 - \frac{|q_{Y_j} \setminus q_{X_i}|}{|q_{Y_j}||q_{X_i}|}$ which yields a score that better captures the prefix similarity. Furthermore, since $d_{acc,q}$ is a local similarity measure, missing short tokens (i.e. X_i with no more than 4 bigrams) could cause lower and biased values for $d_{acc,q}$. Thus, pseudo-counters β_i proportional to $1/|q_{X_i}|$ are added to the similarity measure $d_{acc,q}$. In Figure 1 (b), the values of $d_{acc,2}$ are also reported for strings in Figure 1 (a). Typically, $d_{acc,q}$ increases the score of a strings pair with respect to $d_{light,q}$ in presence of prefixes, missing of short words, exact matches or high similarity between token pairs. The function $d_{acc,q}$ can be generalized to work with records of k fields.

Definition 3.13. *Let R be a relation, let r_x and r_y be two records in R projected on k fields. Let p_i, $1 \leq i \leq k$, be a weight associated to i-th field.*

$$d_{acc_k,q}(r_x, r_y) = \sum_{i=1}^{k} p_i \times d_{acc,q}(r_x[i], r_y[i]) \quad (3)$$

where $p_1 + p_2 + ... + p_k = 1$ *and* $r_x[i]$ *is the projection of* r_x *to the* $i - th$ *field.*

Here, the comparison key is composed of k fields, and $d_{acc,q}$ is computed separately for each field. The contribution of the $i - th$ field depends also on the value of p_i, that can give a higher or lower weight to the $i - th$ attribute in the comparison process. Unfortunately, this function is quadratic with respect to the length of the strings. To overcome this limitation, we introduce a secondary memory indexing technique on the list of distinct q-grams. This index makes similarity computation linear in the number of q-grams of the longest string.

3.2 DDEBIT Indexing

The indexing phase contributes to quickly detect candidate records having an high global similarity (i.e. high values of $d_{light,q}$) and to reduce the running time of $d_{acc,q}$. The key idea is based on the concept of *q-grams array*. Given a table T, a `q-grams array` is a sorted list of all distinct q-grams in T. Prefix q-grams of the form $\#c_1..c_q$ are listed at the beginning, followed by q-grams without special symbols, which are followed by suffix q-grams ending with the symbol $. In our implementation, we use bigrams (i.e. q=2) sorted in a lexicographical order.

Figure 3 reports the pseudocode of the algorithm Create_Index. Given the dataset and the list of fields used for comparisons, the algorithm generates three index files. For each record of the dataset, all distinct q-grams are inserted in the q-grams array. More precisely, in line {1..3}, the q-grams array index is created and stored in secondary memory (dataset.grm). In line {5, 6}, for each element of the q-grams array, an inverted index is generated allocating a bitmap of n bits (n is the size of the dataset). For each record and for each field, the algorithm stores the position of its q-grams (obtained from the q-grams array) in the index file (dataset.idx). Moreover, if a q-gram occurs in the $i - th$ record, the corresponding bitmap is updated (line 13) by setting the $i - th$ bit. Finally, in line {15, 16} bitmaps are stored in a binary file (dataset.bin). Figure 2 (a) shows an example of dataset and corresponding index files.

Concerning the complexity, let m be the number of records, n be the size of the q-grams array and r be the number of q-grams of the longest string. The running time of *Create_Index* algorithm is $O(m*n)$, whereas the secondary memory required by the index files is $O(n)$ for the q-grams array, $O(m*r)$ for saving q-grams ids and $O(m*n)$ for the binary file. Finally, the method requires $O(n*m)$ bits of memory.

3.3 DDEBIT Record Comparisons

The duplicate record detection problem treated in this paper is defined as follows.

Definition 3.31. *Let* B *be a base table and* Q *be a query table. Let* l *be the loose threshold and* t *be the tight threshold. Let* $r_x \in Q$ *and* $r_y \in B$ *be two records*

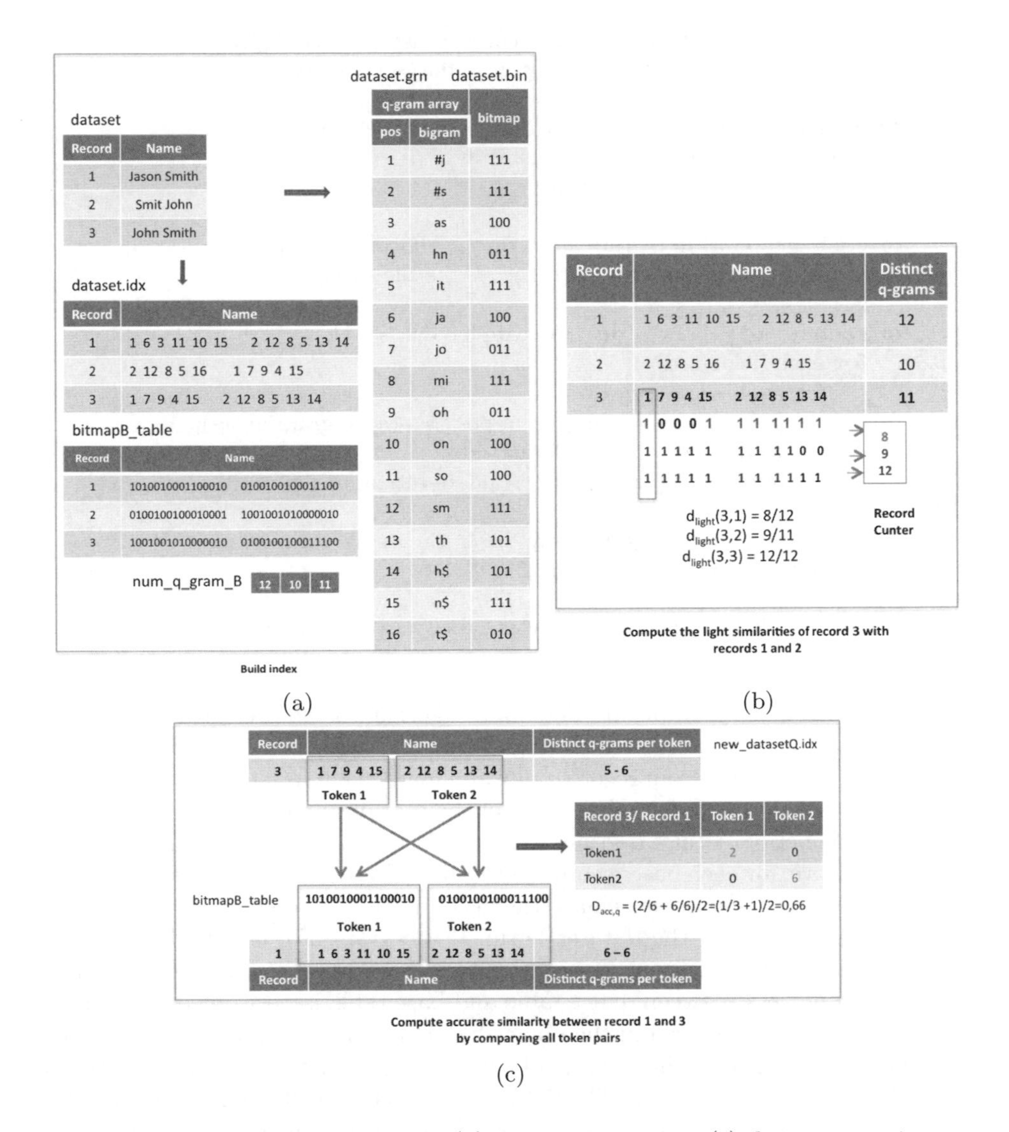

Fig. 2. (a) Indexing construction. (b) d_{light} computation. (c) d_{acc} computation.

projected on k fields. r_x *is considered a duplicate of* r_y*, if* $d_{light,q}(x, y) \geq l$ *and* $d_{acc_k,q}(r_x, r_y) \geq t$.

The novelty introduced by DDEBIT consists in reducing both the number and the cost of the expensive record comparisons using information stored in the index files previously generated for each dataset. More precisely, the inverted indexes stored in the binary files (dataset.bin) are used to quickly compute $d_{light,q}$, while record index files (dataset.idx) are used to reduce the complexity of $d_{acc,q}$. Notice that, when base table B and query table Q do not coincide index files (.grm, .idx, and .bin) are constructed for both B and Q.

```
DDEBIT_Create_Index(D dataset of n records, a list of k fields (f1, f2, ..., fk))
    OUTPUT: Three index files: dataset.grm, dataset.idx, dataset.bin

 1  for each record rec_i in D do // Create file dataset.grm
 2      insert_all_q_grams (rec_i, q-grams_array);
 3  write (q-gram_array, dataset.grm);
 4
 5  for each q-gram q_j in q-gram_array do // Allocate bitmaps
 6      bitmap[q_j] = allocate_bitmap (n); // n is the size of D
 7
 8  for each record rec_i in D do // Create file dataset.idx
 9      for each field f_y do
10          for each q-gram q_j do
11              pos_q_gram = find_q_gram_position (q_j, q-gram_array);
12              write (pos_q_gram, dataset.idx);
13              bitmap[q_j] [rec_i] = 1; // Update bitmap for q_j
14
15  for each q-gram q_i in q-gram_array do // Create file dataset.bin
16      write (bitmap[q_j], dataset.bin);
```

Fig. 3. The Create Index procedure of DDEBIT

The pseudo code of duplicate detection algorithm is showed in Figure 4. In line {1..4}, for each base record, the algorithm (i) retrieves for each field and for each token, all q-grams from record index and loads them in memory by setting a bitmap whose size is equal to the length of the q-grams array, and (ii) maintains the total number of distinct q-grams. The construction of *bitmapB_table* (line 3) is not expensive; for each base record r_b, it is linear in the total number of q-grams of r_b. In line 5, the algorithm remaps the ids of q-grams in datasetQ.idx with the corresponding ids from the datasetB.grm. If a q-gram of query table does not occur in the q-grams array of base table, its id is set to -1. In lines {7..19}, the comparison process between query table and base table is performed. DDEBIT, by using information stored in the binary file, maintains a counter associated to each base record storing the number of common q-grams with the query records (line 10). Such a counter allows to quickly detect base records sharing at least one q-gram with the query record (see Figure 2). We compute (line 14) the $d_{light,q}$ similarity between the query record and each base record whose id corresponds to a value of record_counter greater than 0. If the value of $d_{light,q}$ is above the loose threshold, the algorithm computes $d_{acc_k,q}$ to verify the local similarity. The candidate records are considered duplicated if $d_{acc_k,q}$ is above the tight threshold (see Figures 2 (b),(c) for an example of computation).

Using information stored in the index files, if g is the number of q-grams of the query records, $d_{acc,q}$ requires $O(g)$ comparisons. The reason behind this complexity relies on fact that, for each q-gram i of a query record, the method tests only the i-th bit of the bitmap corresponding to each token of the base record. Thus, the complexity of the method is $O(m*n*g)$, where m is the number of records of the query table, n is the number of records of the base table, and

```
DDEBIT_Duplicate_Detection(datasetB.*: Base table, datasetQ.*: Query table;
a list of k fields (f1, f2, ..., fk); loose threshold l; tight threshold t)
OUTPUT: for each query record, a list of potential duplicated records

1  for each record rec_b in B do //Load base table in memory
2      for each field f_i in rec_b do //Read record ids from file datasetB.idx
3          bitmapB_table[rec_b][f_i]= Allocate_and_set_List_bitmaps();
4      num_q-grams_B[rec_b]=total_num_distinct_q-grams();
5  new_datasetQ.idx = Remap(datasetQ.idx, datasetB.grm);
6
7  for each record rec_q in Q do // Comparison process
8      for each q-gram q_j in rec_q do
9          for each i - th set bit of datasetB.bin[q_j] do
10             record_counter[i]++;
11     num_q-grams_Q[rec_q]=total_num_distinct_q-grams();
12
13     for each i such that record_counter[i]> 0 do //Find candidate pairs
14         d_light= record_counter[i] / max(num_q-grams_B[i],num_q-grams_Q[rec_q]);
15         if (d_light == 1.0 ) then output(Exact match);
16         else if ( d_light >= l ) then
17                 d_acc_k = compute_score(bitmapB_table[i], new_datasetQ.idx[rec_q]);
18                 if (d_acc_k >= t) then
19                     output (Record i and record rec_q are duplicated!);
```

Fig. 4. The Duplicate_Detection procedure of DDEBIT

$g << n$ is the maximum number of q-grams in the records. Moreover, the method requires $O(n*m)$ bits of main memory.

3.4 Application to Clustering

The key idea for detecting duplicates presented in this paper can be used for clustering. Extending a strategy proposed in [16], we introduce a new clustering technique which consists of partitioning the dataset into non-overlapping clusters. Using the similarity measures d_{light} and d_{acc}, the method groups into the same cluster records considered duplicated according to the strategy described in the previous sections. Differently from comparisons process, number of comparisons are slightly reduced. This is due to the fact that a record associated to a cluster will not be used for further comparisons. The complexity is $O(n*g*c)$, where c is the number of clusters, n is the number of records, and $g << n$ is the maximum number of q-grams in the records.

4 Performance Analysis

Experimental analysis was performed on a server HP Proliant DL380 with 4GB RAM, equipped with Linux Debian Operating System. DDEBIT was implemented in C++ language. In all our experiments, q has been set to 2.

Accuracy of DDEBIT measures. We compared DDEBIT with q-grams and positional q-grams [9], Jaro [11], Jaro Winkler [23] and edit distance [13]. To perform comparisons, we used Python implementation of these metrics available in Febrl package[1]. We measured the accuracy of similarity metrics on real datasets concerning information about restaurants, by comparing 331 tuples from Zagat's website with 533 tuples from the Dept. of Health website. We selected the fields `name`, `street`, `city`, and `phone` and computed Recall, Precision and the F1 measure defined as $\frac{2 \times Recall \times Precision}{Recall+Precision}$.

Table 1 shows the accuracy of the metrics related to the best performances for each method. Results clearly show that the proposed method has a better behavior with respect to the other metrics. DDEBIT outperforms q-grams and edit distance since it uses a combination of local and global similarities. Moreover, due to the fact that DDEBIT better captures the similarity in presence of prefixes and missing of short tokens, it also yields an higher value of F1 with respect to Jaro and Jaro Winkler. However, a limitation of DDEBIT is that it does not detect string swaps between attributes neither potential similar strings which differ on an high number of tokens or contain missing values.

Table 1. Comparison among DDEBIT and q-grams, positional q-grams, edit distance, Jaro and Jaro-Winkler

Metric	Tight(Loose) threshold	Real Duplicates	Correct Pred. Duplicates	Predicted Duplicates	Precision	Recall	F1
q-gram(q=2)	0,7	112	107	210	0,51	0,96	0,66
Positional	0,7	112	49	61	0,8	0,44	0,57
Jaro	0,8	112	78	178	0,44	0,70	0,54
Edit	0,75	112	72	140	0,51	0,64	0,57
Jaro Winkler	0,85	112	103	173	0,60	0,92	0,72
DDEBIT	0,8 (0,65)	112	97	106	0,92	0,87	**0,89**
DDEBIT	0,75 (0,6)	112	107	134	0,8	0,96	0,87
DDEBIT	0,7 (0,6)	112	110	181	0,61	0,98	0,75

Efficiency and accuracy of DDEBIT method. By evaluating the indexing construction time on synthetic data[2], we observed a linear trend on both the number of fields (see Figure 5 (a)) and the size of the dataset (see Figure 5 (b)). Concerning the record comparisons time, by performing a self comparison of a base table of 10k and varying number of fields, DDEBIT took a maximum running time of 15,41 sec. (Figure 6 (a)). In Figure 6 (b) we report the average time for query record varying the size of the dataset. The query time increases for larger datasets due to the higher number of candidate records. Figure 7 (a) reports a record comparison between a query table of size 10k and a base table of size 100k. A naive approach (which does not use filtering) performs 10k*100k

[1] http://sourceforge.net/projects/febrl/

[2] We used the generator provided at http://dbgen.sourceforge.net/.

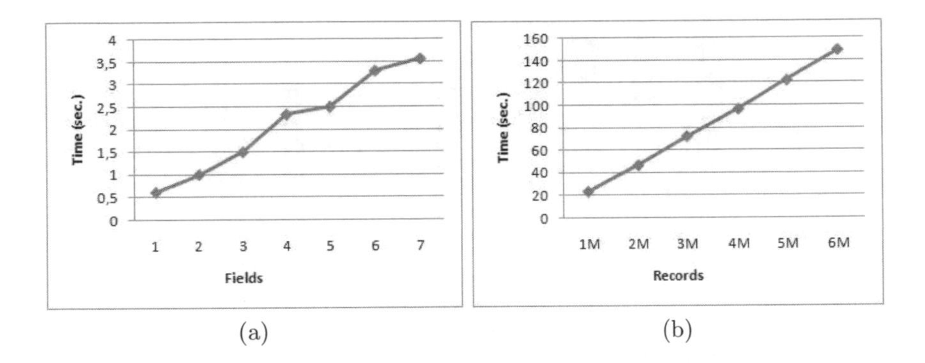

Fig. 5. DDEBIT indexing time. (a) Size of dataset is 100k. (b) Number of fields is 4.

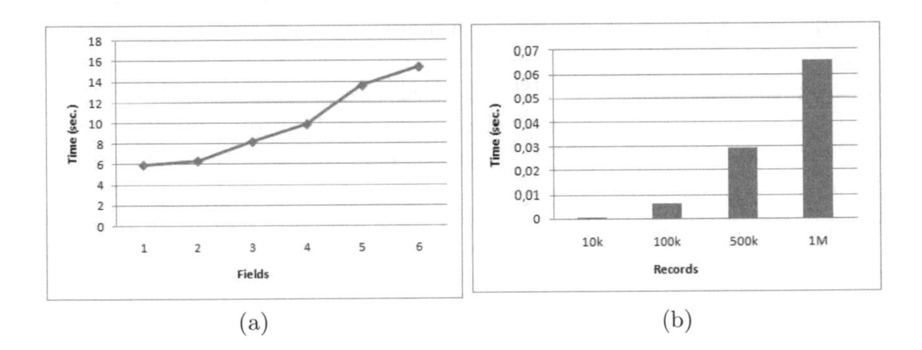

Fig. 6. DDEBIT record comparisons time with light threshold and tight threshold equal to 0.7 and query table of size 10k. (a) Time w.r.t. fields. (b) Number of fields is 4.

comparisons, computing $d_{acc,q}$ on all possible record pairs. $d_{light,q}$ computes an high number of comparisons however the cost of $d_{light,q}$ is lower than $d_{acc,q}$ (common q-grams are retrieved directly from index files). Moreover, it reduces the number of expensive comparisons filtering out non similar record pairs. For example, by setting $k = 3$, $d_{acc3,2}$ performs only 12.294 record comparisons in place of 10k*100k. In Figure 7 (b), DDEBIT drops down the running time up to the 92% with respect to the naive approach. Although the different programming languages, we compared DDEBIT with SB (Standard Blocking with Soundex encoding) [6], QI (Q-gram Indexing) [2], SA and MSA [5] (Similarity Aware and Materialized Similarity Aware using Jaro winkler and 1-gram comparisons metrics)[3]. We generated uniformly distributed synthetic datasets having size ranging from 500k to 1.5M. We computed the index construction time (Figure 8 (a) top) and the average query time (Figure 8 (a) bottom) using a query table of 2k records.

[3] We thank prof. Peter Christen for providing the software.

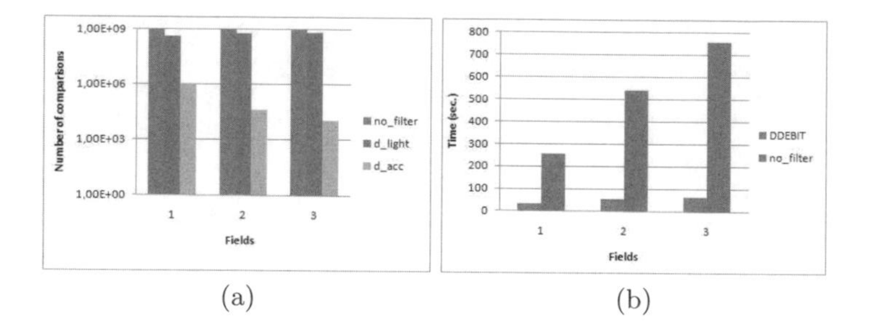

Fig. 7. (a) Number of comparisons using $d_{light,q}$, $d_{acc,q}$ and the naive approach. (b) Running time of DDEBIT vs the naive approach.

We also compared the accuracy of the above methods on synthetic datasets of size 100 with different distributions (uniform, Poisson and zipf) and 4 fields. Each query record is a duplicate of one base record with 1 modification per field and 3 modifications per record. We also used two real data selecting randomly 100 records from *ucdPeopleMatch*[4] dataset which contains prefixes and swap of words within attributes, and *cora*[5] dataset containing citations clustered into groups referring to the same scientific paper.

For real datasets, the number of true matches is slightly increased with respect to the other algorithms (see Figure 8 (b)). Figure 8 (c) shows the Recall of DDEBIT for different values of light and tight thresholds for the cora dataset. Notice that, using proper thresholds, DDEBIT is able to reach a very high Recall. Finally, Figure 8 (d) contains different F1 values varying the weights p_i associated to the attributes, used in $d_{acc_k,q}$. Assigning 1/2 to *title* and 1/6 to the other attributes, DDEBIT yields a very good accuracy with respect to the uniform weights assignment (first row in the table).

Clustering. In order to test the accuracy of DDEBIT clustering, we generated synthetic datasets (using the Febrl generator) having at most 4 modifications per attribute and 15 modifications per record. Each dataset contains one duplicate per record. We used a light threshold equal to 0.6 and a tight threshold equal to 0.65. Results clearly show that F1 (see Figure 9 (a)) is higher using more fields in the comparisons process, due to the fact that $d_{acc,q}$ is highly discriminating and gives a more accurate score using a larger number of fields.

We also tested running time of our clustering approach. Figure 9 (b) shows that, by fixing the number clusters to 10 and varying the size of dataset, DDEBIT has a better running time with respect to fix the percentage of duplicates per record (10%) and varying the number of clusters.

The proposed clustering technique does not take into account, for further comparisons, records already assigned to a cluster. This improves clustering efficiency at cost of inserting some records in wrong blocks. Finally, in Figure 10 we report

[4] http://fingolfin.user.cis.ksu.edu/ERICRAWLER/data/

[5] http://www.hpi.uni-potsdam.de/naumann/projekte/repeatability/datasets/

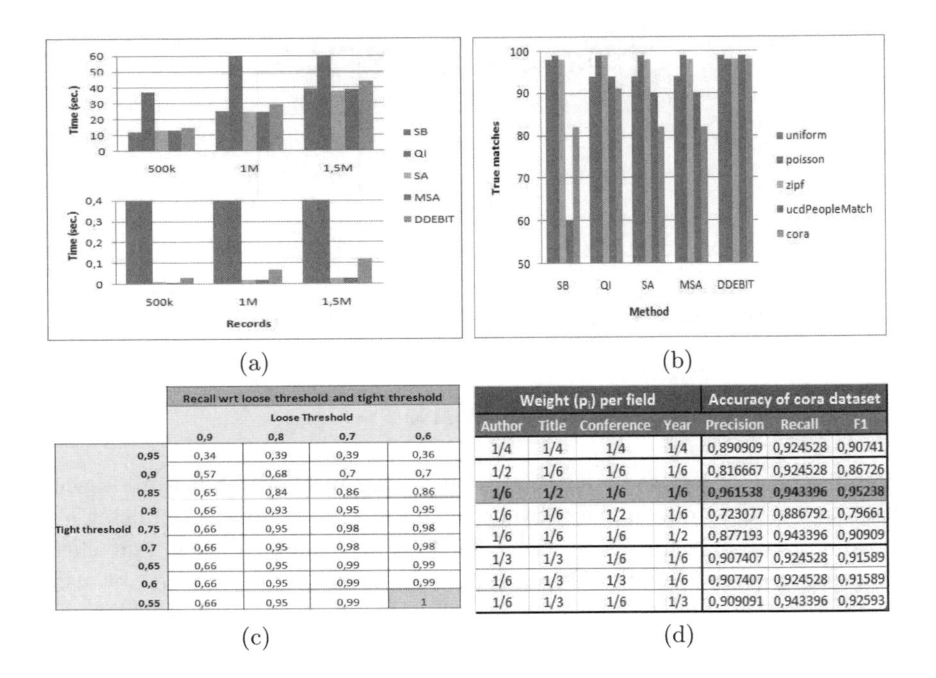

		Recall wrt loose threshold and tight threshold			
		Loose Threshold			
		0,9	0,8	0,7	0,6
Tight threshold	0,95	0,34	0,39	0,39	0,36
	0,9	0,57	0,68	0,7	0,7
	0,85	0,65	0,84	0,86	0,86
	0,8	0,66	0,93	0,95	0,95
	0,75	0,66	0,95	0,98	0,98
	0,7	0,66	0,95	0,98	0,98
	0,65	0,66	0,95	0,99	0,99
	0,6	0,66	0,95	0,99	0,99
	0,55	0,66	0,95	0,99	1

(c)

Weight (p_i) per field				Accuracy of cora dataset		
Author	Title	Conference	Year	Precision	Recall	F1
1/4	1/4	1/4	1/4	0,890909	0,924528	0,90741
1/2	1/6	1/6	1/6	0,816667	0,924528	0,86726
1/6	1/2	1/6	1/6	0,961538	0,943396	0,95238
1/6	1/6	1/2	1/6	0,723077	0,886792	0,79661
1/6	1/6	1/6	1/2	0,877193	0,943396	0,90909
1/3	1/3	1/6	1/6	0,907407	0,924528	0,91589
1/6	1/3	1/3	1/6	0,907407	0,924528	0,91589
1/6	1/3	1/6	1/3	0,909091	0,943396	0,92593

(d)

Fig. 8. (a) DDEBIT vs other techniques on synthetic datasets. Top shows index construction time, bottom query time. (b) True matches (0-100) on different datasets. (c) DDEBIT Recall on cora dataset. (d) DDEBIT accuracy on cora dataset wrt p_i.

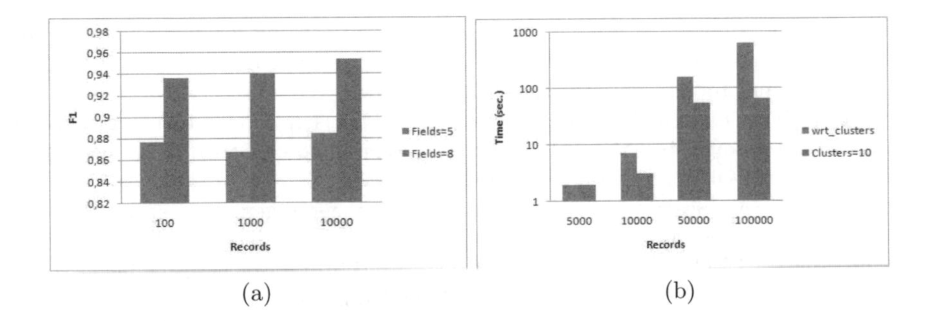

Fig. 9. Clustering. (a) F1 w.r.t. the size of dataset. (b) Running time with fields=5.

a direct comparison between DDEBIT clustering technique and a naive approach that does not filter candidate record pairs. DDEBIT slightly reduces execution time and the number of comparisons. Moreover, accuracy is lower in some cases (records = 1000), because $d_{acc,q}$ introduces some false positives or negatives; but we observed an increased value of Recall in the other cases (records equal to 10k and 20k), due to the fact that $d_{light,q}$ sometimes discards false positives that $d_{acc,q}$ would consider as duplicates.

Records	Time (sec.)		Number of comparisons		Recall	
	DDBEBIT	no_filter	DDBEBIT	no_filter	DDEBIT	no_filter
1000	0,12	324	263	250035	0,998	1
2000	0,33	13,63	1619	1000925	1	1
10000	6,3	457	10007	24816509	0,997	0,9954
20000	19,83	1640	5001	98649409	0,9977	0,9913

Fig. 10. Clustering vs naive approach. Fields=5, light and tight threshold equal to 0.7.

5 Conclusions and Future Work

In this paper we proposed DDEBIT, an indexing method for fast and accurate duplicate record detection. The novelty of this approach consists of using a two-step similarity comparisons process which results more efficient and effective than using a single measure. Moreover, comparisons are optimized by using a q-grams array indexed with bitmaps. Finally, the efficiency of DDEBIT has been confirmed by a new clustering technique based on it. Future work will investigate the memory usage reduction through an efficient bitmap compression and the refinement of similarity metrics to increase accuracy of resulted records.

References

1. Aizawa, A., Oyama, K.: A fast linkage detection scheme for multi-source information integration. In: International Workshop on Challenges in Web Information Retrieval and Integration, pp. 30–39 (2005)
2. Baxter, R., Christen, P., Churches, T.: A comparison of fast blocking methods for record linkage. In: ACM SIGKDD 2003 Workshop on Data Cleaning, Record Linkage, and Object Consolidation, pp. 25–27 (2003)
3. Chatterjee, A., Segev, A.: Data manipulation in heterogeneous databases. ACM SIGMOD Record 20, 64–68 (1991)
4. Chaudhuri, S., Ganjam, K., Ganti, V., Motwani, R.: Robust and efficient fuzzy match for online data cleaning. In: SIGMOD 2003, pp. 313–324 (2003)
5. Christen, P., Gayler, R.: Towards scalable real-time entity resolution using a similarity-aware inverted index approach. Proceedings of AusDM 2008, Glenelg, Adelaide 87, 30–39 (2008)
6. Christen, P., Churches, T.: Febrl: Freely extensible biomedical record linkage Manual (2002)
7. Cohen, W., Richman, J.: Learning to match and cluster large high-dimensional data sets for data integration. In: SIGKDD 2002 (2002)
8. Elmagarmid, A., Ipeirotis, P., Verykios, V.: Duplicate record detection: A survey. TKDE 19 (2007)
9. Gravano, L., Ipeirotis, P.G., Jagadish, H.V., Koudas, N., Muthukrishnan, S., Srivastava, D.: Approximate string joins in a database (almost) for free. In: VLDB 2001, pp. 491–500 (2001)

10. Han, J., Kamber, M.: The data warehouse ETL toolkit: Practical techniques for extracting, cleaning, conforming, and delivering data. John Wiley and Sons, Chichester (2004)
11. Jaro, M.A.: Unimatch: A record linkage system: User's manual. Technical report, U.S. Bureau of the Census, Washington, D.C (1976)
12. Jaro, M.A.: Advances in record linkage methodology as applied to matching the 1985 census of tampa, florida. Journal of the American Statistical Society 84, 414–420 (1989)
13. Levenshtein, V.I.: Binary codes capable of correcting deletions, insertions and reversals. Doklady Akademii Nauk SSSR 163, 845–848 (1965)
14. Hernandez, M., Stolfo, S.: The merge/purge problem for large databases. In: Proceedings of the ACM SIGMOD International Conference on Management of Data (1995)
15. Manber, U., Myers, G.: Suffix arrays: a new method for on-line string searches. SIAM Journal on Computing 22, 935–948 (1993)
16. McCallum, A., Nigam, K., Ungar, L.H.: Efficient clustering of high-dimensional data sets with application to reference matching. In: ACM SIGKDD, pp. 169–178 (2000)
17. Monge, A.E., Elkan, C.P.: An efficient domain-independent algorithm for detecting approximately duplicate database records. In: Proceedings of DMKD 1997, pp. 23–29 (1997)
18. Ramos, J.: Using tf-idf to determine word relevance in document queries. In: Proceedings of the First Instructional Conference on Machine Learning (2003)
19. Smith, T.F., Waterman, M.S.: Identification of common molecular subsequences. Journal of Molecular Biology 147, 195–197 (1981)
20. Sutinen, E., Tarhio, J.: On using q-gram locations in approximate string matching. In: Spirakis, P.G. (ed.) ESA 1995. LNCS, vol. 979, pp. 327–340. Springer, Heidelberg (1995)
21. Ukkonen, E.: Approximate string matching with q-grams and maximal matches. Theoretical Computer Science 92, 191–211 (1992)
22. Ullman, J.: A binary n-gram technique for automatic correction of substitution, deletion, insertion, and reversal errors in words. The Computer Journal 20, 141–147 (1977)
23. Winkler, W.E.: The state of record linkage and current research problems. In: Statistics of Income Division (1999)

Modelling Complex Data by Learning Which Variable to Construct

Françoise Fessant, Aurélie Le Cam, Marc Boullé, and Raphaël Féraud

Orange Labs,
2 avenue Pierre Marzin, 22307 Lannion, France
{francoise.fessant,aurelie.lecam,marc.boulle,
raphael.feraud}@orange-ftgroup.com
http://www.orange.com/en_EN/innovation/

Abstract. This paper addresses a task of variable selection which consists in choosing a subset of variables that is sufficient to predict the target label well. Here instead of trying to directly determine which variables are better, we make use of prior knowledge to learn the properties of good variables and guide the selection towards the most relevant dimensions. For this purpose we assume that a variable can be represented by a set of indicators that describe both the properties of the variable and its potential relationship to the targeting problem. This approach enables the prediction of the relevance of variables without measuring their value on the training instances. We devise a selection methodology that can efficiently search for new good variables in the presence of a huge number of variables and to dramatically reduce the number of variable measurements needed. Our algorithm is illustrated on an industrial CRM application.

Keywords: Variable selection, classification, scoring, CRM.

1 Introduction

Customer Relationship Management (CRM) is a key element of modern marketing strategies. The most practical way to build knowledge on customers in a CRM system is to produce scores to detect churn, propensity to subscribe to a new service, etc. A score (the output of a model) is an evaluation for all target variables to explain. The score is computed using customer records represented by a number of variables or features. Scores are then used by the information system for example to personalize the customer relationship. The rapid and robust detection of the most predictive variables can be a key factor in a marketing application.

An industrial customer platform has been developed at Orange Labs to industrialize the data mining process for marketing purpose. The platform, capable of building predictive models for datasets having a very large number of input variables (thousands) and instances (hundreds of thousands), is currently in use by Orange marketing. Its fully automated data processing machinery includes: data preparation, model building, and model deployment. The system extracts a

T.B. Pedersen, M.K. Mohania, and A M. Tjoa (Eds.): DaWaK 2010, LNCS 6263, pp. 324–335, 2010.

large number of features from a relational database, selects a subset of informative variables and efficiently builds in a few hours an accurate classifier. When the models are deployed, the platform exploits sophisticated indexing structures and parallelization in order to compute the scores of millions of customers, using the best representation. The platform allows building predictive models using two orders of magnitude more exploratory variables than the current state of the art, resulting in a dramatic improvement of performances. Performances of the in-house platform have been benchmarked in an academic context through the recent challenge KDD cup 2009 [1].

Experiments on several marketing campaigns have shown that the improvement of the quality of scoring models is strongly correlated to the number of explicative variables that can be explored. However the processing time associated with data table flattening remains the main limitation to the exploration of even larger data spaces. The variables are very expensive to compute; the evaluation times growing linearly with the number of variables. For the moment, the platform is limited to the analysis of about 20 000 variables for strong industrial time constraints. The efficient exploration of such huge spaces therefore requires the conception of an exploration technique guiding the flattening towards the most promising areas.

This paper presents a methodology for the exploration of a large space of variables consistent with the time constraints. Our idea is to estimate the predictive power of input variables without measuring them and so to avoid the flattening of all variables. A variable is characterized by a set of indicators that describe both the properties of the variable and its potential relationship to the scoring problem. The link between the indicators and the predictive importance of the variables is modelized with a subset of evaluated variables. The learned model is then used to infer the predictive importance of many new variables. Then the set of best variables can be selected for final scoring. In this way, we can explore a large set of variables while measuring only a few of them. What's more we are able to characterize the most important variables and to judge new variables.

We describe the complete methodology of exploration and its evaluation on a raw marketing campaign. The paper is organized as follows: section 2 gives an overall view of the in house Orange customer analysis platform. Section 3 details the methodology of exploration. Experimental results are presented in section 4. We conclude with some further research directions in section 5.

2 Platform Description

Two main steps of the Orange in-house customer analysis platform, data preparation and model building, are described in this section. More about the platform can be found in [2].

2.1 Data Folder

Unlike the current practice of data mining architecture, the explanatory variables are not designed and computed once in a datamart. In our platform architecture, the input data from information system are structured, and stored in a simple relational database called the data folder. The explanatory variables are constructed and selected automatically for each specific marketing project. The data folder model provides a unique view of the available input data sources, normalized according to a star schema:

- The primary table is related to the marketing domain. For customer data analysis, this table contains all the fields directly connected to the customer, such as his name or address,
- The secondary tables have a N-0 relationship with the primary table. Each instance of the primary table may be related to a variable number of instances of a secondary table. For telecommunication data for example, the secondary table contains the list of services, of usages of theses services, the call details.

The star schema offers an efficient trade-of between single table data mining and full multi-relational data mining: it has a large expressiveness, suitable for many data mining problem, and it allows efficiently build aggregated variables from secondary tables. Finally, this star schema allows to design formatted and restricted data extraction languages in order to facilitate automatic control of data extraction.

2.2 Data Extraction

The platform uses a feature construction language dedicated to the marketing domain, to build tens of thousands of features in order to create a rich data representation space. The data extraction functionality of the platform is parameterized using dedicated languages.

- a selection language to filter the instances,
- a construction language to build a flat instance x variables representation from the data folder,
- a preparation language to specify the recoding of the explanatory variables.

These languages are both simple enough to be automatically exploited by the process of variable selection and expressive enough to build a large variety of explanatory variables. Each language expression deals with at most two tables: the primary table plus eventually one secondary table. The join key always belongs to the primary table, and the selection and construction operands exploit the fields of any table, primary or secondary.

We focus on the construction language because it represents one of the sources of prior knowledge exploited in our methodology. A unified framework is used to write each language expression. It is composed of several successive fields (an example is given table 1). The first one is the identification of the variable ("Id"), the second is the type of the variable ("Type" whose values can be numeric or symbolic). The third is the name of the table of origin (the primary table or a secondary one). Fourth item is the name of the operator (several type of operators are used, simple selection with "Get", calculation with "Mean", "Count" or

"Total" and more complex like date and trends). Next item "Operand" identifies the selected field in the table. The four following items correspond to a selection expression. A selection expression is defined by a naming rule "Sel_Id_1", the choice of another field of the table "Sel_Operand_1", one or more selection values "Sel_Value_1", and the choice of a new operator "Sel_TranscodingOperator_1 that can be a ranking operator or a date. The selection expression enables to specify some crosses between several fields of a given table. The language expression can contain from 1 to 4 selection expressions allowing more or less complex crosses. For example, to build the total turnover for several successive quarters for all customers, one single language expression needs to be specified (the expression is illustrated table 1). The table of origin is the secondary table "Photo", the name of the selected field is the operand identifying the turnover "CA". The operator working on the operand is the calculation operator " Total'. The selection expression is defined by the choices of the other field of the table ("M_Photo"), a transcoding operator ("DiffDate") and some values for the selection ([0,1,2] means that the total amount of CA is evaluated on the three last months stored in the data folder). The language expression generates 3 variables of numerical type (the turnover for 3 successive quarters) labelled "CA3M_t1", "CA3M_t2" and "CA3M_t3".

It is then possible to specify up to thousands of variables to construct, using one single expression of the construction language.

Table 1. The expression generates 3 explicative variables about the turnover for 3 successive quarters (CA3M_t1, CA3M_t2, CA3M_t3). It is composed of successive fields: Id, type of the variable (N for numerical in this case), source table name, operator, operand, selection id for variable identification, operand of selection, selection values and trancoding operator. A single expression can contain from 0 to 4 selection id, selection operands, selection values and trancoding operators according to the required complexity.

Id	Type	Table	Operator	Operand	Sel_Id_1	Sel_ Operand_1	Sel_ Value_1	Sel_Trancoding_ Operator_1
CA3M	N	Photo	Total	CA	$-t$	M_ Photo	$[0,1,2]$; $[3,4,5]$; $[6,7,8]$	DiffdateM

2.3 Data Preparation

The platform architecture allows to easily build flat data tables with up to tens of thousands of constructed variables. In order to select the best representation, that is the best subset of informative variables, a robust and efficient variable selection method has been implemented. Explicative variables are individually evaluated by means of a supervised discretization method in the numerical case or by means of an optimal value grouping method in the categorical case. Supervised discretization [3] (or value grouping [4]) is treated as a non parametric model of conditional probability of the output variable given an input variable with the MODL approach (Minimum Optimized Description Length). The discretization is turned into a model selection problem and solved in a Bayesian way. The best discretizations and value groupings are optimized using the bottom-up greedy heuristic described in [3]. One advantage of this filter approach is that non informative variables are discretized in one single interval and can thus be reliably discarded. This approach also quantitatively evaluates the predictive importance of each variable for the target.

2.4 Modelling

The orange in house platform uses the Khiops scoring tool which implements an extension of the naives Bayes classifier (including model averaging) called Selective Nave Bayes classifier. The system has no hyper-parameter to adjust. The tool is designed for the management of large datasets, with hundreds of thousands of instances and tens of thousands of variables, and was successfully evaluated in international data mining challenges. Khiops can be downloaded here: http://www.khiops.com/. Once learned, the model is finally deployed to produce scores for all instances on all the explanatory variables.

3 Predicting the Relevance of a Variable

3.1 Related Work

Our problem can be seen as a problem of variable selection. Classical variable selection task is to choose a small subset of variables that is sufficient to predict the target well. The main motivations for variable selection are computation complexity, reduction of the cost of measurements, improving classification accuracy or problem understanding [5]. The main approaches studied in the literature are filter and wrapper [6]. Filter methods consider the correlation between the input variables and the output variable as a pre-processing step, independently of the chosen classifier. Wrapper methods search the best subset of variables for a given classification technique, used as a black box. Wrapper methods which are time consuming [7] are restricted to the modelling phase of data mining, as a post-optimization of a classifier. Filter methods are better suited for the data preparation phase, since they are time efficient and can be combined with any data modelling approach.

Classical methods of variable selection tell us which variables are better, they don't tell us what characterizes these variables or how to judge new variables which were not measured in the training data. On the basis of these observations Krupka [8] has recently developed another approach to variable selection. Instead of selecting a set of better variables out of a given set, his algorithm learns the relation between some descriptors coming from prior knowledge on initial data and the variable usefulness. This in turn enables him to predict the quality of unseen variables. The scenario is based on an extension of Recursive Feature Elimination [9], a wrapper selection method for linear SVM. Subsets of variables with poor usefulness are successively removed with a recursive process.

Other ideas about the exploitation of prior knowledge about relevance of variables can be found in the literature. For instance, [10] performs transfer learning across tasks, acquiring prior knowledge on one dataset and using it as partial supervision on others. [11] is another example of transfer learning. Our work is based on an idea similar to [8] that consists in exploiting prior knowledge we have on initial variables and linking it to variable relevance. The modelization is completely based on the Khiops tool.

3.2 Acquisition of Prior Knowledge on Variables

As introduced section 2, the platform allows the generation of many variables with very few language expressions. The definition of an expression is composed of several choices: table, variable, operators, operands, values, ... and specifications for the exploitation of the expression, like id for labelling the variable. The language used for the construction of the variables provides the first source of prior knowledge we want to exploit. The initial data are stored in a data folder and this data folder is another source of prior information. For example, we know for a categorical variable details about its modalities (number, frequency) and for a numerical variable the spread of values.

List of descriptors. Each variable has been described by a set of descriptors from these two sources of knowledge. 15 descriptors have been directly retained from the structure of construction of the variable or derived from it:

- Type of the variable (a variable can be categorical or numerical),
- Table name (one of the table of the data folder),
- Operator (the name of the calculation operator: Get, Count, Mean, Trend, ...)
- Type of operator (an operator can be a simple selection or more complex: calculation, date, trend or count),
- Flag for the presence or absence of an operand (yes or no),
- Operand name (the name of one field of the selected table),
- Total number of transcoding operators in the expression (examples of trancoding operators: WeekDay, Diffdate, HourNumber, AscendingRanking, ...),
- Transcoding operator names (vector of 2 dimensions, in our applicative context an expression can have at most 2 items filled),
- Number of transcoding operator in each type (vector of 2 dimensions, a transcoding operator can be a date or a ranking),
- Number of selection operands (a selection operand is a field item of the selected table),
- Names of selection operands (vector of 4 dimensions, in our construction scheme a language expression can have up to 4 items filled in the selection expression),
- Length of the language expression (total number of items in the language expression),
- Flag for the complexity of the expression (yes or no, an expression is considered as complex if at least a part of a selection expression is filled in),

- Number of selection Id (a selection Id is used to label the variable),
- Number of selection values in each type (vector of 6 dimensions, a value can be a single numerical or categorical value, an interval of numerical values, a group of numerical or categorical values, a null value).

The 5 descriptors retained from the initial data in data folder are:

- Operand type (an operand can be numerical, categorical , a date or a time value),
- Number of operands in the table,
- Number of modalities for a categorical operand,
- Entropy for a numerical operand,
- Ratio between the interquartile interval and the median for a numerical operand.

Finally, prior knowledge on an explicative variable is represented by a vector of 30 dimensions.

3.3 Model of Variable Importance

We now define a new supervised problem. The original variables are the instances. The descriptors listed above become the new variables. The target is the predictive importance evaluated by the scoring model. As recalled section 2.3, Khiops analyses each variable independently for the target and return a value that is directly its predictive importance. Khiops is used once again as a classification model to find the required mapping from descriptors to predictive importance. The algorithmic protocol is decomposed into the learning and test steps:

Learning Step: We are able to build a set of N variables from a set of P feature construction expressions. We assume that we evaluate only a subset of these N variables with the scoring platform (it means that only these variables are flattening and a predictive importance is available for each of them). The descriptors associated to this subset of variables correspond to our learning set. We use it to learn the relation between the descriptor values and the variable importance. The problem we learn is not the exact prediction of the importance value but the class of importance (i.e. if the predictive importance value is null (not important) or positive (important)).

Test Step: Based on the previous modelization, the goal is now to generalize to unseen variables. We predict the importance class for the instances of the test set represented by the descriptors of the whole variables including variables that were not part of the training set. This in turn enables us to choose the most relevant variables for the final scoring. The process of variable importance prediction can be summarized as follows:

- Variable and descriptor sets constitution
  Variable set: build variables from a limited number of language expressions
  Descriptor set: extract descriptors for each variable
    15 descriptors based on the language framework
    5 descriptors based on the data stored in the data folder
- Learning of the model of importance
- Generalization on all the constructed variables
  Selection of the most important variables.

4 Experimental Validation

We report in this section practical experiments that have been made on a raw marketing campaign.

4.1 Data Description

For the evaluation, the platform is supplied with data collected on a sample of 30000 customers. The information comes from decisional applications of Orange Company. The goal of the task presented here is to prevent a customer to switch ADSL provider. For this problem we have 24,3% of positive instances. The feature construction language is used to generate 20000 initial explicative variables from 600 feature construction expressions (an example of such expression is given table 1).

4.2 Evaluation Process

The final evaluation concerns the scores produced with the platform. We compared the scores for several sizes of subsets in the model of predictive importance (it means that only the variables corresponding to these subsets are initially flattened and evaluated with the platform). The complete algorithmic protocol is as follows:

- Learning set constitution
  Repeat
    Random selection of a language expression
      Random selection of a variable among those generated by the expression
  Until the expected number of variables is reached
  Evaluation of the importance associated to variables with the scoring platform
  Building of the set of descriptors for the set of variables
- Learning of the model of importance
- Generalization on all the constructed variables
  Selection of the most important variables
- Final scoring with the selected variables
  Scoring evaluation.

4.3 Results

We successively experimented with a sample of 2, 5, 10, 20 and 40 percent of the initial explicative variables. In other words, the model of importance has been built with respectively 400, 1000, 2000, 4000 and 8000 instances (the instances being selected as described section 4.2). The predictive model is evaluated using the area under the ROC curve (AUC) [12] (the higher the criteria, the better, with 1.00 indicating perfect performance).

Table 2 shows for each sample, the number of variables evaluated for the constitution of the learning step, the time of flattening, the AUC of the classifier on the test set, the number of variables that have been classified with a positive predictive importance after generalization and the number of variables really important among them. 1072 variables have been labelled as important when the scoring has been achieved directly with the flattening of all the initial explicative variables. A evaluated variable is tagged as really important if it belongs to this set. 70% of the users are used for the modelization steps, the remaining 30% are kept for the final scoring evaluation.

We observed that less than 4% of the whole variables is considered as important for the targeting by the model. This number regularly increases with the size of the subset used in the learning step. A detailed analysis of the model of importance can help us to characterize good variables (for instance, the descriptors with high level in the model are the name of the operand, the names of the first and second selection operand in the expression and the name of the table).

Only the variables predicted as important are considered now and flattened for final scoring.

We compared the scores produced for the 5 sets of variables predicted as important to those given by the current operational model. The current model requires the direct flattening of 20000 explicative variables.

The performance of a model is measured with the cumulative gain curve. It is a graphical representation of the advantage of using a predictive model to choose which customers to contact. The x-axis gives the proportion of the population with the best probability to correspond to the target, according to the model.

Table 2. Sample parameters for learning the model of importance, flattening time (in minutes), AUC of the classifier on the test set, number of variables classified as important and number of really important variables. It takes 375 minutes to flatten the initial set of 20000 explicative variables.

Sample rate	size of the learning set for importance prediction	flattening time (m)	AUC	nb of variables classified as important	nb of variables really important
2 %	400	35	0.858	284	177
5 %	1000	51	0.850	513	413
10 %	2000	67	0.905	595	439
20 %	4000	121	0.908	604	450
40 %	8000	210	0.913	775	513

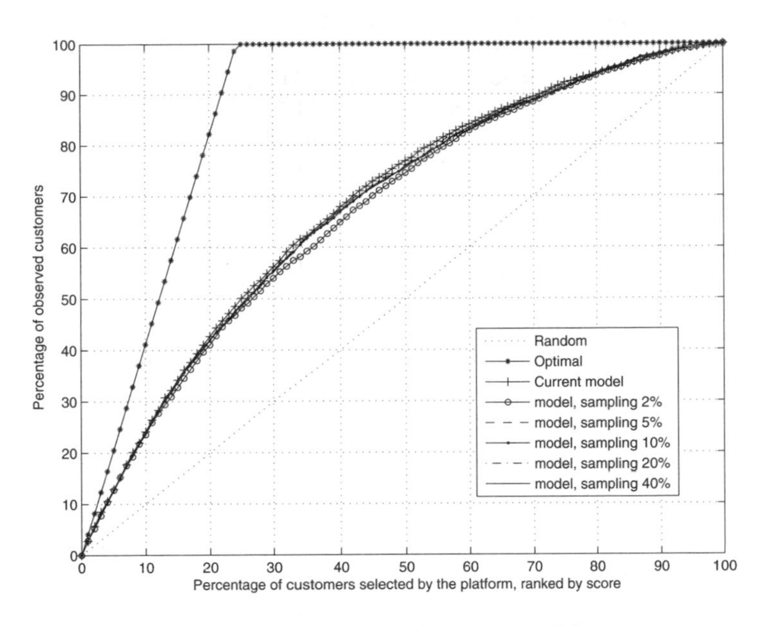

Fig. 1. Lift curves of scoring models

The y-axis gives the percentage of the targeted population reached. The curves are plotted on Figure 1. The diagonal represents the performance of a random model. If we target 20% of the population with the random model, we are able to reach 20% of the fragile customers. With the current model, when 20% of the population is contacted, 40% of the fragile customers is reached.

The curves corresponding to the sampling rates of 5%, 10%, 20% and 40% are almost confused and this remains true for the entire cumulative gain curve. The performance slightly decreases for the sampling at 2%. Numerical results in table 3 complete the previous observations. We give the AUC of the different scoring models. The lowest sampling rate excepted, the scoring based on the variable selection scenario has led to the same scoring accuracy than the actual model. For instance, a sampling of 5% of the initial variables means that 1000 variables among 20000 are first flattened in order to build the model of predictive importance. At the end of the generalization, 513 variables considered as important are retained. In the end the complete scoring process required the evaluation of about 1500 variables. Therefore we can conclude that a reduction of a factor 12 of the number of evaluated variables has been achieved without damage on the final scoring.

The experimental results confirmed the interest of the approach. We obtained similar scoring performances to the actual model with a significant reduction of measurements. A consequence is an important saving of time for the global scoring process. Another point with the method is that we are able to characterize the properties of good variables. An in depth analysis of the 20 best variables kept by the targeting models shows that they share 40% of similar variables

Table 3. Final scoring model performances (AUC)

model (sample rate)	AUC
Current model	0.744
2 %	0.728
5 %	0.735
10 %	0.739
20 %	0.738
40 %	0.740

with the current model. We can notice that efficient scoring can be achieved with several combinations of variables.

5 Conclusion

We have described in this paper a methodology of variable selection whose main idea is to take benefit from prior knowledge on variables to guide the exploration of the input space towards the most promising areas. The approach consists in predicting the quality of variables with measuring few of them. A variable is described by a set of indicators and the link between these indicators and the predictive importance of the variable is modelized. The model is then used to predict the importance of new variables. Only the variables predicted as important are retained and evaluated for final scoring, the other being discarded. The result is a dramatically reduction of the number of variable measurement needed for a similar scoring performance. With this approach, for a given number of variables we can explore more quickly or explore more variables in a fixed duration.

The validity of the approach has been demonstrated on a raw marketing campaign for several thousands of variables. This preliminary work needs to be extended. The exploration of even larger input spaces raises the question of overfitting and the risk that a variable becomes informative by accident. A solution could be a regularization procedure to penalize variables whose computational cost is high. Another research perspective is to combine our methodology with another learning method. A promising example is discussed in [13] where variable selection is formalized as a reinforcement learning problem.

References

1. Guyon, I., Lemaire, V., Boullé, M., Dror, G., Vogel, D.: Analysis of the kdd cup 2009: Fast scoring on a large orange customer database. Journal of Machine Learning Research: Workshop and Conference Proceedings 7, 1–22 (2010)
2. Féraud, F., Boullé, M., Clérot, F., Fessant, F., Lemaire, V.: The orange customer analysis platform. In: Perner, P., Ahlemeyer-Stubbe, A. (eds.) Proceedings of the 10th Industrial Conference on Data Mining. Springer, Heidelberg (2010)

3. Boullé, M.: MODL: a Bayes optimal discretization method for continuous attributes. Machine Learning 65(1), 131–165 (2006)
4. Boullé, M.: A Bayes optimal approach for partitioning the values of categorical attributes. Journal of Machine Learning Research 6, 1431–1452 (2005)
5. Guyon, I., Elisseeff, A.: An introduction to variable and feature selection. Journal of Machine Learning Research 3, 1157–1182 (2003)
6. Kohavi, R., John, G.: Wrappers for feature selection. Artificial Intelligence 97(1-2), 273–324 (1997)
7. Féraud, R., Clérot, F.: A methodology to explain neural network classification. Neural Networks 15, 237–246 (2001)
8. Krupka, E., Navot, A., Tishby, N.: Learning to select features using their properties. Journal of Machine Learning Research 9, 2349–2376 (2008)
9. Guyon, I., Weston, J., Barnhill, S., Vapnik, V.: Gene selection for cancer classification using support vector machines. Machine Learning 46(1-3), 389–422 (2002)
10. Lee, S., Chatalbashev, V., Vickrey, D., Koller, D.: Learning a meta-level prior for feature relevance from multiple related tasks, pp. 489–496 (2007)
11. Helleputte, T., Dupont, P.: Partially supervised feature selection with regularized linear models. In: Bottou, L., Littman, M. (eds.) Proceedings of the 26th International Conference on Machine Learning, Montreal, Omnipress, pp. 409–416 (June 2009)
12. Fawcett, T.: ROC graphs: Notes and practical considerations for researchers. Technical Report HPL-2003-4, HP Laboratories (2003)
13. Gaudel, R., Sebag, M.: Feature selection as a one-player game. In: Proceedings of the second NIPS Workshop on Optimization for Machine Learning, OPT 2009 (2009)

Author Index

Printing: Mercedes-Druck, Berlin
Binding: Stein + Lehmann, Berlin